JAMES S. THOMPSON, M.D.

Department of Anatomy, University of Toronto

MARGARET W. THOMPSON, Ph.D.

Departments of Medical Genetics and
Paediatrics. University of Toronto and The
Hospital for Sick Children, Toronto

genetics in medicine

second edition

W. B. SAUNDERS COMPANY

Philadelphia • London • Toronto

W. B. Saunders Company: West Washington Square
Philadelphia, Pa. 19105

12 Dyott Street
London, WC1A 1DB

833 Oxford Street
Toronto, Ontario M8Z 5T9, Canada

Listed here is the latest translated edition of this book together with the language of the translation and the publisher.

Portuguese: Levrario Atheneu, Rio de Janeiro, Brazil

Genetics in Medicine ISBN 0-7216-8856-X

Print No.: 9 8 7 6 5 4 3

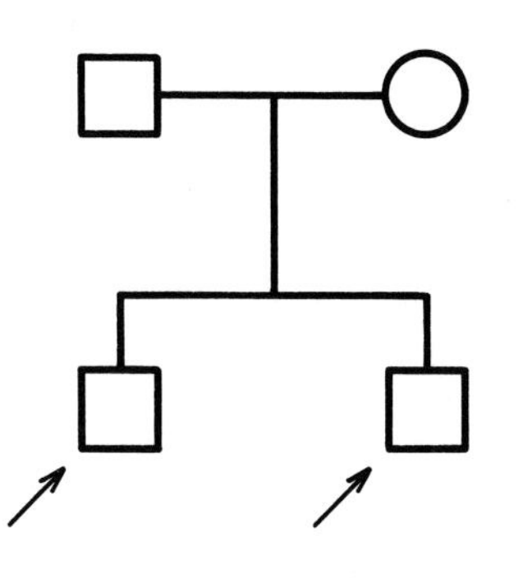

To our own F_1

PREFACE

Since the first edition of this book appeared in 1966, the continuing rapid growth of knowledge in medical genetics has given this field increasing relevance for medical science and practice. Genetics is now regarded as a core subject in the undergraduate medical school curriculum, and an area in which clinicians and other health scientists of many types need to be expert or at least literate.

Because new knowledge pours down upon medical students in an unceasing stream, one of the major problems in teaching medical genetics is the selection of core material that will elucidate the principles of genetics and provide an up-to-date framework of knowledge to serve as a foundation for the student's later training in a variety of related fields. Though we have planned this edition, like the previous one, primarily as a clear, elementary discussion for medical students, we hope it will also be useful to physicians and to other students of the health and life sciences.

In recent years a number of striking advances in research have opened several new and exciting areas of genetic investigation. Among the new topics included in this Second Edition are the banding techniques for chromosome staining, which have at last made possible the identification of individual human chromosomes; human somatic cell genetics, an expanding field that has a great potential in medical practice, but that carries certain risks of genetic manipulation of man, with which we must learn to live; and prenatal diagnosis of genetic disorders, especially by amniocentesis and cell culture.

A major difficulty in teaching and learning genetics is its specialized language. To help the reader familiarize himself with genetic terminology, we have provided problems with answers, and an updated glossary.

It is a difficult task to select from the massive genetic literature that material which will elucidate principles and describe new advances, without exceeding the limits imposed by the multiple pressures of the medical curriculum. Many important and interesting developments have had to be omitted, much to our regret. The selection of material has been facilitated by the availability of several new books, or new editions of older books, concerned with specialized areas of genetics, especially Giblett's *Genetic Markers in Human Blood,* Hamerton's two-volume *Human Cytogenetics,* the third edition of McKusick's *Mendelian Inheritance in Man* and the fifth edition of Race and Sanger's *Blood Groups in Man.* The different series of reviews in medical genetics (*Advances in Human Genetics,*

Annual Reviews of Genetics and *Progress in Medical Genetics*) have also been excellent sources. The work of many individual authors is cited in the appropriate parts of the text, but limitations of space have unfortunately prevented us from acknowledging many of the scientists whose work is quoted.

We are grateful for the contributions of many colleagues in genetics and in medicine, in Toronto and elsewhere, who have been helpful to us in the preparation of this book. In particular we are indebted to those investigators who have provided us with illustrations from their unpublished work, and to the authors and publishers who have granted us permission to use copyright material.

The original illustrations were done by Ella Furness, who also illustrated the first edition, and by Ann Wood and Gary Cousins. In the preparation of the manuscript we received help from many sources, especially from Carol Hutton and Pauline Kowal.

Once again, we thank our friends of the editorial staff of the W. B. Saunders Company for their continuing encouragement, patience and good will.

J. S. THOMPSON

MARGARET W. THOMPSON

CONTENTS

1

Introduction

The place of genetics in medicine was not always as obvious as it is today. Though now we know that genetics has an "axial role in the conceptual structure of biology [and a] ripening yield for the theory and practice of medicine" (Lederberg, 1959), not many years ago genetics was thought to be concerned only with the inheritance of trivial, superficial and rare characteristics, and the fundamental role of genes in basic life processes was not understood.

The discovery of the principles of heredity by the Austrian monk Gregor Mendel in 1865 received no recognition at all from medical scientists and virtually none from other biologists. Instead, his work lay unnoticed in the scientific literature for 35 years. Charles Darwin, whose great work *The Origin of Species* (published in 1859) emphasized the hereditary nature of variability among members of a species as an important factor in evolution, had no idea how inheritance worked. At that time inheritance was regarded as blending of the traits of the two parents, and Lamarck's idea of the inheritance of acquired characteristics was still accepted. Mendel's work could have clarified Darwin's concept of the mechanism of inheritance of variability, but Darwin seems never to have been aware of its significance or even of its existence. Darwin's cousin, Francis Galton, one of the great figures of early medical genetics, also remained ignorant of Mendel's work despite its importance to his own studies of "nature and nurture." Mendel himself, perhaps discouraged by the results of later, less favorably designed experiments, eventually took the course followed by many successful scientists—he abandoned research and became an administrator.

Mendel's laws, which form the cornerstone of the science of genetics, were derived from his experiments with garden peas, in which he crossed pure lines differing in one or more clear-cut characteristics and followed the progeny of the crosses for at least two generations. The three laws he derived from the results of his experiments may be stated as follows:

1. Unit inheritance. Prior to Mendel's time, the characters of the parents were believed to blend in the offspring. Mendel clearly stated

that blending did not occur, and the characters of the parents, though they might not be expressed in the first-generation offspring, could reappear quite unchanged in a later generation. Modern teaching in genetics places little stress upon this law, but in Mendel's time it was an entirely new concept.

2. Segregation. The two members of a *single* pair of genes are never found in the same gamete but instead always segregate and pass to different gametes.

3. Independent assortment. Members of *different* gene pairs assort to the gametes independently of one another.

With the dawn of the new century, the rest of the scientific community was ready to catch up with Mendel. By a curious coincidence, three workers (de Vries in Holland, Correns in Germany and Tschermak in Austria) independently and simultaneously rediscovered Mendel's laws. The development of genetics as a science dates not from Mendel's own paper but from the papers that reported the rediscovery of his laws.

The universal nature of mendelian inheritance was soon recognized, and as early as 1902 Garrod, who ranks with Galton as a founder of medical genetics, could report in alcaptonuria the first example of mendelian inheritance in man. In his paper Garrod generously admitted his debt to the biologist Bateson, who had seen the genetic significance of consanguineous marriage in the parents of recessively affected persons. This is the first clear evidence of the interaction of research by medical and nonmedical geneticists, which has continued to the present day.

A growing understanding of the universal nature of the biochemical structure and functioning of living organisms has brought about an awareness of the crucial role of genes in living organisms. The work of Garrod foreshadowed this knowledge, though in the early years of genetics its fundamental significance was not apparent. The concept was clearly formulated by Beadle and Tatum in 1941 as the "one gene–one enzyme" hypothesis.

The surge of genetic knowledge of the last two decades has had fruitful consequences for clinical medicine. It is estimated that one-third of the children in pediatric hospitals are there because of genetic disorders. This is a great change from the early years of the century, and even from the preantibiotic era. Before the days of immunization, improved nutrition and antibiotics, many children were in the hospital because of infectious diseases or nutritional disorders such as rickets. Today some of the few with infections have genetic defects that impair their resistance, and at least in developed countries most cases of rickets arise not from faulty nutrition but from deleterious genes. Life-saving advances in clinical techniques (transfusion, tube feeding, maintenance of body fluids by intravenous drip) for the management of medical emergencies also play a part in increasing the prevalence of genetic defects. Improvements in surgical procedures have also contributed to the profound alteration that has been effected during the twentieth century.

MAN AS AN OBJECT OF GENETIC RESEARCH

A mouse can complete a generation within two months, a fruit fly within two weeks and a microorganism within 20 minutes; but man has a generation time of at least 20 years. In lower forms it is possible to make test matings to acquire desired information or to test hypotheses; but in man Nature makes the experiment and the investigator can only record the outcome. A mouse can produce scores of offspring in its lifetime, a fruit fly hundreds and a microorganism millions; human families average about three children each. Faced with these formidable obstacles, we must ask ourselves what compensations man has to offset his disadvantages as a suitable animal for genetic research.

Probably most medical geneticists would subscribe to Pope's dictum that "the proper study of mankind is man." Man's fascination with himself has facilitated research into his genetics. Since we consider man so important (in Pope's phrase, "the glory, jest, and riddle of the world"), we have expended more effort upon him than upon some of the more suitable research organisms. Man is far more variable genetically than other living forms, or at least his variants are better explored and documented, and since many of them are deleterious they tend to come to the attention of the medical profession. Though individual families are small and becoming smaller, the total population is very large and becoming even larger; and though we cannot ethically perform experimental matings, we often find that somewhere in the population Nature has performed the experiments for us, or that the new tricks of somatic cell genetics will allow us to approach problems in human genetics from a different angle. A brave beginning has already been made at exploring the genetics of human variation, and this has been so successful that some variants, notably the hemoglobinopathies and the blood groups, have even provided models upon which part of the conceptual structure of genetics has been erected. Some of the many areas in which genetics and medicine mutually illuminate and enrich one another are described in the following chapters.

GENERAL REFERENCES

Boyer, S. H., ed. 1963. *Papers on Human Genetics*. Prentice-Hall, Englewood Cliffs, New Jersey.

Carter, C. O. 1969. *An ABC of Medical Genetics*. Little, Brown and Company, Boston.

Harris, H., and Hirschhorn, K., eds. *Advances in Human Genetics*, Vol. 1, 1970; Vol. 2, 1971; Vol. 3, 1972.

Lenz, W. 1963. *Medical Genetics*. University of Chicago Press, Chicago.

McKusick, V. A. 1969. *Human Genetics*. 2nd ed. Prentice-Hall, Englewood Cliffs, New Jersey. (A *Study Guide* to this book was published in 1972.)

McKusick, V. A. 1971. *Mendelian Inheritance in Man:* Catalogs of Autosomal Dominant, Autosomal Recessive and X-linked Phenotypes. 3rd ed. The Johns Hopkins Press, Baltimore.

Porter, I. H. 1968. *Heredity and Disease*. McGraw-Hill Book Company, New York and Toronto.

Roberts, J. A. F. 1970. *An Introduction to Medical Genetics*. 5th ed. Oxford University Press, Inc., New York and London.

Srb, A. M., Owen, R. D., and Edgar, R. S. 1965. *General Genetics*. 2nd ed. W. H. Freeman and Co., Publishers, San Francisco.

Steinberg, A. G., and Bearn, A. G., eds. *Progress in Medical Genetics*, Vols. 1–9, 1961–1973. Grune & Stratton, Inc., New York.

Stern, C. 1973. *Principles of Human Genetics*, 3rd ed., W. H. Freeman and Co., Publishers, San Francisco.

Sutton, H. E. 1965. *An Introduction to Human Genetics*. Holt, Rinehart & Winston, New York.

Whittinghill, M. 1965. *Human Genetics and Its Foundations*. Reinhold Publishing Corporation, New York.

2

The Physical Basis of Heredity

When a cell divides, the nuclear material is seen to form a number of rod-shaped organelles which are called **chromosomes** (*chromos,* color; *soma,* body) because they stain deeply with cytological stains. Chromosomes are composed of deoxyribonucleic acid (**DNA**) on a framework of protein. Lengths of DNA constitute the **genes**, the units of heredity, of which there are perhaps 100,000 in each human cell. Replication of chromosomes and segregation of the resulting daughter chromosomes occur very precisely, so that each daughter cell has the same chromosome complement and genetic information as the parent cell.

Each species has a characteristic chromosome constitution (**karyotype**), not only with respect to chromosome number, length and shape, but also with respect to the nature and sequence of the genes carried on each chromosome. The genes are arranged along the chromosome in linear order, each gene having a precise position known as its **locus.** Genes whose loci are in the same chromosome are said to be **linked,** or more precisely, **syntenic (in synteny).** The function of the genes is to direct the synthesis of polypeptides, which are the constituents of proteins. The actual DNA composition of a gene may be altered by mutation, so that the gene directs the synthesis of an altered polypeptide (page 46). Alternative forms of a gene which occupy the same locus are called **alleles.** Any one chromosome bears only one allele at a given locus, though in the population as a whole there may be multiple alleles for that locus.

The **genotype** of an individual is his genetic constitution, usually with reference to a single locus; the **phenotype** is the expression of any of those genes as a physical, biochemical or physiological trait. The **genome** is the full set of genes. The comparable term "phenome" is not in common usage, and the term phenotype is often used in a broad sense to describe the expression of the full genome.

Little was known about human cytogenetics until 1956, when Tjio and Levan developed effective techniques for chromosome study and found the human chromosome number to be 46, not 48 as had been believed previously. Since that time the field has developed rapidly; chromosome abnormalities have been shown to be common and important causes of birth defects, mental retardation and pregnancy wastage, and a beginning has been made at charting the location of genes on the human chromosome map.

THE HUMAN CHROMOSOMES

The 46 chromosomes of human cells constitute 23 homologous pairs. The members of a homologous pair match in respect to the genetic information each carries; i.e., they have the same gene loci in the same sequence, but at any one locus they may have either the same or different alleles. One member of each pair is inherited from the father, the other from the mother, at the time of fertilization. Twenty-two pairs are alike in males and females and hence are called **autosomes**. The **sex chromosomes**, which constitute the remaining pair, are different in males and females, and are of major importance in sex determination. Normally the members of a pair of autosomes are microscopically indistinguishable, and the same is true of the sex chromosomes of the female, which are called X chromosomes. In the male the members of the pair of sex chromosomes are different from one another; one is an X, identical to the X's of the female, while the other, which is known as a Y chromosome, is smaller in size than the X and appears not to be homologous to it except possibly with respect to a few genes (see later).

Because females are XX, their reproductive cells all carry an X chromosome; in contrast, males are XY and produce X-bearing and Y-bearing sperm in approximately equal numbers. Hence we speak of human females as the **homogametic sex** and human males as the **heterogametic sex**. This arrangement is characteristic of many living forms, but not of all; for example, in birds the female is the heterogametic sex.

Each parent contributes to the offspring 23 chromosomes, one member of each pair. Each sex cell (gamete), whether ovum or sperm, is said to have the **haploid** (n) chromosome number (*haploos,* single). In man, $n = 23$. The cell formed by the fertilization of the ovum by the sperm, the zygote, has 23 *pairs* of chromosomes, or 46; this is the **diploid** ($2n$) number (*diploos,* double). Almost all human somatic cells are diploid.

MITOSIS

Mitosis is the type of cell division by which the body grows and replaces discarded cells. The cytoplasm at mitosis divides simply by cleaving in half, but the nucleus undergoes a very complicated series of activities which ensures transmission to the two daughter cells of precisely the same chromosomal complement as that of the parent cell. Four stages of mitosis are distinguished: prophase, metaphase, anaphase and telophase. A cell which is not actively dividing is said to be in interphase. The stages of mitosis are shown diagrammatically in Figure 2–1.

During **interphase** (Fig. 2–1A) the chromosomes are elongated and not individually distinguishable. The nuclear material appears granular. The chromosomes are metabolically active (with the exception of the sex chromatin or Barr body, a condensed and inactive piece of chromatin; see page 160). At this stage the chromosomes do not stain differentially. As the cell prepares to divide, they begin to condense by coiling up and

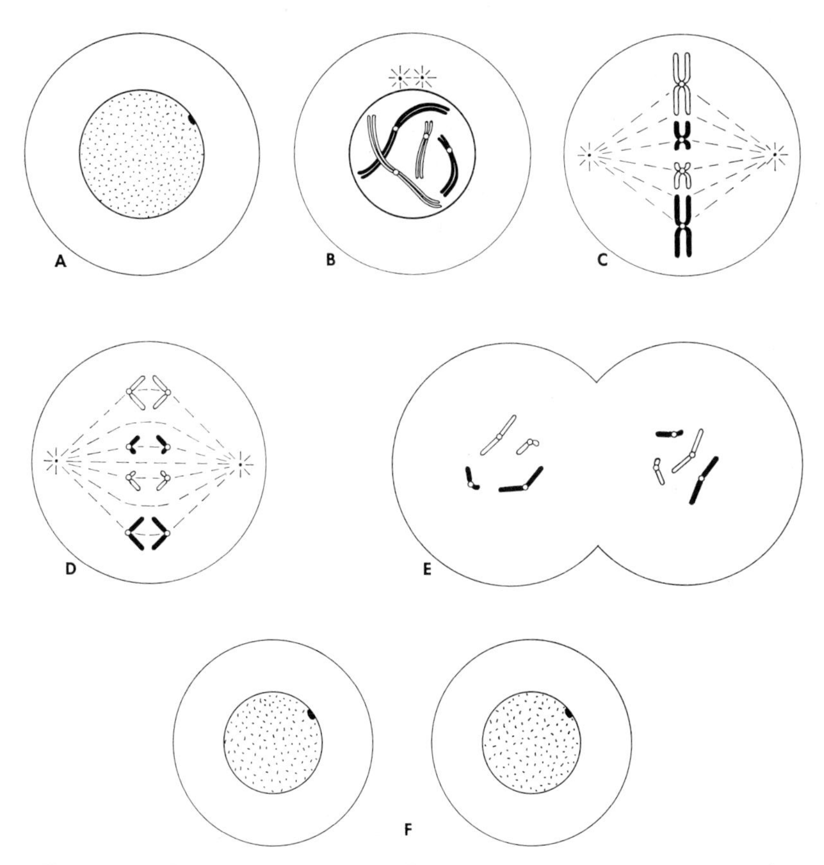

Figure 2–1 *The stages of mitosis. Only two of the 23 chromosome pairs are shown. Chromosomes from one parent are shown in outline; chromosomes from the other parent are in black. For a detailed description, see text. A, interphase; B, prophase; C, metaphase; D, anaphase; E, telophase; F, interphase.*

thus become visible as deeply staining bodies. As soon as the appearance of the nucleus changes and the chromosomes begin to be distinguishable, the cell has entered the first stage of cell division, prophase.

1. **Prophase** (Fig. 2–1B). When the chromosomes have become visible, but before any pattern is discernible in their arrangement, the cell is in prophase. By this time the DNA content of the cell has doubled, and each chromosome can be seen to consist of two long, thin, parallel strands, the **chromatids,** which are held together at one spot, the **centromere** (kinetochore). (Homologous chromosomes are shown in Fig. 2–1 in black and white to signify that one member of each pair is derived from the father, the other from the mother.) The position of the centromere is constant for any one chromosome. Most of the chromosome material stains deeply and is said to be **euchromatic,** but certain areas stain differently, probably because of differential coiling, and these areas are called **heterochromatic.** The types of chromatin associated with euchromatic and heterochromatic areas are euchromatin and heterochromatin.

During prophase the nuclear membrane disappears and the nucleus begins to lose its identity. At the same time, the **centriole,** an organelle just outside the nuclear membrane, duplicates and the two products migrate toward opposite poles of the cell.

2. **Metaphase** (Fig. 2–1C). When the chromosomes, which by now have thickened and become even more densely staining, line up at the equatorial plane of the cell, the cell is in metaphase. This is the stage at which individual chromosomes are most easily studied, since the chromatids have reached their most contracted state and the chromosomes are arranged in a more or less two-dimensional metaphase plate. The chromosomes are connected at their centromeres, with microtubules of protein running in the direction of the centrioles. These microtubules are part of a set of microtubules called the **spindle.** Some microtubules originate from the centromere, others at other microtubule-organizing centers in the general area of the centrioles, but the role of the centriole itself in mitosis is not clear. At metaphase the two members of a chromosome pair have no particular association with each other.

3. **Anaphase** (Fig. 2–1D). The cell enters anaphase when the centromeres divide and the daughter centromeres repel one another, so that the chromatids move apart. The division of the centromeres is parallel to the long axis of the chromatids (Fig. 2–2). The chromatids travel toward

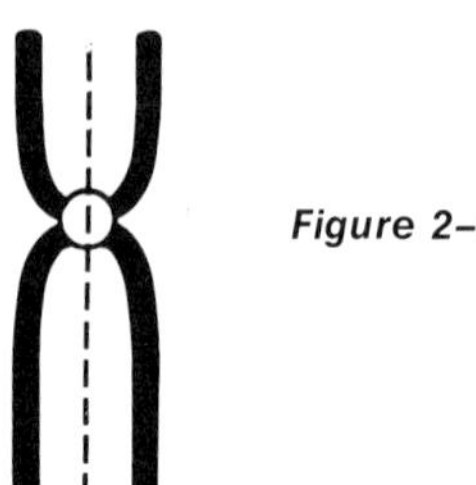

Figure 2–2 *Division of the centromere.*

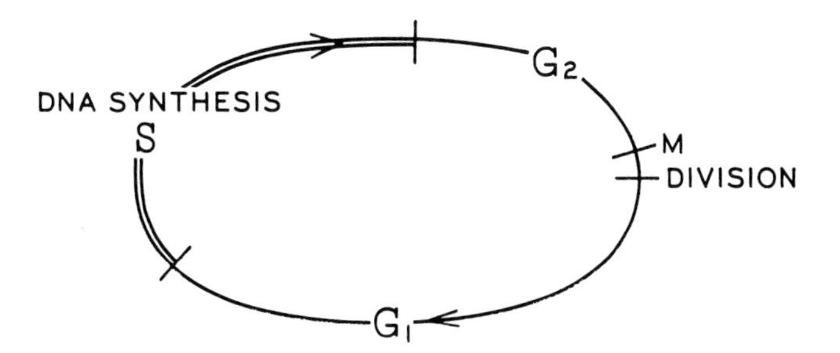

Figure 2–3 *The mitotic cycle. For description, see text. (Modified from Stanners and Till 1960.* Biochim. Biophys. Acta *37:406–419.)*

the poles of the cell as if drawn by the spindle. The centromere precedes the rest of the chromosome so that the two arms trail, giving each chromosome a V or J shape depending on the position of its centromere. The actual mechanism by which chromosomes move along the spindle to the poles of the cell is unknown. When the centromere divides, the chromatids are considered to be **daughter chromosomes.**

4. **Telophase** (Fig. 2–1E). The arrival of the chromosomes at the poles of the cell signifies the beginning of telophase, the final stage of cell division. Concurrently with the onset of telophase, the division of the cytoplasm (cytokinesis) begins by the formation of a furrow near the equatorial plane. Eventually a complete membrane is formed across the cell, which is thereby divided into two new cells, each with the same chromosome complement. Meanwhile, the chromosomes are unwinding and consequently becoming less densely staining. Eventually they no longer stain as individual entities, and the area in which they are located becomes enclosed by a nuclear membrane. Each daughter cell now appears as a typical interphase cell, with the chromosomes actively participating in its metabolism (Fig. 2–1F).

THE MITOTIC CYCLE

It is obvious that a cell must double its DNA content before it divides. The doubling is known to take place during a specific period of interphase. The stages of the mitotic cycle are shown in Figure 2–3. After division, the new cell enters a postmitotic period during which there is no DNA synthesis (G_1 or Gap 1). The period of DNA synthesis (S) follows. There is then a premitotic nonsynthetic period (G_2 or Gap 2), which is ended by the beginning of mitosis (M). Mitosis is of relatively short duration in comparison to the remainder of the cycle. Typical studies of cultured human cells have shown that the complete cycle has a duration of 12 to 24 hours, whereas mitosis requires about one hour.

CHROMOSOME CLASSIFICATION

The human mitotic chromosomes are identified and named, on the basis of size and centromere position, according to the Denver classification, a scheme adopted in 1960 at a meeting of human geneticists in Denver, Colorado. Until the recent development of banding techniques (page 13), it was impossible to classify all the chromosomes exactly into 23 pairs; instead, seven groups had been distinguished, identified by the letters A through G, in order of decreasing length. Only chromosomes 1, 2, 3, 16 and the Y had been individually identifiable. A chromosome set or karyotype, systematically arranged in accordance with the Denver classification, is shown in Figure 2–4.

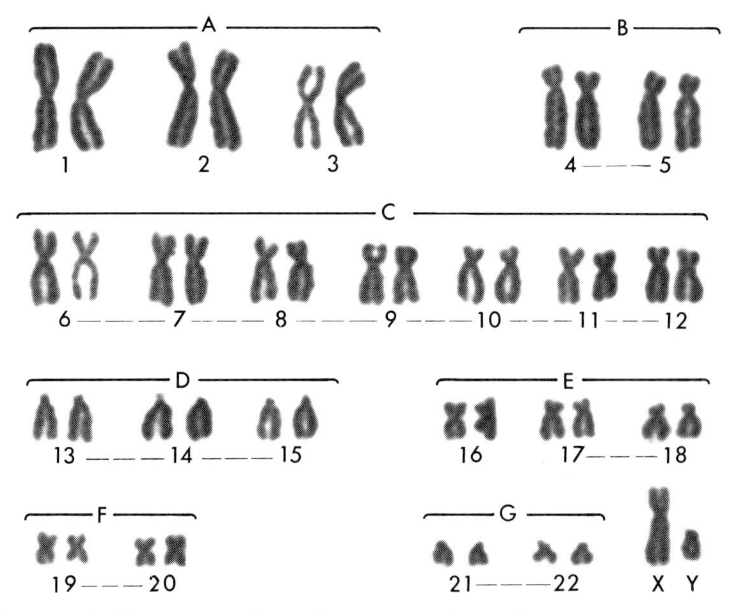

Figure 2–4 *Human metaphase chromosomes from a leucocyte culture, arranged in a standard classification known as a karyotype. (Courtesy of D. H. Carr.)*

By 1966, much knowledge had accumulated concerning numerical and structural aberrations of the chromosomes, and at a conference in Chicago it was agreed to add to the original system a set of symbols which would designate certain normal and unusual chromosome features. The following brief list includes some of the more commonly used symbols. A full list appears in Hamerton (1971).

/	Diagonal line to indicate mosaicism; e.g., 46/47 indicates that the subject is a mosaic with 46–chromosome and 47–chromosome lines.
+ and −	Signs indicate presence of an extra (+) or missing (−) chromosome or part of a chromosome.

p Short arm of chromosome.
q Long arm of chromosome.
r Ring chromosome.
t Translocation.

As an example of the use of these symbols, consider a male who has 45 chromosomes, including a translocation of the long arm of a G chromosome to the long arm of a D chromosome. The symbols 45,XY,D–,G–,t(DqGq)+ describe the karyotype. Aberrations of the chromosomes are discussed in Chapters 6 and 7.

By 1971, when a standardization conference was held in Paris, the development of the new banding techniques using quinacrine fluorescence or Giemsa staining (described in the following section) had made it necessary to enlarge the standard system of nomenclature to include banding patterns. The great advantage of the banding techniques is that they make it possible to identify each human chromosome individually.

Preparation of Chromosomes for Study

Cells for chromosome analysis must be ones that will grow and divide rapidly in culture. The most readily accessible cells that meet this requirement are the white cells of the blood. (Mature red cells have no nuclei and therefore no chromosomes.) To prepare a short-term culture, a sample of peripheral blood is obtained and mixed with heparin to prevent clotting. It is then centrifuged at carefully regulated speed so that the white cells form a distinct layer. Cells of this layer are collected, placed in a suitable tissue culture medium and stimulated to divide by the addition of a mitogenic (mitosis-producing) agent, phytohemagglutinin (an extract of red bean). The culture is then incubated until the cells are dividing well, usually about 72 hours.

When the cultured cells are multiplying, a very dilute solution of colchicine is added to the medium to interfere with the action of the spindle and prevent the centromeres from dividing. Because colchicine stops mitosis at metaphase, cells in metaphase accumulate in the culture. A hypotonic solution is then added to swell the cells and to separate the chromatids while leaving their centromeres intact. The cells are fixed, spread on slides and stained, ready for microscopic examination and photography. Individual chromosomes may be cut from the photographic print, matched in pairs and mounted in accordance with the Denver classification. This process is usually called karyotyping and the completed picture is a karyotype.

Chromosome cultures derived from peripheral blood have the disadvantage of being short-lived. Long-term cultures can be derived from other tissues, such as skin, which is the easiest to obtain. A skin biopsy is a minor surgical procedure. The sample, which must include dermis, grows in culture. The chief cell type in skin culture is the fibroblast, an elongated, spindle-shaped cell capable of continuous multiplication *in vitro* for many generations. These cells can be used for chromosome analysis and for a variety of biochemical and histochemical studies. Fetal cells obtained by the procedure of amniocentesis (page 346) can be cultured by a similar technique.

When the chromosomes of a cell have been karyotyped, it is possible to determine whether they are normal in number and structure. The chromosome number is found simply by counting, but a structural abnormality may be difficult to detect, and only quite large alterations can be identified; small ones remain below the limit of analysis. The new banding techniques for chromosome study (see later) have greatly increased the probability of identifying specific chromosomal aberrations.

Classification of Chromosomes by Centromere Position

The location of the centromere, or primary constriction, is a constant feature of each chromosome. Human chromosomes can be classified by centromere position into three groups. If the centromere is central the chromosome is metacentric (or mediocentric); if the centromere is somewhat off-center the chromosome is submetacentric (submediocentric); and if the centromere is near one end the chromosome is acrocentric (Fig. 2–5). In man, chromosomes 1, 3, 16, 19 and 20 are metacentric, or nearly so; the D(13–15) and G(21, 22 and Y) chromosomes are acrocentric; and other chromosomes are submetacentric. A fourth type, telocentric, with the centromere at one end of the chromosome, does not occur in man.

Satellited Chromosomes

The human acrocentric chromosomes (except the Y) have small masses of chromatin attached to their short arms by narrow stalks (secondary constrictions). These masses, which are called **satellites,**

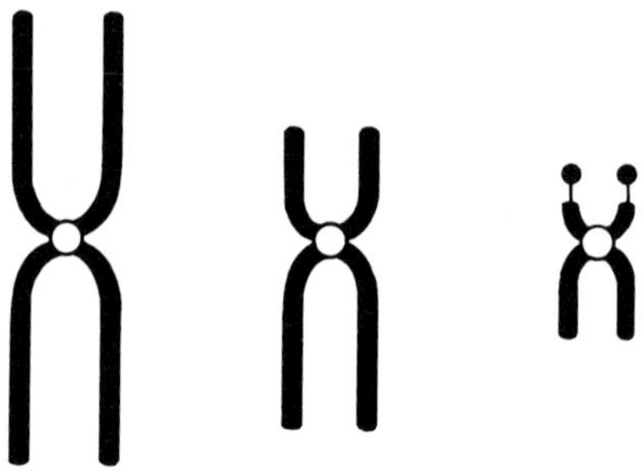

Figure 2–5 *Diagrammatic representation of the three types of human chromosomes. Left to right: metacentric, submetacentric, and acrocentric. Note the satellites on the short arms of the acrocentric chromosome.*

appear to be composed of tightly coiled chromosomal material attached to the rest of the chromosome by a stalk of less tightly coiled chromatin. In metaphase spreads, satellited chromosomes are often seen in **satellite association** (Fig. 2–6). Satellites are too irregular in appearance and staining properties to be used in the identification of individual chromosomes. They appear to play a role in the organization of the nucleoli, which are intracellular organelles composed largely of ribonucleic acid of the ribosomal type (rRNA).

Autoradiography

Identification of chromosomes by autoradiography takes advantage of the fact that different chromosomes replicate at different times during the S stage of the cell cycle. Tritiated thymidine is added to a cell culture, usually late in S, so that the chromosomes which are replicating at that point incorporate the radioactive label (Fig. 2–7). Autoradiography is useful in identification of the individual chromosomes of the B, D, E and G groups. This technique has also revealed that in females one X chromosome replicates later than the other and forms the sex chromatin (Barr body) (page 160).

Banding Techniques

Since 1970, new techniques have been developed which make it possible to identify each chromosome pair individually by its characteristic pattern of cross bands, as shown by special staining methods.

Quinacrine Fluorescence. Caspersson and his colleagues (1970a) were the first to demonstrate that when chromosomes are stained with

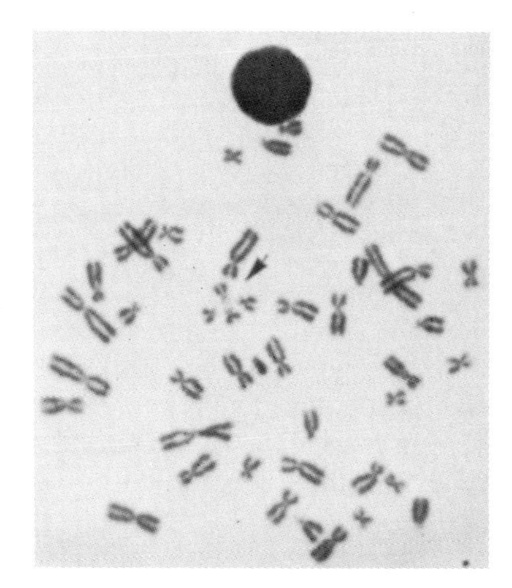

Figure 2–6 *Metaphase cell showing satellited chromosomes in satellite association. (From Hamerton, J. L. 1971.* Human Cytogenetics. *Vol. I,* General Cytogenetics. *Academic Press, New York and London, by permission.)*

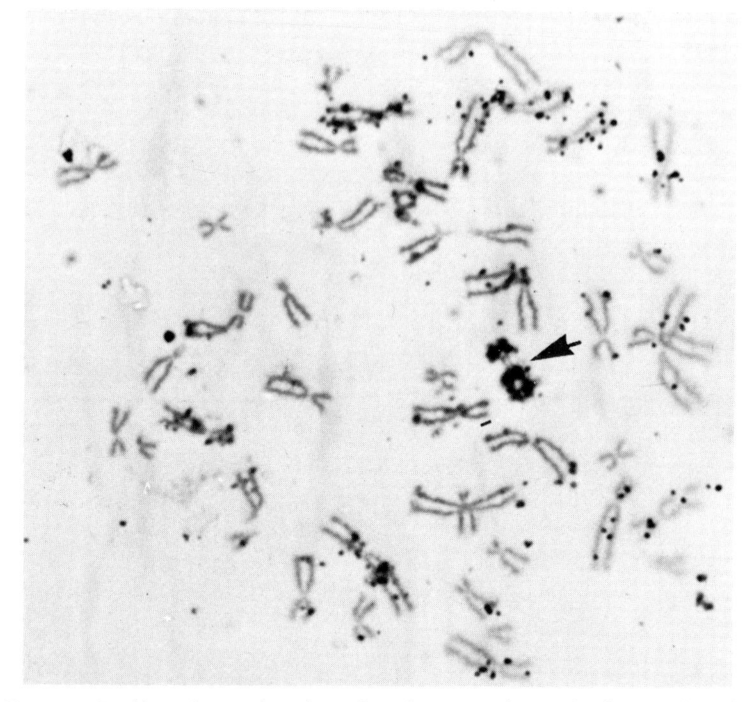

Figure 2–7 *Metaphase plate from lymphocyte culture of a human female after autoradiography. The late-replicating X chromosome is indicated by an arrow. (From Hamerton, J. L. 1971.* Human Cytogenetics. *Vol. I,* General Cytogenetics. *Academic Press, New York and London, by permission.)*

quinacrine mustard or related substances and examined by fluorescence microscopy, each pair stains in a specific pattern of dark and light bands (Q bands) (Fig. 2–8).

In particular, part of the long arm of the Y chromosome fluoresces intensely and can even be identified in interphase cells (Fig. 2–9). Thus, by examining interphase cells by conventional light microscopy for the Barr body and by fluorescence techniques for the Y chromosome, the sex chromosomes of interphase cells can often be determined, though complete karyotyping is more reliable.

Giemsa Banding. When chromosomes are treated with a protein-denaturing agent such as trypsin and then stained with Giemsa stain, they take up stain in a pattern of dark and light bands (G bands) very similar but not identical to Q bands (Fig. 2–10). Giemsa banding is simpler and less expensive than fluorescent banding and provides much the same information, so it will probably become more widely used.

Applications. The banding techniques have greatly broadened the usefulness of chromosome analysis in human and medical genetics. Now

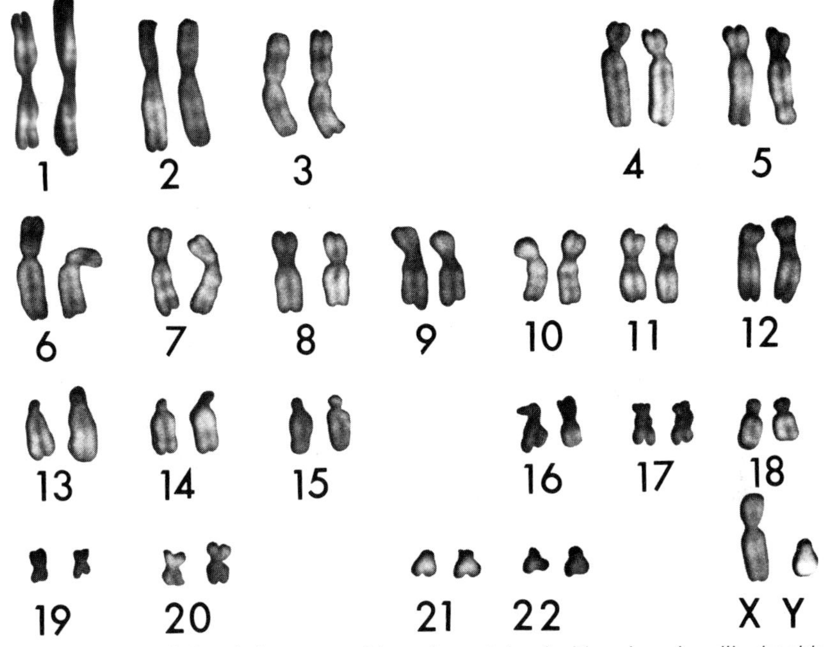

Figure 2–8 *Q bands in a normal karyotype stained with quinacrine dihydrochloride and photographed under fluorescent light. Each chromosome can be individually identified. (Photomicrograph courtesy of Linda J. Stevens.)*

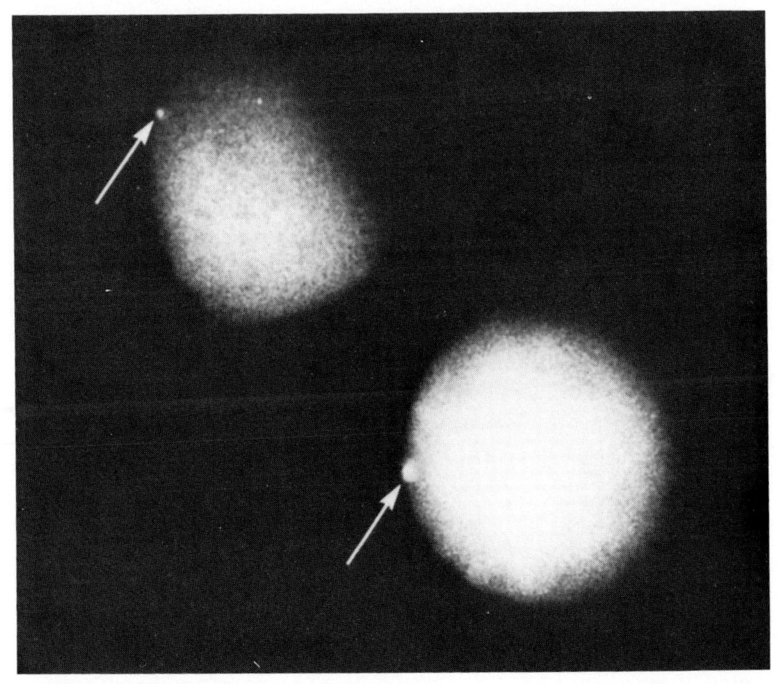

Figure 2–9 *Fluorescent Y chromosome in interphase cells. (Photomicrograph courtesy of Linda J. Stevens.)*

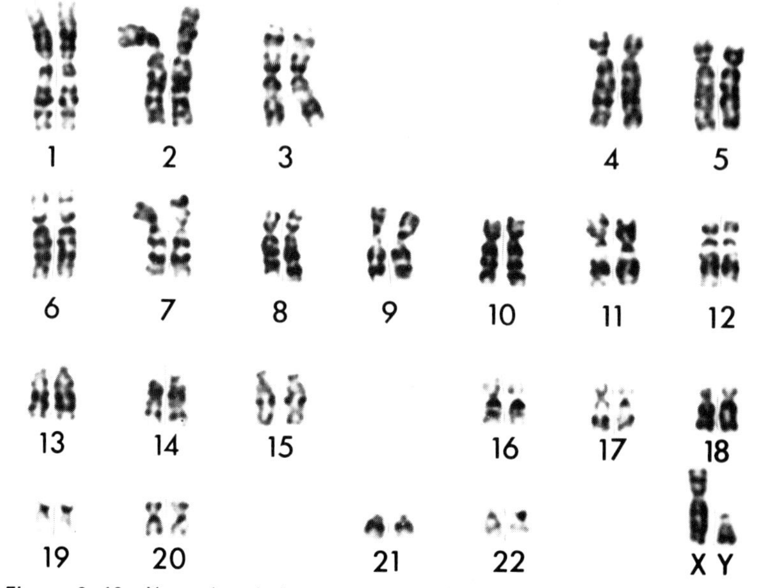

Figure 2–10 *Normal male karyotype with Giemsa banding. Each chromosome can be individually identified. (Photomicrograph courtesy of R. G. Worton.)*

that chromosomes can be individually identified, chromosomal rearrangements can be more readily recognized and the chromosomes involved can be specifically identified. As a consequence, mapping of the genes on the human chromosomes is facilitated.

One of the first applications of Q banding was the demonstration that the so-called Philadelphia chromosome (a G chromosome with deletion of part of the long arm, found in bone marrow cells of patients with chronic myelogenous leukemia) is not the same chromosome that is involved in Down's syndrome (no. 21), but the other G chromosome, no. 22 (Caspersson et al., 1970b).

Minor individual differences in banding patterns are not unusual, so it is sometimes possible to trace individual chromosomes through families. In an interesting exploitation of these differences, Uchida and Lin (1972) described quinacrine fluorescence studies of a triploid infant who survived for a few hours after birth. In this baby, the Q banding pattern showed that the extra chromosome set had probably come from the father in a diploid sperm, produced by failure of the first meiotic division in spermatogenesis. This is a very rare event.

MEIOSIS

Meiosis is the special type of cell division by which gametes are produced. Each daughter cell formed by meiosis has only half the number of chromosomes present in the parent cell, including one representative of each chromosome pair. This is in contradistinction to mitosis, in which each daughter cell is identical to the parent cell in chromosomal constitution. Some of the stages distinguished in meiosis are shown diagrammatically in Figures 2–11 and 2–12.

There are two successive meiotic divisions. In the first, the **reduction division**, homologous chromosomes pair during prophase and then segregate from one another at anaphase, each chromosome's centromere remaining intact. At the second meiotic division, as in ordinary mitosis, the centromere of each chromosome divides and the chromatids are drawn to opposite poles.

FIRST MEIOTIC PROPHASE

In prophase of the first meiotic division, five stages are distinguished:

1. **Leptotene** (Fig. 2–11A) is characterized by the first appearance of the chromosomes as thin threads. Although the DNA has duplicated prior to this stage, the threads still appear single on microscopic examination.

2. **Zygotene** (Fig. 2–11B) is characterized by pairing (synapsis) of the homologous chromosomes. The two members of a pair lie parallel to each other in point-for-point association to form bivalents. *Pairing of homologues does not occur in mitosis.*

3. **Pachytene** (Fig. 2–11C) is the stage of chromosomal thickening. The chromosomes coil tightly and stain more darkly. The bivalents (paired chromosomes) are in close association and each chromosome is now seen to consist of two chromatids, so that each bivalent consists of four strands.

4. **Diplotene** (Fig. 2–11D) is recognizable by a longitudinal separation that begins to appear between the two components of each bivalent. The centromeres remain intact, so that although the two chromosomes of each bivalent separate, the two chromatids of each individual chromosome remain together. During the longitudinal separation, the halves of each bivalent are seen to be in contact in several places called **chiasmata** (only one is shown in Fig. 2–11D), which may mark where homologous chromosomes have exchanged material between their chromatids.

5. **Diakinesis,** the final stage of prophase, is characterized by even tighter coiling of the chromosomes, which in consequence become still more deeply staining and shorter.

FIRST MEIOTIC METAPHASE, ANAPHASE AND TELOPHASE

Metaphase begins, as in mitosis, when the nuclear membrane disappears and the chromosomes move to the equatorial plate (Fig. 2–11E).

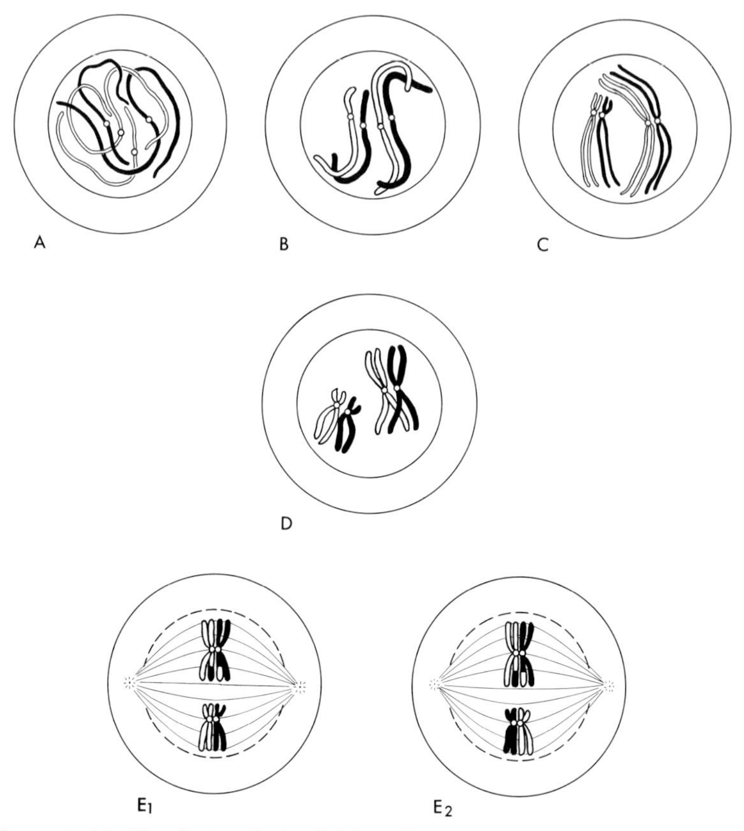

Figure 2–11 *The first meiotic division. Only two of the 23 chromosome pairs are depicted. Chromosomes from one parent are shown in outline; chromosomes from the other parent are in black. A, leptotene; B, zygotene; C, pachytene; D, diplotene;* E_1 *and* E_2*, metaphase;* F_1 *and* F_2*, early anaphase;* G_1 *and* G_2*, late anaphase,* H_{1a} *and* H_{1b}*,* H_{2a} *and* H_{2b}*, telophase. One possible combination of the two chromosome pairs is shown in* E_1*,* F_1*,* G_1*,* H_{1a} *and* H_{1b}*; the alternative combination is shown in* E_2*,* F_2*,* G_2*,* H_{2a} *and* H_{2b}*.*

Illustration Continued on Opposite Page

At anaphase the two members of each pair separate, one member going to each pole (Fig. 2–11F and 2–11G). The bivalents assort *independently of one another* so that the chromosomes received originally as a paternal set and a maternal set are now sorted into mixtures of paternal and maternal chromosomes, *but with one representative of each pair going to each pole.* The separation of homologous chromosomes is the physical basis of segregation, and the random recombination of maternal and paternal chromosomes in the gametes is the basis of independent assortment; thus, the behavior of the chromosomes at the first meiotic division provides the physical basis for mendelian heredity. The parallel between

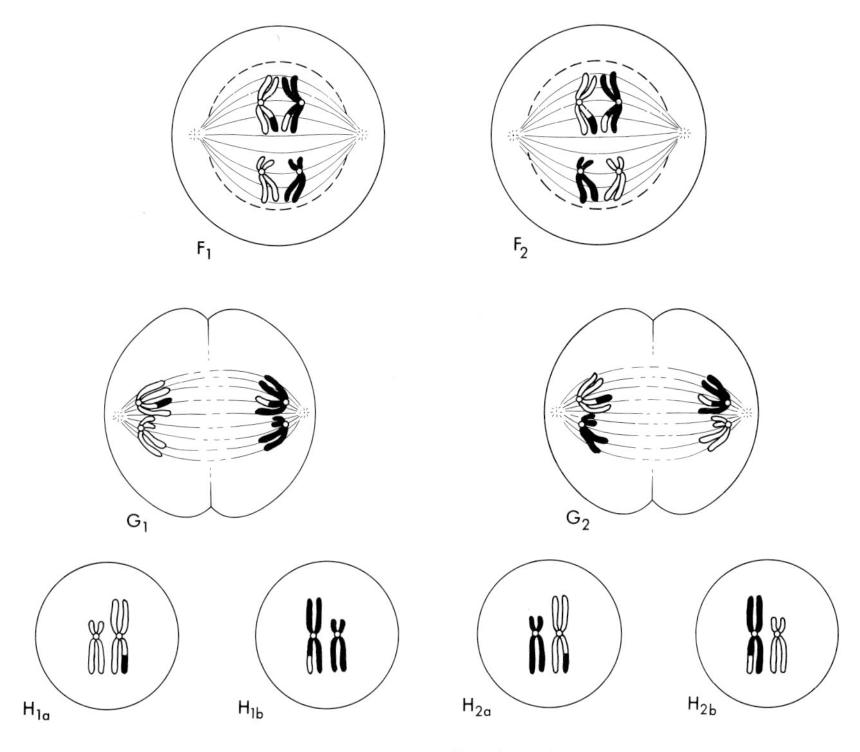

Figure 2–11 *Continued.*

the behavior of genes and chromosomes in heredity was first noted by Sutton in 1903, soon after the rediscovery of Mendel's laws.

Since the centromeres remain intact, by telophase the two chromatids of any one chromosome have passed to the same pole (Fig. 2–11H).

SECOND MEIOTIC DIVISION

The second meiotic division follows upon the first, without DNA replication. It resembles an ordinary mitotic division in that the centromeres divide and the chromatids pass to opposite poles to produce two identical daughter cells (Fig. 2–12A through C).

Thus, the end result of meiosis is the production of four daughter

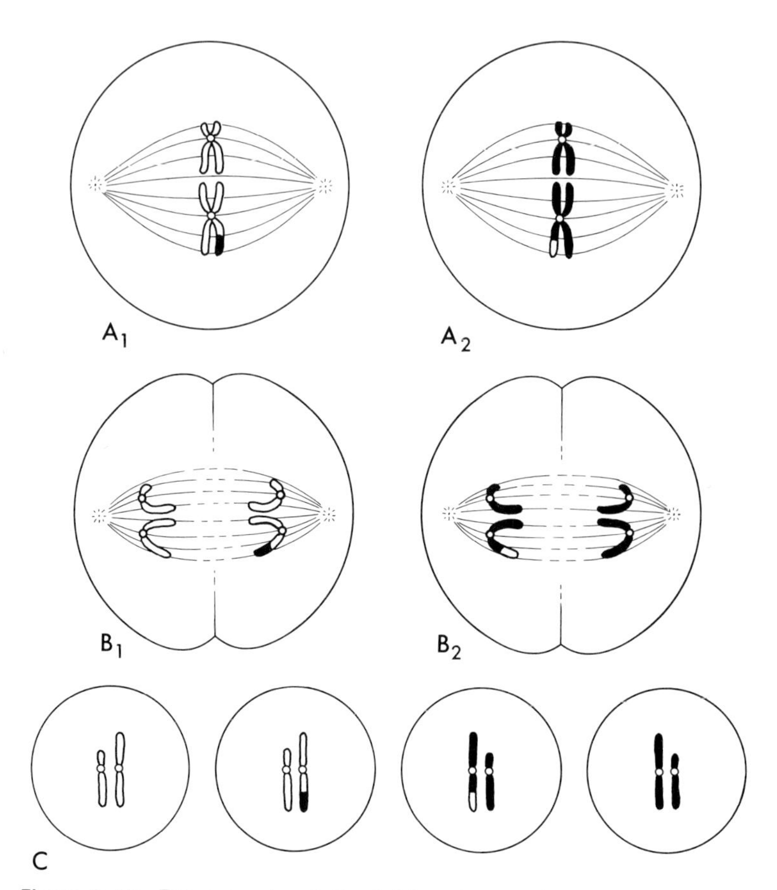

Figure 2–12 *The second meiotic division. A, metaphase; B, anaphase; C, telophase.* A_1 *and* A_2 *represent* H_{1a} *and* H_{1b} *of Figure 2–11.*

cells, formed by only one doubling of the chromosomal material. Each daughter cell is haploid; i.e., it contains only 23 chromosomes, which are unpaired.

CHIASMATA AND CROSSING OVER

Characteristic of meiosis are the chiasmata (singular, chiasma) which are seen to hold paired homologous chromosomes together from the diplotene stage through metaphase. Chiasmata are believed to mark the positions of **crossovers,** sites where homologous chromosomes have exchanged segments by breakage and recombination (Fig. 2–13). Only two chromatids, one from each member of a homologous chromosome pair, take part in any one crossover event. As meiosis proceeds, the chiasmata move to the ends of the chromosomes (terminalization) until eventually the bivalents are held together only at the ends. There are

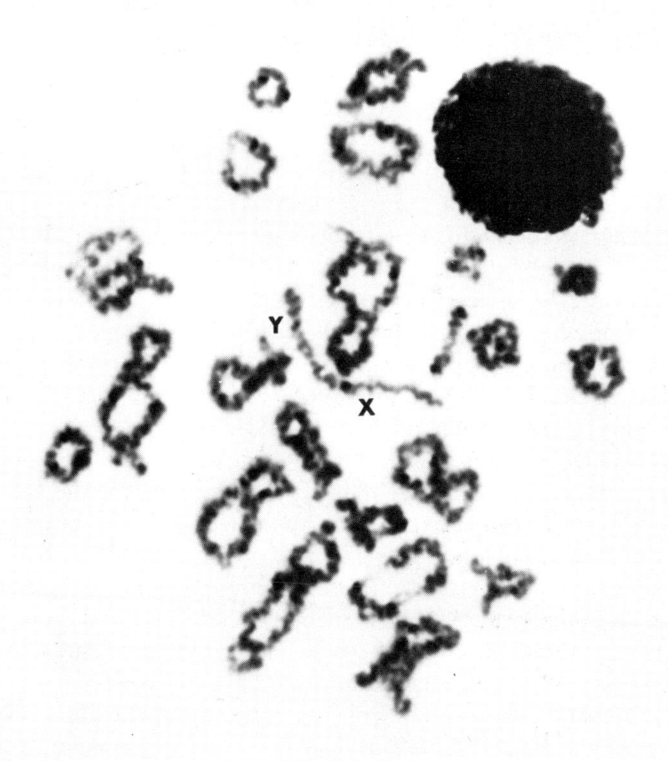

Figure 2–13 Meiosis in the human male. Note chiasmata in the bivalents, centromeres of individual chromosomes and terminal association of the X and Y. (Photomicrograph courtesy of A. Chen.)

approximately 55 chiasmata per set of meiotic chromosomes in males and 50 in females. The number increases with age in males but decreases with age in females. Though females have fewer chiasmata per cell, the frequency of recombination of linked genes in females is about 1.4 times the frequency in males. (See page 282.)

Linkage and Recombination

Genes which have their loci on different chromosomes, or far apart on the same chromosome, assort independently to the gametes and thus to the next generation. But if two genes have their loci close together on the same chromosome, they are likely to be transmitted together, not independently, and the loci are said to be *linked.* The inheritance of linked genes is therefore an exception to Mendel's law of independent assortment.

When the members of a pair of linked genes are separated by crossing over, the new gene combinations formed are called **recombinants**, as distinguished from the parental combinations. (See also Chapter 11, p. 284.)

HUMAN SPERMATOGENESIS

All stages of spermatogenesis occur in the seminiferous tubules of the testes of the mature male. At the periphery of the tubule are the earliest cells of the series, the spermatogonia. Clermont and Leblond (1959) have identified in the monkey five different stages of spermatogonial development, which progress from the most primitive (A1) to the most advanced (B3) by a series of mitoses. The final stage, by mitosis, forms **primary spermatocytes**.

Heller and Clermont (1963, 1964), using radioactive thymidine in human volunteers, have shown that spermatogenesis in the human is like that in lower animals, except that the stages differ in duration. The primary spermatocyte undergoes the first meiotic division to produce two **secondary spermatocytes**, each with 23 chromosomes; each secondary spermatocyte quickly undergoes a second meiotic division to form two **spermatids**, each with 23 chromosomes. The spermatids mature without further division to become **spermatozoa**.

Since human material is difficult to obtain, the exact duration of the first and second meiotic divisions has not been determined; but it is known that the development of primary spermatocytes from the pre-leptotene stage into mature sperms takes about 48 days, and that the two meiotic divisions are completed within 16 to 32 days. The total time from primitive spermatogonium to mature sperm is probably about 74 days (Fig. 2–14).

In rats, severe radiation (approximately 300 r) produces temporary sterility, apparently as a result of its effect upon the spermatogonia. Cells which have developed beyond the spermatogonium stage continue to mature, but the sperm count gradually drops to zero. The spermatogonia slowly recover, and spermatogenesis then proceeds until, by 65 days after irradiation, mature sperm are again found (Partington et al., 1962).

The recovery of the spermatogenic process is much slower in man than in animals. Oakes and Lushbaugh (1952) record the case of a man who received an accidental exposure of about 390 r. Following initial radiation sickness, his sperm count dropped to zero and spermatogenesis was shown by biopsy to be completely suspended. It was not until eight months after the accident that the spermatogonia began to recover and not until the second year that sperms could be found in the ejaculate. After three and one-half years the patient's sperm count reached 17

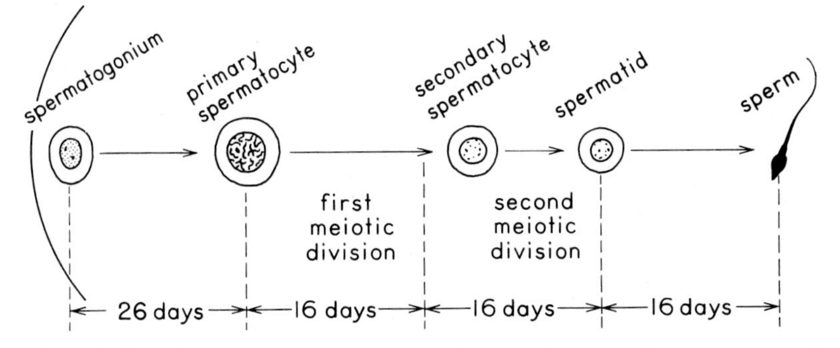

Figure 2–14 *Spermatogenesis in man. The total duration of spermatogenesis is believed to be 74 ± 5 days including the several mitotic divisions which occur before meiosis and culminate in the formation of primary spermatocytes. Primary spermatocytes, by meiosis I, produce secondary spermatocytes, which immediately undergo meiosis II to form spermatids. Spermatids, without further division, mature into sperm which are released from the germinal epithelium. The period from the beginning of meiosis I to the release of mature sperm is about 48 days. (Based upon data of Heller and Clermont, 1963, 1964.)*

million per cc. (60 million is considered normal), and about four years after the accident he became a father. Clearly, recovery is much slower in humans than in animals, but it can nevertheless be very extensive and, at least in this case, complete enough for reproduction.

Male meiotic chromosomes can be studied in material obtained by testicular biopsy. Figure 2–13 shows a human male cell in metaphase of meiosis. Note the 22 pairs of autosomal bivalents, and the X and Y, which are paired in tandem by the ends of their short arms.

HUMAN OOGENESIS

The ova develop from **oogonia**, which are located in the cortical tissue of the ovary. Each oogonium is the central cell in a developing follicle, and by about the third month of intrauterine development the oogonia of the embryo start to differentiate into **primary oocytes**. By the time of birth the primary oocytes have already reached prophase of the first meiotic division and the DNA has replicated (Ohno et al., 1962). The primary oocytes remain in "suspended prophase" (dictyotene) for many years until sexual maturity is reached. Then, as each individual follicle starts to mature, the suspended first meiotic division resumes, and it is completed at about the time of ovulation. This may be 12 to 50 years after the beginning of meiosis.

The primary oocyte completes meiosis in such a way that, while each daughter cell receives 23 chromosomes, one receives most of the cytoplasm and becomes the **secondary oocyte**, while the other becomes a **polar body**. The second meiotic division commences almost immediately and proceeds as the ovum passes into the Fallopian tube. This division is not completed until after fertilization, which usually occurs before the ovum reaches the uterus. The second meiotic division gives rise to the mature **ovum** and a second polar body. The polar bodies are believed to be incapable of forming embryos, though rare exceptions may occur.

The differences between oogenesis and spermatogenesis have genetic significance. The long duration of meiotic prophase in females may be causally related to the increasing risk of meiotic nondisjunction (failure of paired chromosomes to disjoin) with increasing maternal age, whereas the opportunity for error in the numerous replications of the genetic information that take place during spermatogenesis in adult males may account for the fact that fathers of new autosomal mutants are often above the average paternal age.

FERTILIZATION

At fertilization, a sperm penetrates the ovum completely and its head rounds up to form the **male pronucleus**. Normally only a single sperm enters the ovum. At this time the second meiotic division is being completed, and the **female pronucleus** is being formed. The two pronuclei lose their membranes, and the combined total of 46 chromosomes replicates as at any mitosis to form two 46-chromosome daughter cells. The new zygote has now begun the long series of divisions which will eventually produce the mature individual.

EXPERIMENTAL STUDIES OF HUMAN REPRODUCTIVE CELLS

Fertilization of a human egg *in vitro* has already been accomplished. The next step, reimplantation of a fertilized egg in the uterus to continue the egg's development, has not yet been reported for human eggs, but should be quite feasible. These techniques could be of practical value in permitting some sterile women to bear children and might even make possible the correction of genetic defects at a one-cell or few-cell stage by genetic engineering. It is far from certain, however, that the techniques themselves would not be teratogenic. The long-range social and ethical problems associated with this kind of research are currently of great concern to medical scientists.

GENERAL REFERENCES

Hamerton, J. L. 1971. *Human Cytogenetics*. Vol. 1, *General Cytogenetics*. Academic Press, Inc., New York and London.

Swanson, C. P., Merz, T., and Young, W. J. 1967. *Cytogenetics*. Prentice-Hall, Inc., Englewood Cliffs, New Jersey.

PROBLEMS

1. At a certain locus an individual is heterozygous, having the genotype A/A'.
   (a) With respect to this locus, what are the genotypes of his gametes?
   (b) When does segregation of the alleles A and A′ take place:
      (1) if there is no crossing over between this locus and the centromere of the chromosome?
      (2) if there is a crossover between this locus and the centromere?

2. Assume that a man is heterozygous at two loci on different chromosomes, having the genotype A/A' B/B'. With respect to these loci, what genotypes are possible in his spermatozoa, and in what proportions?

3. How many different genotypes are possible in the ova of a woman who is:
   (a) heterozygous at three independent loci?
   (b) heterozygous at five independent loci?
   (c) heterozygous at 23 independent loci?

4. Assume that a man inherits the linked genes ABCD from his father and the genes A′B′C′D from his mother.
   (a) At how many of these loci is the man heterozygous?
   (b) What are the possible genotypes of his spermatozoa?
   (c) Which of these genotypes are parental combinations?
   (d) Which genotypes represent recombinations (by crossing over) of the parental alleles?

3

Molecular Genetics

Molecular genetics is concerned with the reinterpretation of mendelian genetics in molecular terms. It deals with the genetic material, **deoxyribonucleic acid** (DNA); the **replication** of DNA to produce more DNA; the **transcription** of DNA into **ribonucleic acid** (RNA); and the **translation** of RNA into protein in the form of polypeptide chains. The DNA–RNA protein relationship, summarized in the following diagram, is often called the **central dogma** of molecular genetics.

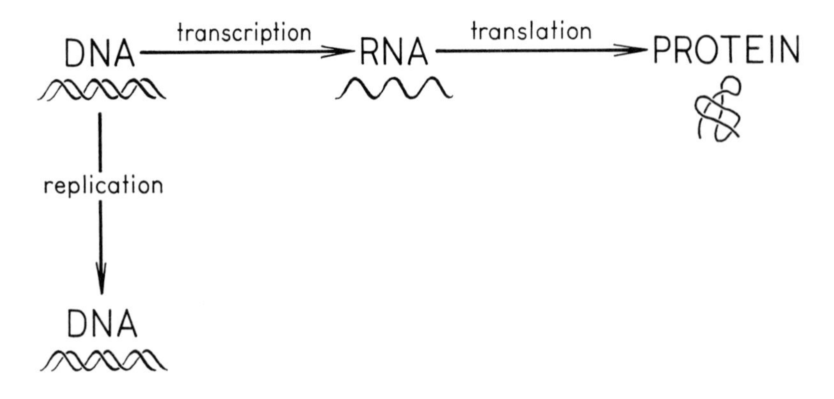

Much of our present knowledge of molecular genetics has come from studies with microbial organisms, and no attempt will be made here to provide a detailed account of this large and rapidly growing field. Excellent references are those of Hartman and Suskind (1969) and Watson (1970). This chapter will merely review some basic facts, and spotlight some recent developments of particular significance in medical genetics.

Genetic information is coded in DNA by the sequence of the four nitrogenous bases which form part of the DNA molecule. The **genetic code** may be compared to a four-letter "alphabet" (the four bases), from which three-letter "words" (sequences of three bases) are spelled out. Each three-base sequence (**triplet**) codes for a specific amino acid. A linear series of triplets specifies the linear order in which amino acids are arranged into a polypeptide chain. A **gene** may be defined as a sequence of triplets which contains the code for one polypeptide.

The polypeptide may in itself be a complete protein (e.g., myoglobin), but in many cases two or more polypeptides combine to form a protein (e.g., hemoglobin).

NUCLEIC ACIDS

The nucleic acids, DNA and RNA, are macromolecules (polymers) with three types of components:

1. Sugar: a five-carbon sugar (pentose). In DNA the sugar is **deoxyribose**; in RNA it is **ribose.**

2. Phosphate.

3. Base: a nitrogen-containing base, which may be a **purine** or a **pyrimidine**. In DNA, the two purine bases are **adenine** (A) and **guanine** (G), and the two pyrimidines are **thymine** (T) and **cytosine** (C). In RNA, uracil (U) replaces thymine. The bases are capable of pairing by hydrogen bonds. Normally A pairs only with T (A = T) in DNA, and G pairs only with C (G ≡ C) in both DNA and RNA. The U of RNA pairs with A (A = U).

A base, a sugar and a phosphate combine to form a **nucleotide.** (A nucleoside is composed of a base and a sugar only.) Nucleotides combine to form nucleic acids, which are **polynucleotides.** A polynucleotide consists of a backbone of alternating sugar and phosphate from which the bases project. Each base is attached to a sugar.

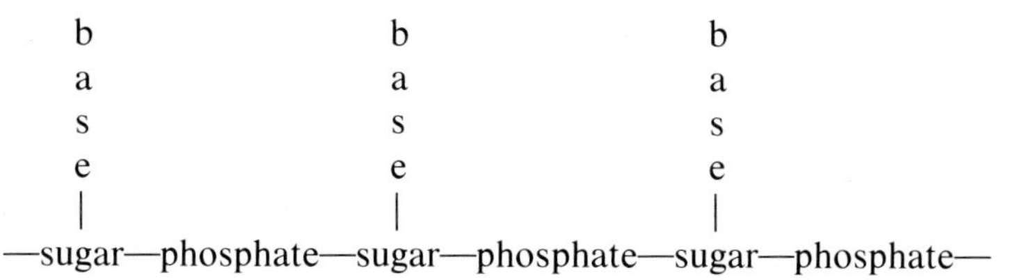

We will abbreviate this to:

```
 b   b   b                  s = sugar
 |   |   |                  p = phosphate
-s-p-s-p-s-p-               b = base
```

```
 b
 |
```

Each –s–p-unit is one nucleotide. Since there are four different bases, there are four different nucleotides. In DNA the number of nucleotides containing A is always equal to the number containing T, and the number containing G is always equal to the number containing C. However, the A + T/G + C ratio varies in the DNA of different organisms within rather wide limits (0.5 to 2.5). In man the ratio is approximately 1.4.

DNA

The 1/1 ratio of A/T and of G/C was one piece of evidence that led Watson and Crick to propose their famous model of the structure of the DNA molecule. Another important observation was made by Wilkins and his group, who showed by X-ray diffraction studies that the DNA molecule has a spiral shape and contains more than one polynucleotide chain. For their pioneering work in this field, Watson, Crick and Wilkins shared a Nobel Prize (1961).

The Watson-Crick model of DNA is easier to draw than to describe, and the illustration that appeared in the original paper is reproduced in Figure 3–1. The molecule is composed of two long polynucleotide chains twisted to form a double spiral. The chains run in opposite directions and are held together by hydrogen bonds between A of one chain and T of the other, or between G of one chain and C of the other. There are about 10 nucleotide pairs per complete turn of the double chain. Imagine a long rubber ladder twisted around its long axis to form a double helix. Each lengthwise piece of the ladder represents alternating molecules of

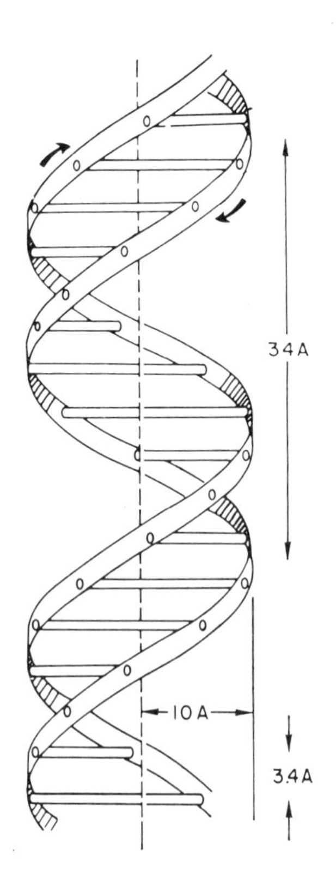

Figure 3–1 *Model of the DNA molecule proposed by Watson and Crick (1953). A = angstrom units. (From* Nature, *171:737–738.)*

sugar and phosphate and each rung represents a pair of bases. Part of the molecule, unwound, has the following structure:

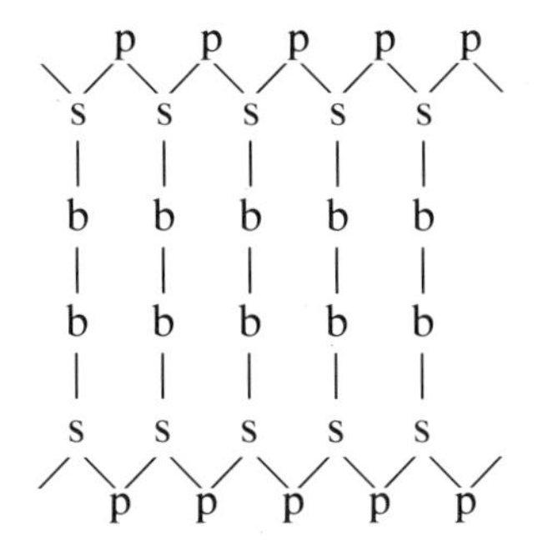

The base pairs, A = T and G ≡ C, can be in any sequence. For example:

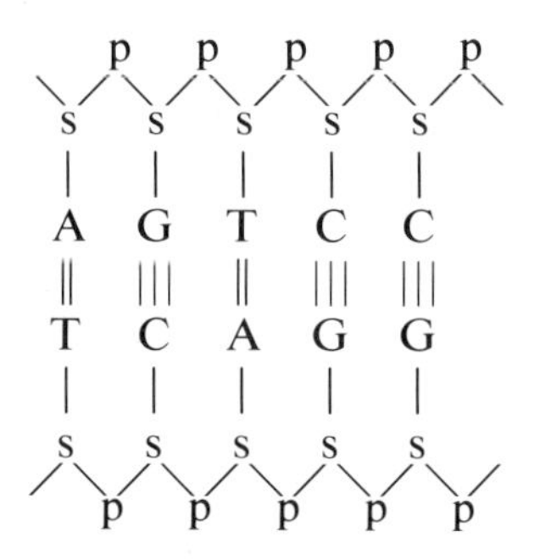

A is bound to T by two hydrogen bonds; G is bound to C by three hydrogen bonds. Because of the configuration that would be necessary if A tried to pair with G (i.e., one bond on G unsatisfied), this pairing occurs only rarely and by error. Similarly, pairing of C and T (the two pyrimidines) is very unusual.

Since A = T and G ≡ C pairing is obligatory, the parallel strands must be complementary to each other; thus if one strand reads A G T C C A, the complementary one reads T C A G G T. An important consequence of the complementary nature of the two strands is that a DNA molecule can readily replicate by separation of the strands followed by formation of two new, complementary strands (see later).

Almost infinite variations are theoretically possible in the arrangement of the bases along a polynucleotide chain. In any one position there are four possibilities; thus there are $4 \times 4 \times 4 = 64$ possible triplet combinations of bases, or codons.

THE GENETIC CODE

One of the two functions of DNA is to direct the synthesis of **polypeptides**, which are molecules built up of amino acids in a specific linear sequence (primary structure), held together by peptide linkages. The properties of the polypeptide depend upon the order of the amino acids in its molecule.

The "one gene–one enzyme" hypothesis proposed by Beadle and Tatum some 30 years ago is interpreted in modern terms as "one gene–one polypeptide chain." A gene is a part of a linear polynucleotide molecule, and a polypeptide (with respect to its primary structure, the amino acid sequence) is also a linear molecule. The sequence in which amino acids are incorporated into a polypeptide chain is dictated by the order of the corresponding triplets of bases in *one* of the two polynucleotide chains of the DNA molecule. The DNA molecule and the polypeptide molecule are said to be *colinear*.

The genetic code shown in Table 3–1 was worked out through experiments using synthetic polynucleotides. The first synthetic polyribonucleotide used as a messenger RNA was poly U (polyuracil, a sequence of nucleotides in which all the bases are U). Poly U directs the formation of a polypeptide chain composed exclusively of phenylalanine, thus demonstrating that UUU codes for phenylalanine. Since there are only 20 amino acids and 64 possible codons, most amino acids are specified by more than one codon; hence, the code is said to be *degenerate*. For instance, as Table 3–1 shows, the base in the third position of a triplet can often be either purine, either pyrimidine or, in some cases, any one of the four bases, without altering the coded message. One of the amino acids, leucine, can be coded for by six different codons.

CHARACTERISTICS OF THE GENETIC CODE

1. The code is a **triplet code**; three adjacent bases, forming a codon, code for one amino acid.

2. The code is highly **degenerate**; several codons may code for the same amino acid.

3. The code is **nonoverlapping**; any one base is part of only one codon.

4. The codons for a particular gene are in sequence, with **no spacers** between successive codons.

5. The code is (with a few possible exceptions) **universal**; the same amino acids are coded for by the same codons in all organisms studied, from bacteria to man.

6. The order of the codons in a gene **dictates the order** of specific amino acids in a polypeptide chain.

7. Three of the 64 codons (UAA, UAG, UGA) indicate the **termination** of a gene. The manufacture of a polypeptide stops when the "reading" of the RNA reaches one of these codons.

8. The codon AUG **initiates** a gene, but the same codon codes for

TABLE 3–1 THE GENETIC CODE*

First Base	Second Base: *U*	Second Base: *C*	Second Base: *A*	Second Base: *G*	Third Base
U	UUU phe	UCU ser	UAU tyr	UGU cys	U
	UUC phe	UCC ser	UAC tyr	UGC cys	C
	UUA leu	UCA ser	UAA stop	UGA stop	A
	UUG leu	UCG ser	UAG stop	UGG try	G
C	CUU leu	CCU pro	CAU his	CGU arg	U
	CUC leu	CCC pro	CAC his	CGC arg	C
	CUA leu	CCA pro	CAA gln	CGA arg	A
	CUG leu	CCG pro	CAG gln	CGG arg	G
A	AUU ile	ACU thr	AAU asn	AGU ser	U
	AUC ile	ACC thr	AAC asn	AGC ser	C
	AUA ile	ACA thr	AAA lys	AGA arg	A
	AUG met	ACG thr	AAG lys	AGG arg	G
G	GUU val	GCU ala	GAU asp	GGU gly	U
	GUC val	GCC ala	GAC asp	GGC gly	C
	GUA val	GCA ala	GAA glu	GGA gly	A
	GUG val	GCG ala	GAG glu	GGG gly	G

Abbreviations for amino acids:

ala alanine
arg arginine
asn asparagine
asp aspartic acid
cys cysteine

gln glutamine
glu glutamic acid
gly glycine
his histidine
ile isoleucine

leu leucine
lys lysine
met methionine
phe phenylalanine
pro proline

ser serine
thr threonine
try tryptophan
tyr tryosine
val valine

Other abbreviation:
stop termination of a gene

*Codons are shown in terms of messenger RNA. The corresponding DNA codons are complementary to these.

methionine. How a single codon fulfills two such different functions is not yet understood.

EVIDENCE THAT DNA IS THE GENETIC MATERIAL

The role of DNA in genetic transmission has been worked out mainly by experiments with microorganisms. Microbial genetics has many advantages, including the exceedingly short generation time, the ease of manipulation and the direct relationship between the gene and its effect. The chief disadvantages are that cells of microorganisms do not differentiate as do those of higher organisms, and that it is uncertain to what extent the genetic mechanisms known in microorganisms apply to man.

BACTERIAL TRANSFORMATION: THE GRIFFITH EXPERIMENT

The pneumococcus (*Diplococcus pneumoniae*) is the causative organism of one of the most common types of pneumonia. It can exist in two phenotypes:

1. The S form: smooth and encapsulated; **virulent.**
2. The R form: rough and unencapsulated; **benign.**

There are several varieties of the S form (designated IIS, IIIS and so forth). By mutation, each can give rise to an R form. IIS can mutate to IIR, and IIIS can mutate to IIIR. These mutations change the virulent S forms to the relatively benign R forms. Occasional reverse mutations of R to S forms have also been observed.

In 1928 Griffith showed that mice infected with live, nonvirulent IIR pneumococci would die if also given heat-killed IIIS bacteria (which, if alive, would alone have killed the mice). Live IIIS pneumococci were recovered from the dead mice. The IIIS could not have arisen by mutation of IIIR, since no IIIR was present. Instead, the IIR had been **transformed** to IIIS. The transformation was permanent and could be transmitted to the progeny (Fig. 3–2).

$$\text{Live IIR} + \text{dead IIIS} \rightarrow \text{Live IIR and live IIIS}$$

BACTERIAL TRANSFORMATION: THE AVERY-MacLEOD-McCARTY EXPERIMENT

In 1944 Avery and his group made it clear that the substance which transformed the live IIR to IIIS was DNA. They did this by adding to

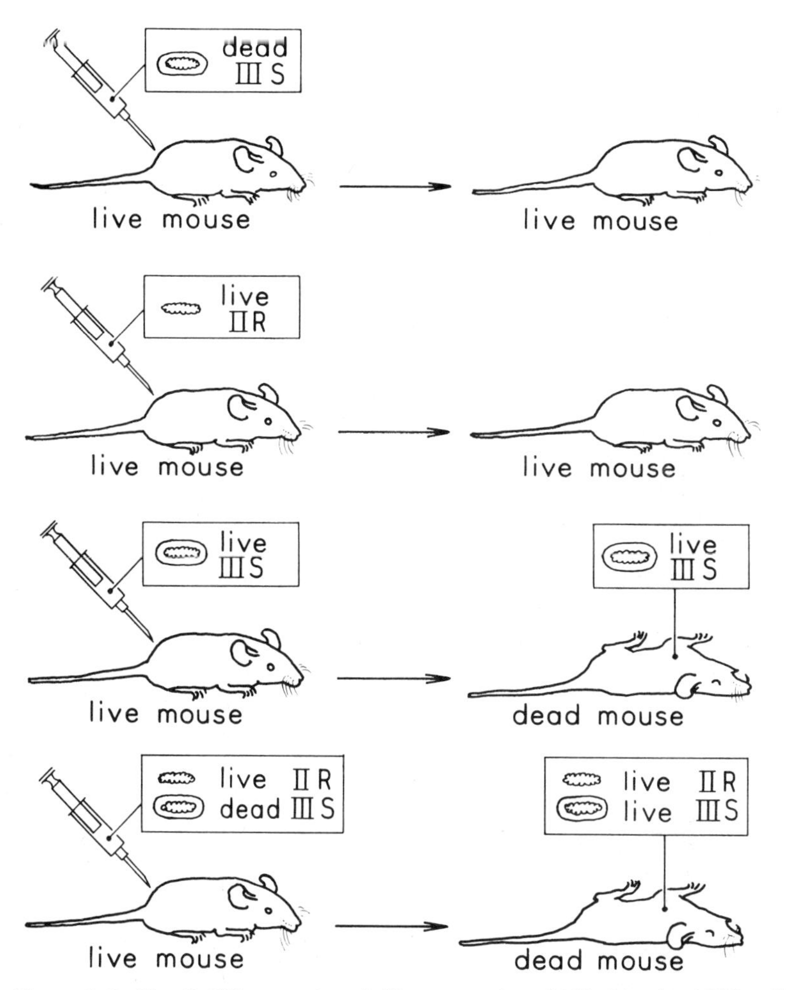

Figure 3–2 *The Griffith experiment. The mouse is not killed by dead IIIS or live IIR pneumococci given alone, but is killed by live IIIS. Live IIR may be partially transformed by dead IIIS to produce some live IIIS, which kills the mouse.*

IIR, not dead IIIS but instead an extract containing DNA from IIIS. Some of the mice injected with this mixture died, and live IIIS was recovered from them. The control experiment, using an identical extract to which the enzyme DNase had been added to destroy the DNA, showed that removal of the DNA from the extract removed its ability to transform IIR to IIIS. Thus, the part of the extract responsible for the transformation of IIR to IIIS was DNA. The interpretation was that some DNA of the IIIS type had been incorporated into the genetic material of IIR, bringing about a permanent transformation. In effect, a genetic change had been produced by DNA:

$$\text{IIR} + \text{DNA of IIIS} \rightarrow \text{IIR} + \text{IIIS}$$

BACTERIOPHAGE REPRODUCTION: THE HERSHEY-CHASE EXPERIMENT

Additional evidence that DNA is the genetic material came in 1952 from Hershey and Chase's work with bacteriophage (phage, for short). This organism is a small bacterial virus, and the one in question, T4 phage, is one of a number which infects the common colon bacillus *Escherichia coli* (*E. coli,* for short). A phage has a head which contains DNA in a protein coat, and a tail by which it attaches to the bacterium it attacks. When T4 phage infects a culture of *E. coli,* a phage particle becomes attached to a bacterium, injects material into it, takes over its metabolism and reproduces. The bacterial cell lyses, and new phage particles are released to continue the cycle (Fig. 3–3).

Hershey and Chase designed an experiment to demonstrate which part of the phage was important in the reproduction of the phage. *E. coli* was grown in two kinds of culture, one containing radioactive phosphorus (P^{32}), which specifically labels DNA but not protein, the other containing radioactive sulfur (S^{35}), which specifically labels protein but not DNA.

T4 phage was allowed to grow on the two kinds of culture, so that two types of phage were produced, one with label in its DNA and the other with label in its protein coat. When the DNA-labeled phage was

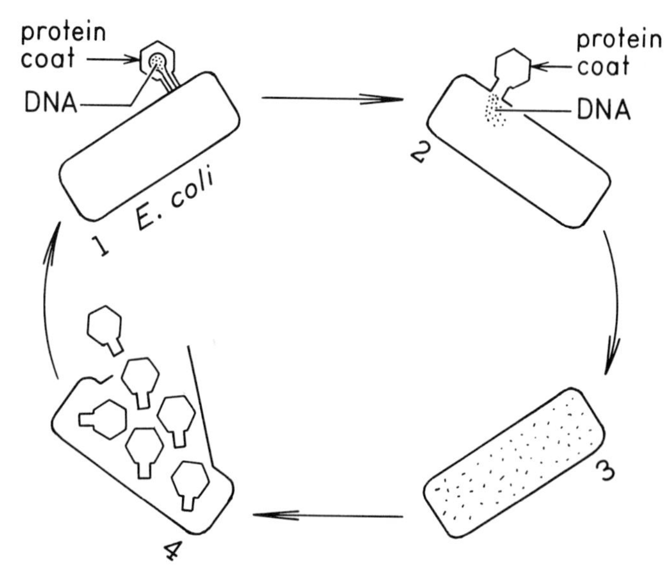

Figure 3–3 *Bacteriophage attacking an* E. coli *bacillus: (1) The phage particle attaches itself to the bacterial cell; (2) phage DNA is injected into the bacterium; (3) the DNA spreads through the cell; and (4) the bacterium then produces more phage particles, which are released when the cell lyses.*

then grown on unlabeled *E. coli*, the label appeared *inside the E. coli*; but when the protein-labeled phage was used, the label remained *outside* the bacillus. In other words, the material which entered the cell and allowed the phage to reproduce was DNA; protein took no part in the reproductive process.

REPLICATION OF DNA

DNA must be able to replicate with great accuracy so that the genetic information does not become garbled during its transmission to the next generation. Watson and Crick suggested a simple mechanism for this replication. The two chains unwind and separate, and each serves as a template on which the missing partner can be reconstructed from nucleotides present in the cell, these nucleotides being united by the action of DNA polymerase. Because the two threads are complementary, each will rebuild a second thread identical to the one from which it has been separated, and the end result will be two complete molecules, each identical to the original (Fig. 3–4).

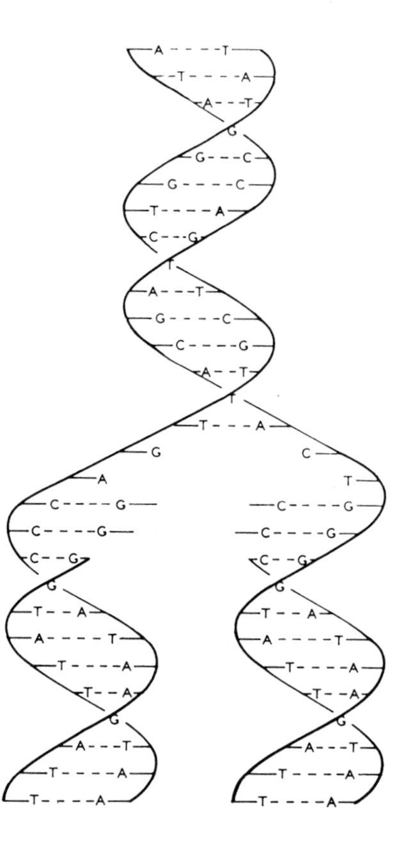

Figure 3–4 *Replication of DNA. Note that the original molecule (top) unwinds and the halves separate. The new half-molecules of DNA are formed on the old halves by A = T and C ≡ G pairing. The result is two complete molecules, each identical to the original molecule. (From Sutton 1961.* Genes, Enzymes and Inherited Diseases. *Holt, Rinehart and Winston, Inc., New York.)*

REPAIR OF THE DNA MOLECULE

Though formerly it was believed that breaks in the DNA strand would be lethal because they would inevitably lead to failure of replication of the molecule, it is now known that repair of the DNA molecule can take place. If one strand of the double helix is broken, for example, by X-rays or ultraviolet (UV) irradiation, the damaged area may be excised and a correct sequence of nucleotides may be inserted into the damaged chain, on the template provided by the unbroken strand. The significance of repair synthesis is that it provides a degree of protection against chromosome breakage by such agents as radiation, and thus may serve an evolutionary function.

Xeroderma pigmentosum is a genetic disorder, inherited as an autosomal recessive, in which cells lack completely or almost completely the ability to perform the first step in repair of damage to DNA caused by UV irradiation (Cleaver, 1968). As a consequence, the skin of affected individuals is sensitive to sunlight, and may develop cancers.

SEMICONSERVATIVE REPLICATION: THE EXPERIMENT OF MESELSON AND STAHL

Confirmation of the Watson-Crick hypothesis of DNA replication came from an experiment, by Meselson and Stahl (1958), which made use of a heavy isotope of nitrogen, N^{15}. Bacteria were allowed to reproduce in a medium in which the only source of nitrogen was N^{15}, so the DNA of the bacteria became labeled with heavy nitrogen. They were then transferred to a medium containing ordinary N^{14} and allowed to reproduce for two further generations. The proportion of N^{15} and N^{14} in the bacterial DNA was measured after each generation. The consequences are shown in Figure 3–5. After one generation of replication in the N^{14} medium, the weight of the "hybrid" DNA was halfway between that of DNA containing N^{14} and that of DNA containing N^{15}. After a second generation, a 3:1 distribution of light to heavy DNA was found. These observations were to be expected if, at the first replication, each strand marked with N^{15} replicated itself using the only N available, which was N^{14}. Half the strands would then be light and half heavy. At the second replication the strands would have separated again and each would have replicated a second time, again using only N^{14}. This time one "heavy" and three "normal" DNA strands would result from each single DNA strand that had existed two generations earlier. Since replication involves the preservation of the old DNA strands and the formation by each of a new strand, it is said to be **semiconservative**.

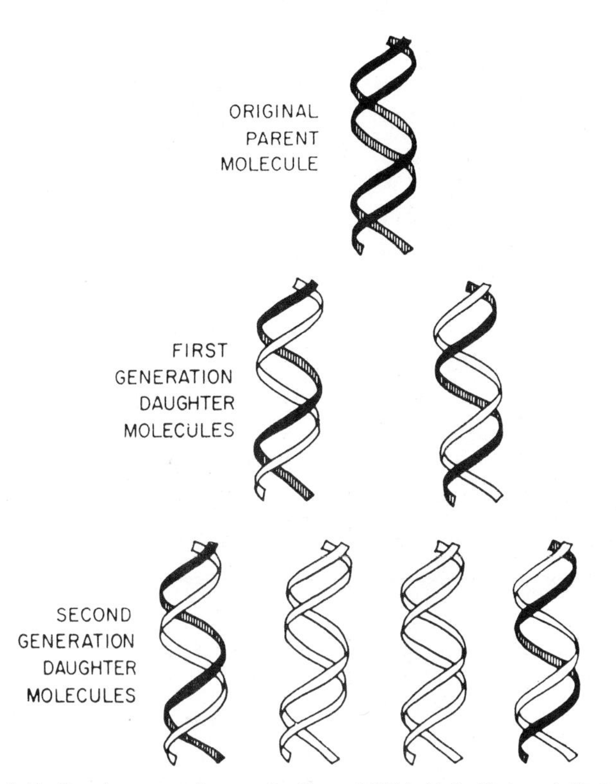

Figure 3–5 *Semiconservative replication of DNA. Note that each first-generation daughter molecule consists of one old and one new strand of DNA. The same is true for second (and succeeding) generations, but only the labeled strands are detectable. For further description, see text. (From Meselson and Stahl 1958.* Proc. Nat. Acad. Sci. *44:671–682.)*

ORGANIZATION OF DNA IN CHROMOSOMES

As we have seen, DNA has a linear structure, and genes are known to be located in a linear fashion along the chromosome. It is therefore tempting to speculate that the chromosome consists of a number of strands of DNA running parallel to the long axis of the chromosome. Unfortunately, there are difficulties in this model. For example, if the double helix of DNA simply stretched in a straight line the length of the chromosome, the longest human chromosome in interphase cells would have to be many centimeters long.

The exact arrangement of DNA in the chromosome is still a matter of speculation. One possibility, suggested by Taylor (1963), is shown in Figure 3–6. More recently (1970) Ris and Kubai have reviewed the

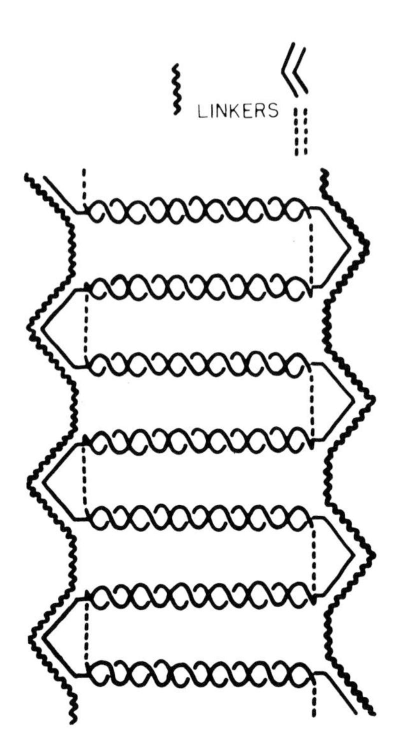

Figure 3–6 A possible arrangement of DNA in the chromosome. (From Taylor 1963. Molecular Genetics. Part I. Academic Press, Inc., New York and London, by permission.)

current knowledge concerning it. Much of the DNA in the mammalian chromosome is probably redundant, yet it must all be packaged in such a way that it is readily accessible. DNA in its familiar double helix apparently combines with histone, so that each fraction forms about 50 per cent of the chromosome. The DNA–histone (DNH) complex forms a basic fiber about 100Å thick, in which the DNH is coiled upon itself, and the whole complex is maintained in this coiled condition by an unstable "cross-bridging." If the bridging is broken, the combined DNH fiber stretches out, the histone retaining its properties and the DNA becoming accessible for replication or transcription. In the intact nucleus the DNH actually appears to be in fibers about 250Å wide, representing either parallel 100Å fibers coiled together, or a single supercoiled 100Å fiber.

THE STRUCTURE AND FUNCTION OF RNA

The genetic code is contained in DNA in the chromosomes of the cell nucleus, but polypeptide synthesis takes place in the cytoplasm, in association with cytoplasmic organelles called **ribosomes**. The link between DNA and polypeptide is RNA.

RNA is a single-strand molecule, very similar to one strand of the double-strand molecule of DNA. All types of RNA are synthesized on a DNA template by direct **transcription** of the DNA code into a complementary RNA code.

Transcription depends upon RNA polymerase, a complicated molecule, part of which recognizes the specific DNA triplet that signals the start of a gene. The polymerase "reads" the DNA gene, meanwhile forming a complementary molecule of RNA. When the polymerase reaches a terminator codon, the reading stops and the completed RNA molecule is released. Figure 3–7 shows electron micrographs of this process in the oocytes of *Triturus*.

MESSENGER, TRANSFER AND RIBOSOMAL RNA

Three kinds of RNA take part in protein synthesis: **messenger** RNA, **transfer** RNA and **ribosomal** RNA. Messenger RNA transmits the genetic information from the DNA molecule to the cytoplasm; transfer RNA acts as the "adaptor" which brings amino acids into place on the growing polypeptide chain; and ribosomal RNA serves a nonspecific function in connection with the ribosomes, which are found in large numbers in cells actively engaged in synthesizing protein.

Messenger RNA

Messenger RNA (mRNA) is formed in the nucleus upon a template composed of one of the two strands of the DNA double helix, and is structurally complementary to this strand. It passes from the nucleus to the cytoplasm, where it is found in association with ribosomes. The mRNA code, transcribed from the DNA of the chromosomes, prescribes the sequence in which amino acids are incorporated into a polypeptide.

Transfer RNA

Transfer RNA (tRNA), sometimes called sRNA because it is soluble, or adaptor RNA because of its structure, has a molecule much smaller than that of mRNA. Its function is to transfer amino acids from the cytoplasm to their specific places along the mRNA template. These amino acids must first be activated by special **amino acid activating enzymes**, one for each amino acid. An activating enzyme recognizes a special site on the amino acid molecule and also a special "recognition site" on the tRNA molecule. At another site on the tRNA molecule is an **anticodon**, which recognizes a specific codon on the mRNA chain.

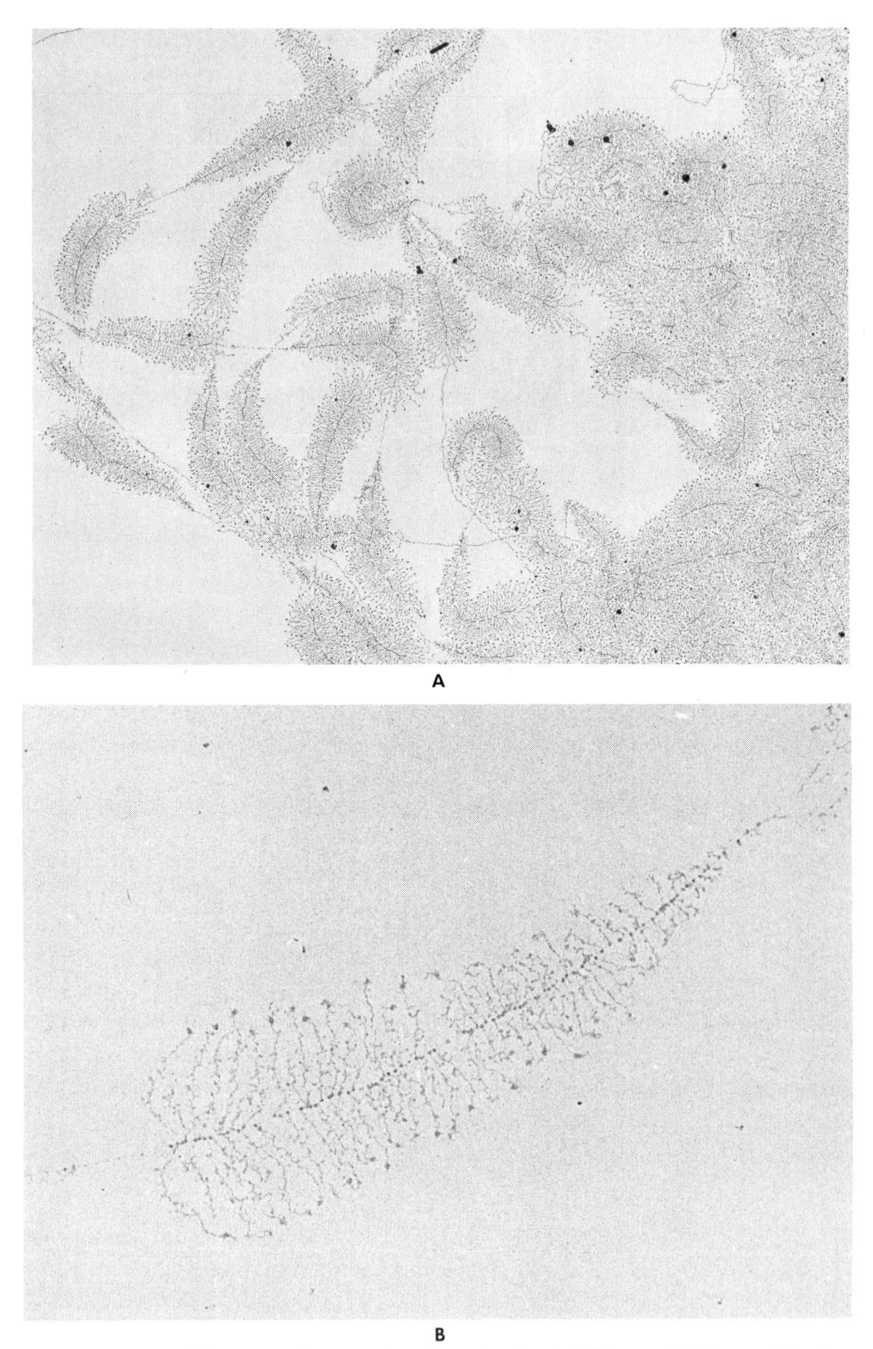

Figure 3–7 *Electron micrographs of synthesis of RNA on a DNA template in a "lampbrush" chromosome of an oocyte of the newt* Triturus, *in which only a small part of the DNA is unwound and functional.* A, *DNA molecules (long strands) each with many RNA molecules in the process of transcription projecting from it.* B, *One DNA strand enlarged to show RNA polymerase (black dots) with RNA molecules projecting laterally from each unit of polymerase. Note gradual increase in length of RNA molecules as transcription proceeds along the DNA template. (From Miller and Beatty 1969.* J. Cell Physiol. *74, Suppl. 1:225–232, by permission).*

Since several codons may code for one amino acid, one amino acid may be carried by several different tRNA molecules. It is not yet clear whether alternative codons are equally efficient, and whether the alternative tRNA molecules are available in adequate quantities. The AA-tRNA (amino acid–transfer RNA) unit is placed in the appropriate position on the linear mRNA molecule by the matching of codon and anticodon (Fig. 3–8).

Ribosomal RNA

Ribosomes consist of protein and a nonspecific RNA (**rRNA**) in about equal proportions. The role of the ribosome is to adhere to a

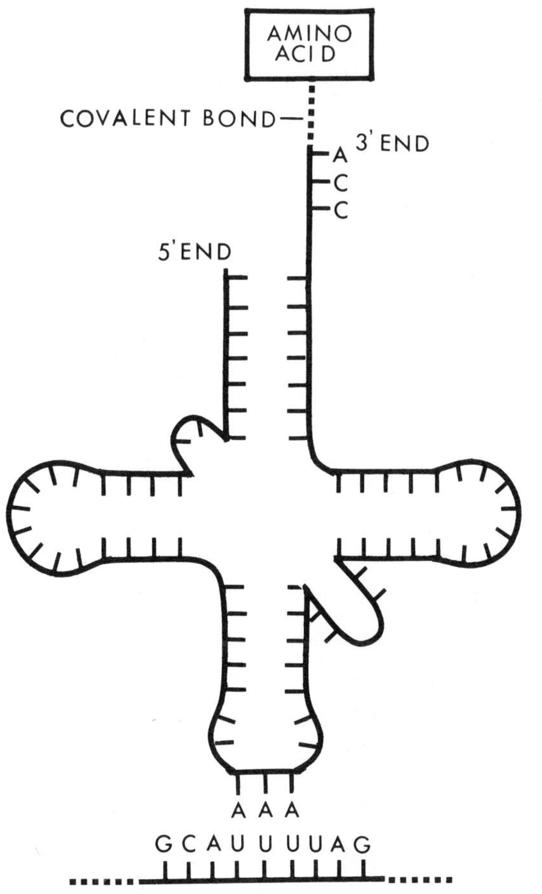

Figure 3–8 *Schematic representation of a molecule of transfer RNA. For details, see text.*

"sticky spot" on the mRNA molecule and then proceed along the mRNA, "reading" the code and bringing AA-tRNA units into line at the proper codons. Peptide bonds are then formed between the amino acids by an enzyme of the ribosome. Once the peptide bond is formed, the polypeptide chain begins to "pull away" from the ribosome (Fig. 3–9). The ribosome takes about 10 seconds to read the length of a mRNA molecule, and a single mRNA molecule may have several ribosomes reading it at once.

Preparations of rabbit reticulocytes are useful in the study of protein synthesis because reticulocytes are engaged in making only one protein, hemoglobin. In these preparations the ribosomes can be seen to cluster in groups (polyribosomes or polysomes) (Fig. 3–10).

FORMATION OF A POLYPEPTIDE CHAIN

The successive steps in the formation of a polypeptide chain, schematically shown in Figure 3–11, can now be summarized:

1. The two strands of DNA in the double helix dissociate in the area of the gene to be transcribed. Dissociation of the DNA strands appears to require no specific enzyme. Only one of the two DNA strands acts as a template for RNA synthesis.

2. Under the influence of RNA polymerase, a molecule of mRNA forms on the DNA template.

3. The mRNA moves into the cytoplasm.

4. In the cytoplasm, specific enzymes activate amino acid molecules and bring them into association with tRNA to form AA-tRNA complexes. Each enzyme activates only one kind of amino acid and attaches it to a specific tRNA, which has in some position an anticodon complementary to the appropriate codon of mRNA.

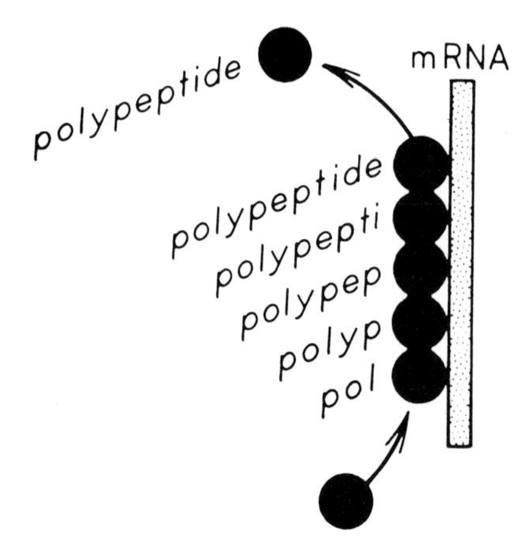

Figure 3–9 *Schematic representation of the formation of polypeptide chains by ribosomes. As the ribosome (black circle) moves along the mRNA molecule, it "reads" the "message" of the mRNA and, with the assistance of tRNA, translates it into a polypeptide chain. When the ribosome reaches the end of the mRNA strand, it releases the completed polypeptide, "jumps off" and seeks another mRNA molecule to translate.*

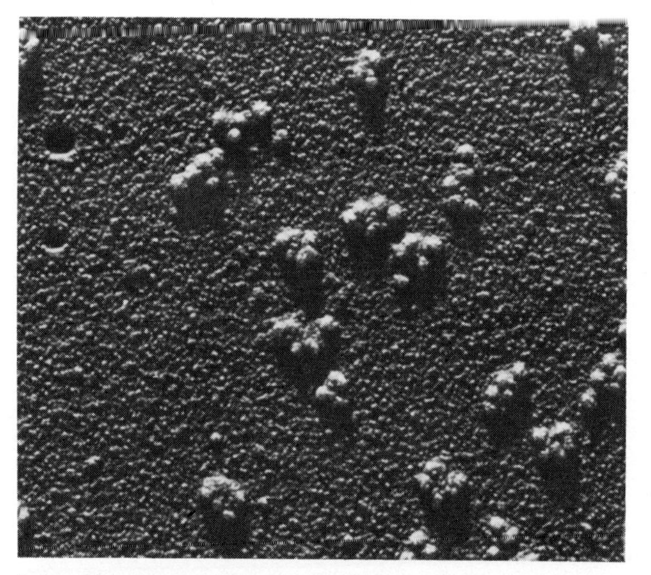

Figure 3–10 Polyribosomes as seen in electron micrograph of rabbit reticulocyte preparation. (Electron photomicrograph courtesy of A. Rich.)

5. Ribosomes, which are approximately half protein and half non-specific RNA, adhere to "sticky points" on the mRNA molecule and move along the molecule. Under their influence AA-tRNA complexes are lined up in the correct order along the mRNA chain, commencing at an AUG codon.

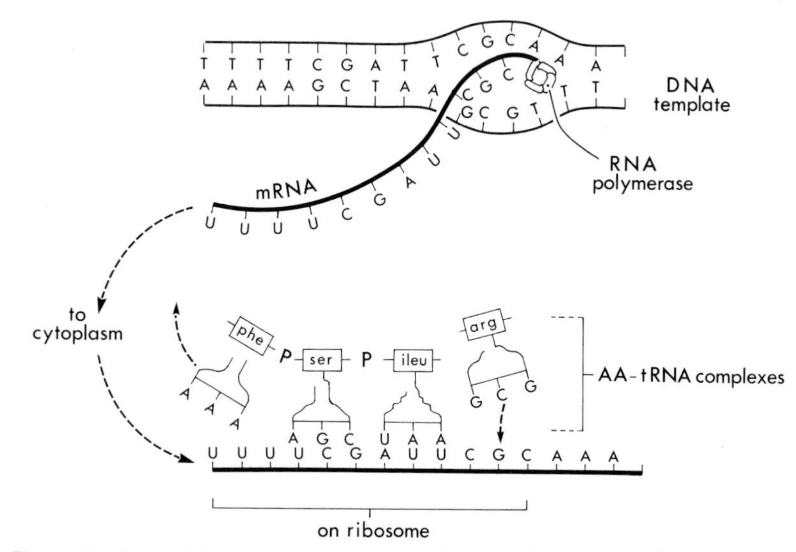

Figure 3–11 An interpretation of protein biosynthesis. For description, see text.

6. An enzyme in the ribosome causes the formation of peptide bonds between the amino acids of successive AA-tRNA complexes.

7. When the peptide bond has formed, the AA-tRNA complex breaks up and the tRNA is free to unite with another activated AA.

8. As it is formed, the molecule of polypeptide swings away from the mRNA strand. In about 10 seconds the ribosome reaches a terminator codon at the end of the gene, "jumps off" and becomes available to form another polypeptide.

GENETIC CONTROL MECHANISMS

Not all the genes of the genome of a given cell are active in protein synthesis at the same time, and those genes that are active may be synthesizing their respective proteins in widely varying amounts. This is true both of bacteria and of multicellular organisms such as man. The factors that determine which proteins a particular cell will synthesize, and in what quantities, are not fully known. The problem is of great importance, not only in regard to understanding the functioning of single cells in molecular detail, but also with respect to elucidation of the control of development and differentiation in multicellular organisms.

Regulation of protein synthesis is exerted partly by genes and partly by factors in the external milieu. Our present understanding is based chiefly upon studies with bacteria, and may not apply to man.

The generally accepted concept of genetic control mechanisms was originated by Jacob and Monod (1961) and is known as the Jacob-Monod model. This model (Fig. 3–12) involves the interaction of two kinds of units, **regulatory genes** and **operators** (as distinct from the **structural genes**, which code for the structural proteins and enzymes of the organism).

REGULATORY GENES

Regulatory genes exert a negative control over protein synthesis by coding for **repressors**, specific protein molecules which act through the cytoplasm to repress the synthesis of specific proteins or groups of related proteins. Repressors inhibit protein synthesis by binding to their corresponding operators (see below), thus forestalling messenger RNA transcription and consequently preventing protein synthesis.

Some repressors cannot function just as they are synthesized but must first combine with some other substance that will give the repressor the right conformation to bind with its operator. A substance which binds with a repressor to activate it is called a **corepressor**.

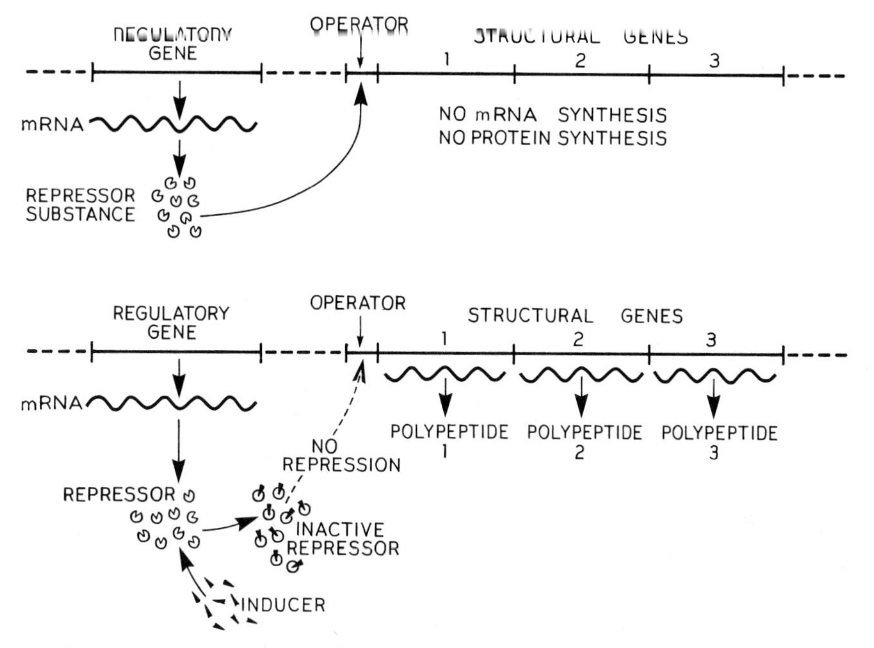

Figure 3–12. *An interpretation of the regulation of protein synthesis. For description, see text.*

Because a repressor inhibits synthesis of its particular protein, it must be inactivated if synthesis of the protein is to proceed. A variety of substances can act as **inducers**, specific molecules which combine with the repressor molecules to prevent them from binding with their specific operators.

OPERATORS

The inhibition of transcription of a gene or of a sequence of related genes depends upon a site at one extremity of the region called an operator. (An operator is a relatively short sequence of nucleotides. It is not regarded as a "whole" gene but rather as a site within a gene or gene complex; however, operators can undergo mutation.) The operator, together with the adjacent structural gene or genes controlled by it, constitutes an **operon**. Unlike regulatory genes, the operator affects only the genes in its own operon, not those on the homologous chromosome.

We will not go further into the complexities of the control of protein synthesis here. Even for bacteria, an understanding of the process is incomplete; indeed, the further knowledge advances, the more subtle and varied the mechanisms involved appear to be.

At present there is no good example of the Jacob-Monod type of genetic control in man. The thalassemias (page 102), in which normal hemoglobin chains are made in abnormal quantities, may possibly represent disorders of the control of hemoglobin synthesis.

THE GENE AND GENE MUTATION

The classic concept of the gene has been that it is a unit with three special properties:

1. It is the unit of **structure**, within which recombination cannot occur.
2. It is the unit of **function**, having one primary function.
3. It is the unit of **mutation**.

It is now evident that none of these terms is satisfactory in its original meaning. A gene is a segment of DNA forming the code for a particular polypeptide. Mutation involves an alteration in the code at one particular site, i.e., involving one nucleotide (see below). A single gene, with hundreds of nucleotides, has hundreds of possible mutational sites. Recombination (crossing over) between homologous DNA strands occurs between successive nucleotides, again at many possible sites in the DNA strand. Even the concept of the gene as a functional unit cannot be accepted uncritically, because there are many examples of complex loci, units which are transmitted as single genes but have more than one recognizable primary effect.

MUTATION

In general terms, a mutation is any sudden heritable change in DNA. This definition is broad enough to include alterations of the structure of whole chromosomes, but in actual practice the term is often restricted to point mutations. A point mutation involves the substitution of one base for another within a triplet. If this substitution alters the codon so that it codes for a different amino acid, the mutation results in synthesis of an altered polypeptide chain with an amino acid substitution at a position colinear with the site of the mutation in the corresponding DNA codon. We shall return to this topic in connection with abnormal hemoglobins (page 93).

Referring again to the genetic code in Table 3–1, note that not every base substitution will lead to an amino acid substitution, especially if it is the third base of the triplet that is substituted. The table also indicates the kinds of substitutions that can occur. By a single mutation, an amino acid can be replaced only by one that appears in the same row or column as itself. Any other substitution would require two separate mutational events.

Because the genetic message is read in groups of three bases, insertion or deletion of a single nucleotide can alter the "reading frame" and result in a sequence of amino acids quite different from the original and completely lacking in biological activity. Such mutations are called "nonsense mutations," as distinguished from the "mis-sense" mutations produced by single amino acid substitutions. Mutation to a chain-terminating codon at some site in a gene could result in the synthesis of a short, unstable, biochemically inactive polypeptide. On the other hand, mutation of a chain-terminating codon could produce a polypeptide longer than the normal one, with a "gibberish" sequence at the end.

Other mechanisms by which alterations in the DNA code can arise are discussed elsewhere (see unequal crossing over, page 99; and nonhomologous crossing over with partial gene duplication or deletion, pages 101 and 232).

GENERAL REFERENCES

Hartman, P. E., and Suskind, S. R. 1969. *Gene Action.* 2nd ed. Prentice-Hall, Inc., Englewood Cliffs, New Jersey.

Watson, J. D. 1968. *The Double Helix.* Atheneum Publishers, New York.

Watson, J. D. 1970. *Molecular Biology of the Gene.* 2nd ed. W. A. Benjamin, Inc., New York.

PROBLEMS

1. State the DNA triplets corresponding to the following mRNA triplets:
   A G C
   A A U
   U C U
   C G A
   G G U

2. A portion of a DNA molecule has the base sequence ACA.
   (a) What is the base sequence of the portion of mRNA formed on this template?
   (b) A tRNA molecule will bring a certain amino acid into position on the polypeptide chain for which this mRNA molecule carries the code. What is the base sequence of the part of the tRNA molecule that "recognizes" this mRNA triplet?
   (c) In terms of Table 3–1, what amino acid is involved?

4

Patterns of Transmission of Genes and Traits

In medical practice, the chief significance of genetics is its role in the etiology of a large number of defects. Virtually any trait is the result of the combined action of genetic and environmental factors, but it is convenient to distinguish between (1) those disorders in which defects in the *genetic* information are of prime importance, (2) those in which *environmental* hazards (including hazards of the intrauterine environment) are chiefly to blame and (3) those in which a *combination* of genetic constitution and environment is responsible.

Broadly speaking, **genetic disorders** are of three main types:

1. **Single-gene defects.**
2. **Chromosome disorders.**
3. **Multifactorial traits.**

Single-gene defects are caused by mutant genes. The mutation may be present singly, i.e., on only one chromosome of a pair (matched with a normal allele), or in "double dose," i.e., at the same locus on each of a pair of homologous chromosomes. Hence, the cause of the defect is a single major error in the genetic information. If the gene is expressed when it exists in single dose, it is said to be **dominant**; if it must be in double dose to show an effect, it is said to be **recessive.** If the gene is on an autosome, it is **autosomal**; if it is on the X chromosome, it is **X-linked.**

In chromosome disorders, which are discussed in Chapters 6 and 7, the cause of the defect is not a single mistake in the genetic blueprint but instead is developmental confusion arising from an abnormal number or arrangement of chromosomes, which upsets their normal balance. For example, in Down's syndrome the presence of an extra chromosome

21 produces a characteristic disorder, even though all the genes on the extra chromosome may be quite normal. Usually chromosome disorders do not run in families, but there are exceptions.

Multifactorial inheritance (also known as quantitative or polygenic inheritance) is discussed in Chapter 11. Here again there is no one major error in the genetic information but rather a combination of small variations which together can produce a serious defect. Multifactorial disorders tend to cluster in certain families, but do not show the clear-cut pedigree patterns of single-gene traits.

Before describing the patterns of single-gene ("mendelian") inheritance, we must first introduce some terms which have special connotations in genetics.

The family member who first brings a family to the attention of an investigator is the **propositus** (proband, index case). **Sibs** (or siblings) are brothers or sisters, of unspecified sex. The parent generation is designated the P_1, and the first generation offspring of two parents is the F_1, but these terms are used more frequently in experimental genetics with inbred lines of plants or animals than in human genetics.

Recall that genes at the same locus on a pair of homologous chromosomes are **alleles**. In more general terms, alleles are alternative forms of a gene. When both members of a pair of alleles are identical, the individual is **homozygous** (a homozygote); when they are different, the individual is **heterozygous** (a heterozygote or carrier). The term **compound** is used to describe a genotype in which two different *mutant* alleles are present, rather than one normal and one mutant. These terms (homozygous, heterozygous and compound) can be applied either to an individual or to a genotype.

An allele which is always expressed, whether homozygous or heterozygous, is **dominant**; an allele which is expressed only when it is homozygous is **recessive.** Strictly speaking, it is the **trait** (phenotypic expression of a gene) rather than the gene itself which is dominant or recessive, but the terms **dominant gene** and **recessive gene** are in common use.

The distinction between dominant and recessive genes is not absolute. In heterozygotes, both members of a pair of alleles are expressed, regardless of whether the expression is phenotypically distinguishable. (The exception to this general rule is that certain mutations code for chain-terminating triplets, so their products are short, unstable polypeptide chains.) By definition a recessive has no detectable expression in a heterozygote, but some genes are defined as recessive simply because when heterozygous they are not phenotypically evident *under the conditions of analysis.* If the phenotype is examined in a different way, the expression of the gene may be detectable. For example, if the looked-for phenotypic expression is a manifest disease, a gene may be regarded as recessive; but, if the study is on the biochemical level, the "recessive" gene may be detectable even when heterozygous. Later we will mention a number of examples of recessive genes that are detectable in heterozygotes.

Disorders in structural (nonenzymatic) proteins are often inherited as dominants, whereas changes in enzyme proteins are usually recessive. The explanation is that in heterozygotes an abnormal structural protein would be formed in all cells in which the gene determining its abnormal structure is active, and consequently would affect the phenotype; whereas the gene for an abnormal enzyme usually would produce no obvious phenotypic effect in heterozygotes, because the margin of safety in enzyme systems allows normal function even if only one of a pair of alleles is producing the normal enzyme.

Family data can be summarized in a **pedigree chart**, which is merely a shorthand method of classifying the data for ready reference. The symbols used in drawing up a pedigree chart are shown in Figure 4–1. Variants of the symbols in Figure 4–1 are acceptable, but the ones shown are all in common use. Special symbols may be invented to demonstrate special situations. By convention, gene symbols are always in italics, and usually a capital letter is used for a dominant gene and the same letter in lower case is used for a corresponding recessive allele; alter-

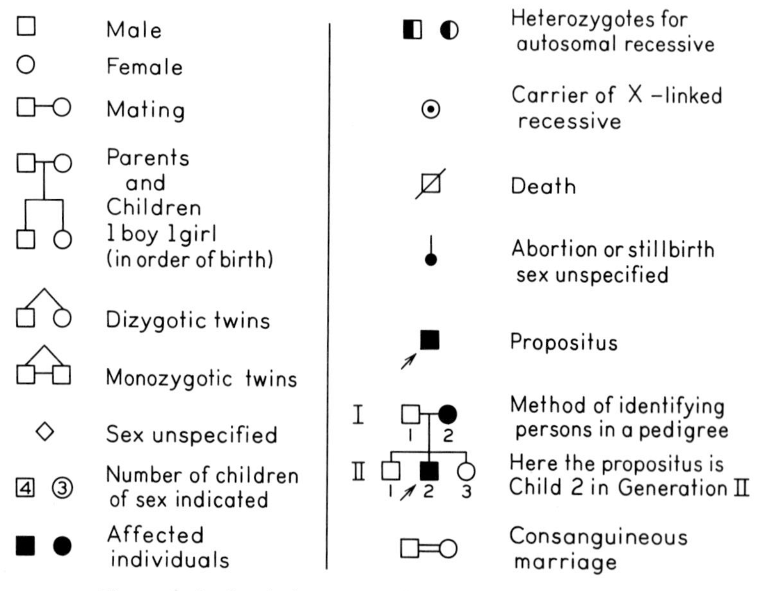

Figure 4–1 *Symbols commonly used in pedigree charts.*

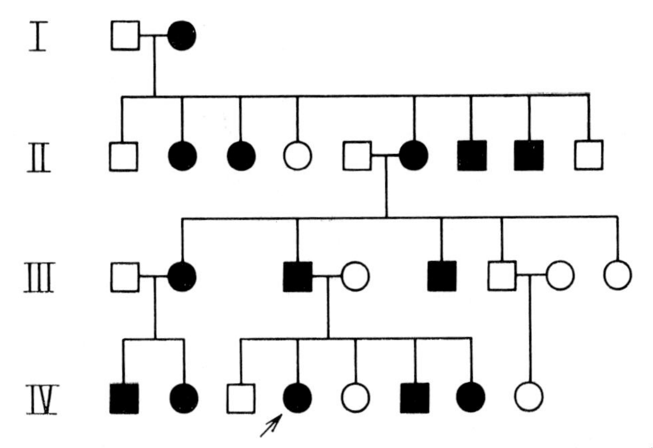

Figure 4–3 *Pedigree of dentinogenesis imperfecta, an autosomal dominant disorder of dentine formation. (Courtesy of N. Levine.)*

of a family. The condition of the teeth and gums of one of the affected children is shown in Figure 4–4.

Let us use the symbol D for the dominant gene for dentinogenesis imperfecta, and d for its normal allele. Then an affected person has the genotype D/d and a normal person has the genotype d/d. Fifty per cent of the progeny of $D/d \times d/d$ matings are D/d and 50 per cent are d/d.

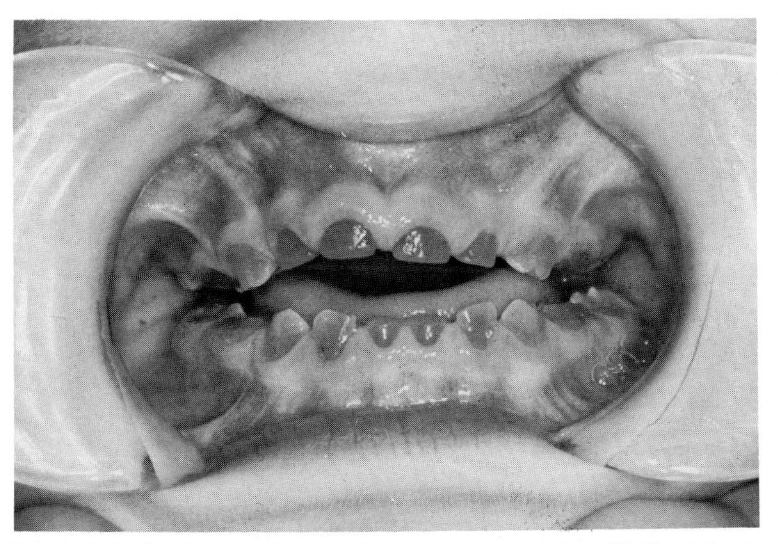

Figure 4–4 *Dentinogenesis imperfecta in the propositus (IV-4) of the family shown in Figure 4–3 (Courtesy of N. Levine.)*

The *D/d* parent forms *D* and *d* gametes in equal numbers. The *d/d* parent forms *d* gametes only:

PROGENY OF *D/d* × *d/d* MATING

		Normal Parent	
		d	*d*
Affected Parent	*D*	*D/d* Affected	*D/d* Affected
	d	*d/d* Normal	*d/d* Normal

The Homozygous Dominant Genotype

A dominant gene, by formal definition, has the same expression in the heterozygous state as in the homozygous state, but in actual practice homozygotes for rare dominants are seldom encountered. This is because homozygotes are produced only by the mating of two heterozygotes, and it is statistically unlikely that two persons, both affected by the same rare dominant trait, will marry. Homozygotes theoretically form one-quarter of the offspring of the marriage of two heterozygous parents:

PROGENY OF *D/d* × *D/d* MATING

	D	*d*
D	*D/D* Homozygous affected (usually rare)	*D/d* Heterozygous affected
d	*D/d* Heterozygous affected	*d/d* Homozygous normal

A D/D person may be distinguishable from a D/d person by his progeny. Since a D/D person has only D genes, he can produce only affected children. However, in a small human family, by chance a D/d with a d/d spouse might have only D/d children, though theoretically half of his children could be d/d. Thus, in human families it is very difficult to identify a homozygote with certainty, even if he or she is the offspring of two heterozygous parents.

Even though the definition of a dominant gene implies that its phenotypic expression is the same in homozygotes as in heterozygotes, experience shows that rare dominants may actually have more severe effects in homozygotes than in heterozygotes; their effects in homozygotes may even be lethal. In mice, many genes that produce little or no effect in heterozygotes are lethal or sublethal in homozygotes. One such gene is the "dominant spotting" allele W. Mice of genotype W/w are normal, but the W gene is expressed in the coat color as a large white patch of fur on the belly and a small white blaze on the head. Homozygous W/W mice are black-eyed white mice which are severely anemic; though born alive, they usually die within a few days (Russell, 1954).

Figure 4–5 demonstrates how much more severe a dominant anomaly may be in homozygotes than in heterozygotes. In this family both parents, who are first cousins, have a mild skeletal anomaly affecting the hands and feet. Their children, who are presumably homozygous for the

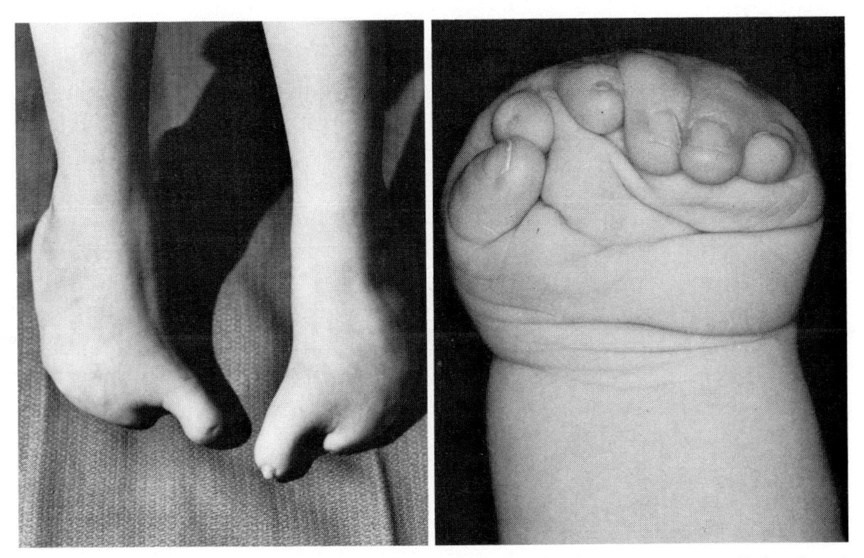

Figure 4–5 *A probable example of homozygosity for an autosomal dominant. Feet and hand of a child whose first-cousin parents had minor congenital malformations of the hands and feet. (From Edwards and Gale 1972.* Amer. J. Hum. Genet. *24:464–474, by permission.)*

abnormal gene, have very severe malformations of the limbs and urogenital anomalies and are retarded.

Since a heterozygote has one normal allele and a homozygote does not, it is not surprising to see a more severely abnormal phenotype in homozygotes; the observation demonstrates that the classification of genes as "dominant" and "recessive," though both alleles are functional, is a convenient label rather than an essential distinction between dominance and recessivity.

Criteria for Autosomal Dominant Inheritance

The criteria for "diagnosis" of autosomal dominant inheritance may be summarized as follows:

1. The trait appears in every generation, with no "skipping."
2. The trait is transmitted by an affected person to half his children (on the average).
3. Unaffected persons do not transmit the trait to their children.
4. The occurrence and transmission of the trait are not influenced by sex; i.e., males and females are equally likely to have or to transmit the trait.

AUTOSOMAL RECESSIVE INHERITANCE

A trait transmitted as an autosomal recessive is expressed only in a person who receives the recessive gene *from both parents* and so is homozygous for it. Typically the trait appears in some of the sibs of the propositus but not in relatives outside the sibship, although other relatives are sometimes involved, especially in large kindreds. Nearly 800 clinical disorders with autosomal recessive inheritance have been recognized (McKusick, 1971).

The most common autosomal recessive disease in white children is **cystic fibrosis.** In cystic fibrosis there are abnormalities of several exocrine secretions, including pancreatic and duodenal enzymes, sweat chlorides and bronchial secretions. The thick, viscid mucus produced by the bronchi is particularly serious, since it makes the affected children very susceptible to pneumonia. The loss of chlorides in the sweat can be severe enough to cause heat prostration in warm weather.

Cystic fibrosis affects perhaps one child in 2000 births. About one person in 22 is a heterozygous carrier. (The mathematical relationship of gene frequency, trait frequency and carrier frequency is discussed in Chapter 11.) In a hypothetical family of four children produced by two carrier parents, one child will be homozygous normal, two children will be heterozygous normal and one child will receive a "double dose" of the abnormal gene and be affected. In other words, if c represents the

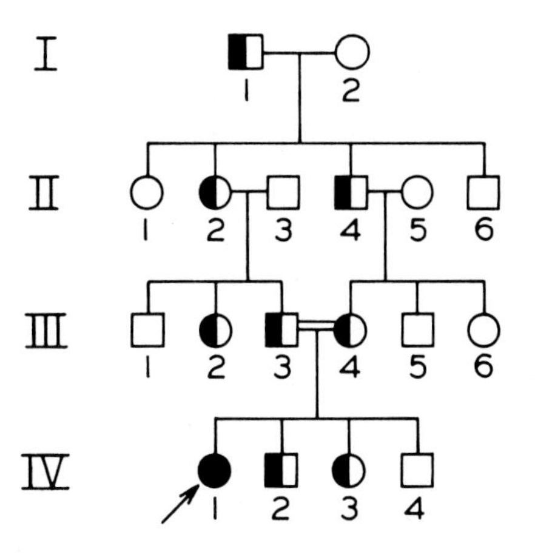

Figure 4–6 *Stereotype pedigree of autosomal recessive inheritance, involving a cousin marriage (III 3–4). Note that both parents of the propositus, who is homozygous for the recessive gene, are heterozygous, having inherited the gene from their common ancestor I–1.*

cystic fibrosis gene and C its normal allele, the genotypic ratio of the offspring is $1\ C/C : 2\ C/c : 1\ c/c$, and the phenotypic ratio is 3 normal : 1 affected (see also Table 4–1, fourth line).

Because children with recessive traits usually have phenotypically normal (but heterozygous) parents, the only families that can be recognized and studied are those in which there is already at least one affected child. Families in which no child is affected merge with the general population and are not ascertained. This **bias of ascertainment** creates a statistical problem in that the proportion of affected children in the sibships that can be ascertained will be well above the theoretical one-fourth, unless the sibship size is very large (about 12). We will return to this problem and ways of dealing with it in Chapter 11. A stereotype pedigree of autosomal recessive inheritance, involving a cousin marriage, is shown in Figure 4–6.

Consanguinity and Recessive Inheritance

A carrier of a recessive gene can have affected children only if the other parent is also a carrier. The risk that a carrier of cystic fibrosis will marry another carrier is one in 22 (the incidence of carriers in the population). However, since rare recessive genes are passed down in families and so are concentrated in family groups, the risk that a carrier will marry another carrier is usually higher by at least one order of magnitude if he marries a near relative than if he marries at random (i.e., without regard for the genotype of the spouse).

For example, consider the chance that a carrier of cystic fibrosis who marries a first cousin will have children homozygous for the gene. Figure 4–7 sets out the various probabilities involved. In brief, the risk

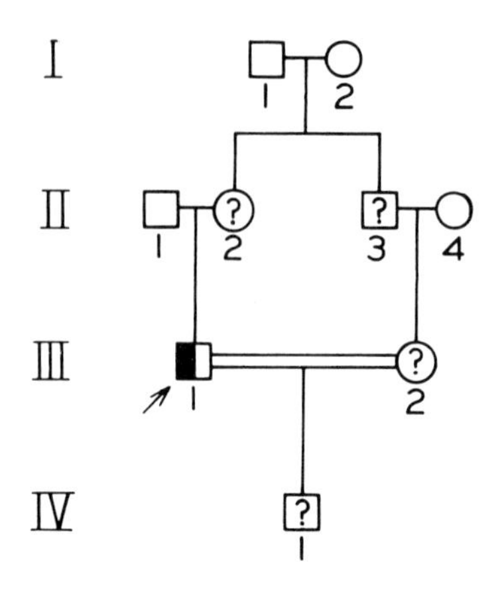

Figure 4–7 *The probability that a carrier of cystic fibrosis (III–1) who marries a first cousin will have an affected child.*

1. *If III–1 is a carrier, the chance II–2 is a carrier is* $\frac{1}{2}$.
2. *If II–2 is a carrier, the chance II–3 is a carrier is* $\frac{1}{2}$.
3. *If II–3 is a carrier, the chance III–2 is a carrier is* $\frac{1}{2}$.

Therefore *if III–1 is a carrier, the chance that III–2 is also a carrier is* $\frac{1}{2} \times \frac{1}{2} \times \frac{1}{2} = \frac{1}{8}$.

4. *If III–1 and III–2 are both carriers, the chance that IV–1 is homozygous for that gene is* $\frac{1}{4}$. Therefore *if III–1 is a carrier who marries a first cousin, the chance that their first child will be homozygous for that gene is* $\frac{1}{4} \times \frac{1}{8} = \frac{1}{32}$.

that a first cousin of a carrier is also a carrier because of inheritance of the gene from a common ancestor is one in eight. The mathematical aspects of consanguinity are treated in additional detail in Chapter 11.

The effect of **inbreeding** in producing homozygous recessives is the biological basis of the prohibition of cousin marriage in many societies. For a condition as common as cystic fibrosis, the chance that a carrier who marries a cousin will have an affected child is only a little greater (about three times as high, on the basis of our incidence figure) than the general population risk. For less common conditions, the risk is higher. Some examples are given in Table 4–2, which compares the incidence of recessive traits of three different frequencies for children of cousin marriages with the incidence in the general population. Note the inverse relationship between the frequency of a recessive trait in the general

TABLE 4–2 EFFECT OF COUSIN MARRIAGE ON THE INCIDENCE OF A RECESSIVE CONDITION*

	Affected Individuals		
Frequency of Recessive Gene	*A. In General Population*	*B. Among Children of Cousin Marriages*	***Ratio B/A***
0.2	0.04	0.05	1.25
0.02	0.0004	0.0016	4.00
0.002	0.000004	0.00013	32.25

*After data of Li, C. C. 1963. *Amer. J. Med.* 34:702.

population and the frequency of consanguinity among the parents of affected children; for example, if a trait has a frequency of only four per million, it will appear 32 times as often in the children of cousin marriages as in the general population.

Rare Recessives in Genetic Isolates

There are many small groups in which the frequency of certain rare recessive genes is quite different from that in the general population. Such groups may have become **genetically isolated** from their neighbors by geographic, religious or linguistic barriers.

In Ashkenazi Jews in North America, the gene for **Tay-Sachs disease** is very common. Tay-Sachs disease is an autosomal recessive neurological degenerative disorder which develops at about six months of age. Affected children become blind and regress mentally and physically. The disease is usually fatal in early childhood. The frequency is 100 times as high in Ashkenazi Jews (one in 3600) as in other populations (one in 360,000).

When a recessive reaches such a high frequency, cousin marriage is no longer a striking feature of pedigrees of the trait. This is because the gene frequency is so high that a carrier who marries another member of the same group is almost as likely to marry another carrier as he would have been had he married a close relative. Consequently, among Ashkenazi Jews the parents of affected children are usually not closely consanguineous, whereas in other North Americans the consanguinity rate in the parents of Tay-Sachs patients is high.

There are many other examples of rare recessives in genetic isolates. **Tyrosinemia,** a very rare and lethal hepatic disease of early infancy, has been recognized in 51 French-Canadian children in 29 families in the isolated Lac St. Jean-Chicoutimi region of Quebec in recent years. It is virtually unknown elsewhere. Laberge (1969) has been able to trace both parents of each affected child to a couple who were in Quebec City by 1644. Probably, one member of this couple was a carrier of the tyrosinemia gene, which has been passed down 10 or more generations to "meet itself" in double dose in some of the remote descendants. The carrier frequency in the isolate is of the order of one in 30, and, as expected, the parents of the affected children are not closely consanguineous.

Genetic Counseling in Consanguineous Marriages

Are cousin marriages unwise? There is considerable disagreement among geneticists as to whether the extra risk of defective offspring is significant. A commonly quoted statistic is that for Japanese data: the proportion of abnormal progeny was about 2 per cent for random marriages and about 3 per cent for consanguineous marriages (Neel and Schull, 1962). This finding is interpreted by some as an inconsequential increase, but from the standpoint of the individual family it does mean that the risk of defective children is increased by half. The high frequency of cousin marriage observed among the parents of congenitally deaf children requiring special educational facilities (Sank, 1963) is one measure of the personal and social cost of consanguineous marriages. (Incidentally, we might add that life insurance companies do not consider the offspring of cousins to be any different from normal insurance risks.)

If cousin parents have a child who suffers from a recessively inherited disease, they are thereby proven to be carriers, with a one-quarter chance of having a similarly affected child at any later pregnancy. Furthermore, first cousins may share more than one deleterious recessive gene; in fact, they have one-eighth of their genes in common, and their progeny will, on the average, be homozygous at $1/16$ of all their gene loci.

Consanguineous marriages more distant than those of first cousins have a correspondingly lower risk of producing affected offspring. For example, second cousins have only $1/32$ of their genes in common and third cousins only $1/128$. For individuals less closely related than second cousins, the risk of defective offspring is near enough to the general population risk to be of little or no genetic consequence. Nevertheless, if a child has a rare or undiagnosed defect and his parents are even remotely consanguineous, it is a reasonable hypothesis that his defect has autosomal recessive inheritance, with a one in four recurrence risk for later-born sibs.

Matings closer than first-cousin marriages are not legal anywhere in North America (with the exception of marriages of double first cousins, which are not proscribed although the relationship of double first cousins is as close as the uncle-niece relationship). However, parent-child and brother-sister matings do take place and do produce children who usually become public wards and must be assessed as suitable candidates for adoption. Children produced by such close inbreeding are homozygous at one-fourth of their gene loci and have a correspondingly high risk of being homozygous for some deleterious gene. Very few studies of children of incest have been reported. Carter (1967) in a series of 13 cases found four to have been severely abnormal, definitely or probably because of homozygosity for recessives.

Not only do cousin marriages have an increased risk of producing children homozygous for deleterious recessive genes, but they also have a higher than normal capacity for producing children with common congenital malformations or other conditions whose manifestation depends upon several genes. This aspect is discussed under polygenic inheritance (Chapter 11).

Carrier Identification

Though by definition recessive genes are not expressed in heterozygotes, many recessives which are not clinically significant in heterozygotes may nevertheless have a detectable phenotype if an appropriate carrier test is used. For cystic fibrosis no carrier test is generally available as yet, though recent advances suggest there will be one shortly. In a number of other recessive disorders, useful carrier tests have been developed (page 339). Carrier identification is helpful in genetic counseling. An unaffected sib of a recessively affected child has a two-thirds chance of being a carrier; uncles and aunts each have a one-half chance; and first cousins have a one-fourth chance. Because only carrier × carrier matings can produce affected offspring, these individuals often wish to know definitely whether they do or do not have the abnormal gene, and, if so, whether their mates or prospective mates are also carriers. Identification and investigation of carriers can also add to our understanding of the mechanism of specific diseases, as exemplified by the hemoglobinopathies (page 93).

The Offspring of Homozygous Recessives

To indicate the types of offspring a homozygous recessive can have, we can return to our original example, the nontaster trait. Table 4–1 shows three matings involving at least one nontaster parent: a *t*/*t* person may have a *T*/*T*, *T*/*t* or *t*/*t* spouse. In brief:

Parents	**Progeny**	
	Genotypes	*Phenotypes*
t/*t* × *T*/*T*	All *T*/*t*	All taster
t/*t* × *T*/*t*	½ *T*/*t*, ½ *t*/*t*	½ taster, ½ nontaster
t/*t* × *t*/*t*	All *t*/*t*	All nontaster

Note that the *t*/*t* × *T*/*t* mating, in which an affected individual has an apparently normal spouse but half of the children are affected, mimics the pattern of inheritance of a rare dominant trait. Recessive inheritance which resembles dominant inheritance because an affected person marries a carrier is termed **quasidominant inheritance.** This pattern cannot always be firmly distinguished from ordinary autosomal dominant inheritance, but a pedigree of such a trait over more than two generations will probably reveal that in each generation the characteristics of recessive inheritance are present; in particular, the trait will usually appear only in the parent sibship and the offspring sibship, not elsewhere in the kindred, and consanguinity may be present in the parents and/or in one set of grandparents.

Criteria for Autosomal Recessive Inheritance

1. The trait characteristically appears only in sibs, not in their parents, offspring, or other relatives.
2. On the average, one-fourth of the sibs of the propositus are affected; in other words, the recurrence risk is one in four for each birth.
3. The parents of the affected child may be consanguineous.
4. Males and females are equally likely to be affected.

CODOMINANCE AND INTERMEDIATE INHERITANCE

Up to this point we have discussed only the pedigree patterns produced by genes that are fully dominant or fully recessive in their expression. These are not the only relationships possible between pairs of alleles; it has already been pointed out that dominance and recessivity are relative concepts, and that when a gene is said to be recessive in heterozygotes, the reason may be our inability to recognize an effect, rather than the total absence of an effect.

If both alleles of a pair are fully expressed in the heterozygote, the genes are said to be **codominant.** Many examples of codominance are provided by the various blood group systems. For example, a person of blood group AB has both A and B antigens on his red cells. The allelic genes *A* and *B* are therefore codominant.

If the heterozygote is different from both homozygotes, the genes concerned are said to show **intermediate inheritance.** A good example is provided by sickle cell anemia. Heterozygotes for the abnormal allele do not have the severe sickle cell anemia found in homozygotes, but a proportion of their red cells do show the sickling phenomenon; in other words, heterozygotes are intermediate between normal homozygotes and sickle cell homozygotes, and are said to have the sickle cell trait. (The use of the term *trait* to indicate a heterozygous phenotype, as in *sickle cell trait,* is a specialized clinical usage. In human genetics, *trait* is used interchangeably with *phenotype* to indicate the visible or detectable expression of a gene.)

The difference between codominance and intermediate inheritance is often indistinct. If we look at the hemoglobin of normal individuals and that of heterozygotes and homozygotes for the sickle cell anemia gene, we see that heterozygotes do not have an "intermediate" type of hemoglobin but have both normal hemoglobin A and sickle cell hemoglobin (hemoglobin S). Therefore, these genes are more properly codominant.

MULTIPLE ALLELES

All the examples used so far in this chapter have involved only a single pair of alleles, usually one "normal" and the other "abnormal." This simple situation is not the only possibility, for at many loci more than two different alleles are known. When for a single locus more than two alternative alleles exist in the population, the alleles are called **multiple alleles.**

The classic example of multiple allelism is provided by the series of alleles that determines the ABO blood groups. The best known alleles of the series are designated O, A^1, A^2 and B. For simplicity we will disregard A^2 and consider only the three main alleles. On this basis, the relationships of genes, genotypes and phenotypes are as follows:

Genes	***Genotypes***	***Phenotypes (Blood Groups)***
O, A, B	*O/O*	O
	A/A *A/O*	A
	B/B *B/O*	B
	A/B	AB

Note that the gene *O* is recessive to genes *A* and *B*, which are codominant. If *A* and *B* are present, both corresponding antigens are formed. *O* is an **amorph,** that is, a gene which has no effect and leaves the substrate (H substance) unaltered.

It was originally believed that the ABO blood groups were determined by genes at two independent loci, i.e., by two pairs of nonallelic genes. Alleles can be distinguished from nonalleles by analysis of family data, since alleles segregate in the progeny, whereas nonalleles assort randomly. The progeny of an AB × O mating is always A or B, never AB or O. In other words, gene *A* and gene *B* are never both transmitted from the parent to the child, but always segregate; thus they must be allelic.

It is not always easy in human genetics to prove whether two rare genes are alleles or whether they are at independent loci. One reason is that families suitable for analysis are not often seen. Furthermore, even if two genes do not assort independently, they may be *linked* rather than *allelic*. The critical test for distinguishing linkage from allelism is that recombination can occur between the loci of linked genes (by crossing over in meiotic prophase), but not within a locus, i.e., not between alleles. However, as we shall see, there are exceptions even to this rule.

X-LINKED INHERITANCE

Genes on the sex chromosomes are distributed unequally to males and females within kindreds. This inequality produces characteristic and readily recognized patterns of inheritance, and has led to the identification of many "sex-linked" conditions in man.

"Sex-linked" genes may be X-linked or Y-linked, but for all practical purposes only X linkage has any clinical significance. Apart from genes essential for male sex determination, the Y chromosome appears to have few loci, or at least few loci at which segregation has been recognized. Genes on the Y show **holandric** inheritance; that is, they are passed down rather like the family surname, in the male line exclusively, by an affected man to all his sons and to none of his daughters. From time to time pedigrees showing holandric inheritance have been reported, but few of these have withstood searching examination (Stern, 1957). At present there is only one well-substantiated Y-linked anomaly, the "hairy pinna" trait (Dronamraju, 1960). Apart from this one example, the terms sex linkage and X linkage may be used synonymously, but most medical geneticists prefer to use X linkage because it is more specific.

The distribution of X-linked traits in families follows the course of the X chromosome carrying the abnormal gene. Since females have a pair of X chromosomes but males have only one, there are three possible genotypes in females but only two in males. One convenient way to indicate that a gene is X-linked is this: let X_H represent a dominant gene *H* on the X, and X_h its recessive allele *h*. Expressed in symbols, the following are the possible combinations in males and females:

MALES	FEMALES
$X_H Y$	$X_H X_H$
$X_h Y$	$X_H X_h$
	$X_h X_h$

A male has only one representative of any X-linked gene, and so is said to be **hemizygous** rather than homozygous or heterozygous. A female can be homozygous or heterozygous.

An important difference between autosomal and X-linked inheritance is that whereas both members of a pair of autosomal alleles are genetically active, in females only one member of the pair of X chromosomes is active; the second X remains condensed and nonfunctional, appearing in interphase cells as the **sex chromatin** (Barr body). Thus, in females, as in males, there is only one functional X, but in heterozygous females it is a chance matter whether her paternal or her maternal X is the functional one in a given cell. Consequently, though we distinguish X-linked "recessive" and "dominant" patterns of inheritance, it must be borne in mind that females heterozygous for either dominant or recessive X-linked mutant genes have the mutant as the only functional allele at that locus in about half their body cells. (See Lyon hypothesis, page 163.)

X-LINKED RECESSIVE INHERITANCE

The inheritance of recessive genes on the X chromosome follows a well-defined pattern. A trait inherited as an X-linked recessive is expressed by all males who carry the gene: but females are affected only if they are homozygous. Consequently, X-linked recessive diseases are practically restricted to males and are rarely if ever seen in females.

Hemophilia (classical hemophilia, hemophilia A) is an X-linked recessive disease in which the blood fails to clot normally because of a deficiency in antihemophilic globulin. The clinical features, which include severe arthritis as a consequence of internal hemorrhages into the joints, are secondary to the clotting defect. The incidence is about one in 10,000 male births. Its hereditary nature was recognized in ancient times, and it has since achieved notoriety by its occurrence among descendants of Queen Victoria, who was a carrier.

We have used the symbol X_h to represent a recessive gene h on the X chromosome, and X_H to represent its dominant allele. To demonstrate the pedigree patterns of X-linked recessive inheritance, these symbols will now be used to denote the hemophilia gene and its normal counterpart.

An affected male has the genotype X_hY, and a normal female has the genotype X_HX_H. The offspring of these parents can be demonstrated by the checkerboard method:

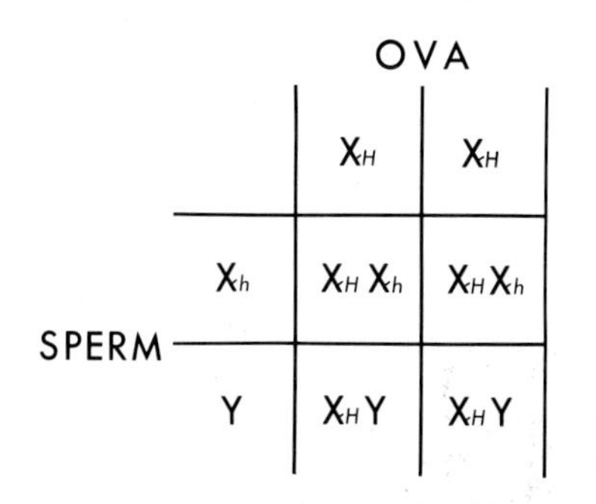

Daughters: 100 per cent heterozygotes (carriers)
Sons: 100 per cent normal

As the checkerboard shows, an affected male does not pass the gene to any of his sons, but he passes it to all of his daughters, who are carriers. If a carrier daughter marries a normal male, four genotypes are possible in the offspring, in equal proportions.

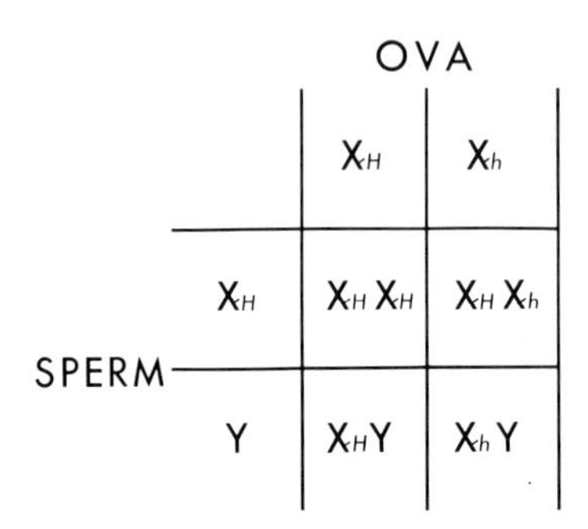

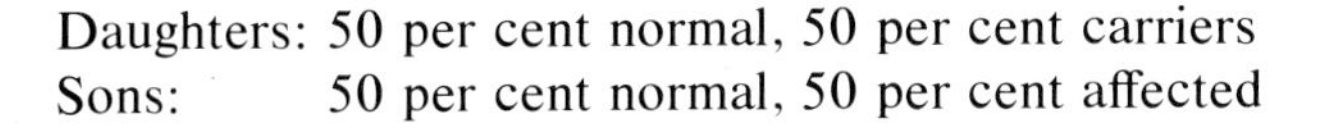
Daughters: 50 per cent normal, 50 per cent carriers
Sons: 50 per cent normal, 50 per cent affected

Note that the X-linked recessive trait of the maternal grandfather, which did not appear in any of his own progeny, now reappears among his grandchildren. X-linked recessives are passed from an affected male through all his daughters to half their sons. Figure 4–8 is a stereotype pedigree to show the characteristics of X-linked recessive inheritance.

Note also that half the daughters of carriers are carriers. By chance, an X-linked trait may be transmitted through a series of carrier women before it makes its appearance in an affected male. **Duchenne muscular dystrophy** (pseudohypertrophic muscular dystrophy) is an X-linked disease which affects young boys. It is usually apparent by the time the

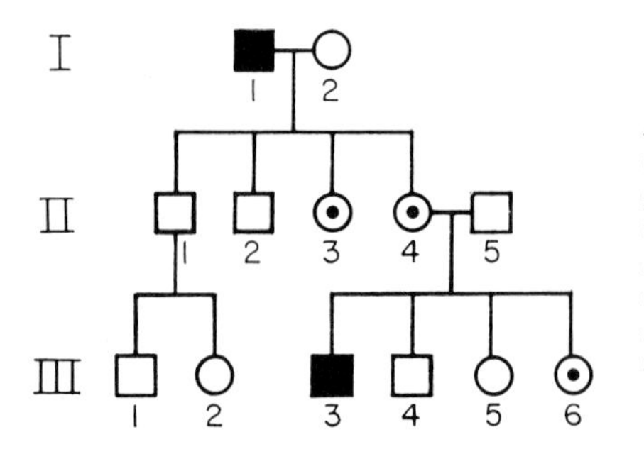

Figure 4–8 *Stereotype pedigree of X-linked recessive inheritance. Note that the affected grandfather I–1 transmits the gene to each of his daughters (who are therefore carriers). They in turn can transmit it to half of their sons, who are affected, and half of their daughters, who are carriers. Male-to-male transmission does not occur.*

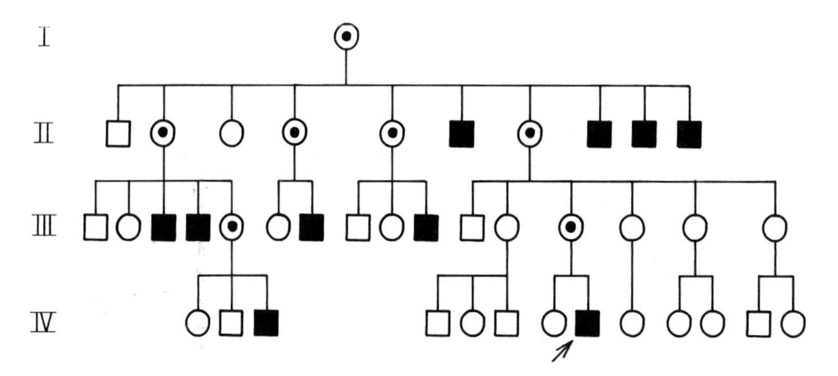

Figure 4–9 Pedigree of Duchenne muscular dystrophy, an X-linked recessive disorder in which affected males do not reproduce.

child begins to walk, and progresses inexorably, so that the child is confined to a wheel chair by about the age of 10, and is unlikely to survive his teens. This disorder is a **genetic lethal,** in that its nature prevents its transmission by affected males; but it is spread by carrier females, who themselves rarely show any clinical manifestation of muscular dystrophy. A representative pedigree of Duchenne muscular dystrophy is shown in Figure 4–9. (For further discussion of the genetics of this disorder, see page 341.)

For convenience, the mating types, the gametes produced and the progeny expected on the basis of X-linked inheritance are shown in Table 4–3.

Criteria for X-Linked Recessive Inheritance

1. The incidence of the trait is much higher in males than in females.
2. The trait is passed from an affected man through all his daughters to half their sons.
3. The trait is never transmitted directly from father to son.
4. The trait may be transmitted through a series of carrier females; if so, the affected males in a kindred are related to one another through females.

X-LINKED DOMINANT INHERITANCE

Whereas X-linked recessive traits typically occur in males only, X-linked dominants are approximately twice as common in females as in males. (For further discussion of this point, see Chapter 11.) The chief

TABLE 4–3 X-LINKED INHERITANCE: MATING TYPES, GAMETES AND EXPECTED PROPORTIONS OF PROGENY FOR A PAIR OF X-LINKED ALLELES X_H AND X_h

MATING TYPES	GAMETES		PROGENY	
	OVA	*SPERM*	*GENOTYPES*	*PHENOTYPES*
$X_H X_H \times X_H Y$	X_H, X_H	X_H, Y	$X_H X_H$ $X_H Y$	all normal
$X_H X_h \times X_H Y$	X_H, X_h	X_H, Y	X_H, X_H $X_H X_h$	daughters: ½ normal, ½ carriers
			$X_H Y$ $X_h Y$	sons: ½ normal, ½ affected
$X_h X_h \times X_H Y$	X_h, X_h	X_H, Y	$X_H X_h$	daughters all carriers
			$X_h Y$	sons all affected
$X_H X_H \times X_h Y$	X_H, X_H	X_h, Y	$X_H X_h$	daughters all carriers
			$X_H Y$	sons all normal
$X_H X_h \times X_h Y$	X_H, X_h	X_h, Y	$X_H X_h$ $X_h X_h$	daughters: ½ carriers, ½ affected
			$X_H Y$ $X_h Y$	sons: ½ normal, ½ affected
$X_h X_h \times X_h Y$	X_h, X_h	X_h, Y	$X_h X_h$ $X_h Y$	all affected

Only a few genetic disorders exhibit the X-linked dominant pattern. An example is hypophosphatemia, also called vitamin D–resistant rickets. In rare X-linked dominants, affected females (heterozygotes) are twice as frequent as affected males (hemizygotes) but usually have a milder form of the defect.

Criteria for X-Linked Dominant Inheritance

1. Affected males transmit the trait to all their daughters and to none of their sons.

2. Affected females who are heterozygous transmit the condition to half their children of either sex. Affected females who are homozygous transmit the trait to all their children. Transmission by females follows the same pattern as an autosomal dominant. In other words, X-linked dominant inheritance cannot be distinguished from autosomal dominant inheritance by the progeny of affected females, but only by the progeny of affected males.

3. In rare X-linked dominant disorders, affected females are twice as common as affected males but are likely to express the condition in a milder form.

VARIATION IN THE EXPRESSION OF GENES

The pattern of inheritance shown by any trait which is determined by a rare gene at a single locus can often be readily recognized if the trait segregates sharply, i.e., if the normal and abnormal phenotypes can easily be distinguished. In ordinary experience, however, the abnormal persons in a kindred may not show an obvious pattern in their relationships, even though the incidence of the abnormality is clearly higher within the kindred than in the general population. We will now consider some of the factors that can affect gene expression and so lead to confusion in interpretation of pedigree data.

PENETRANCE AND EXPRESSIVITY

A mutant gene may not always be phenotypically expressed, or if it is expressed, the degree of expression of the trait may vary widely in different individuals. **Penetrance** applies to a gene's ability to be expressed at all; **expressivity** refers to the degree of expression, i.e., whether mild, moderate or severe.

When the frequency of expression of a trait is below 100 per cent, that is, when some individuals who have the appropriate genotype fail to express it, the trait is said to exhibit **reduced penetrance.** For example, in Figure 4–11, the mutant gene must be carried by II–4, since she transmits

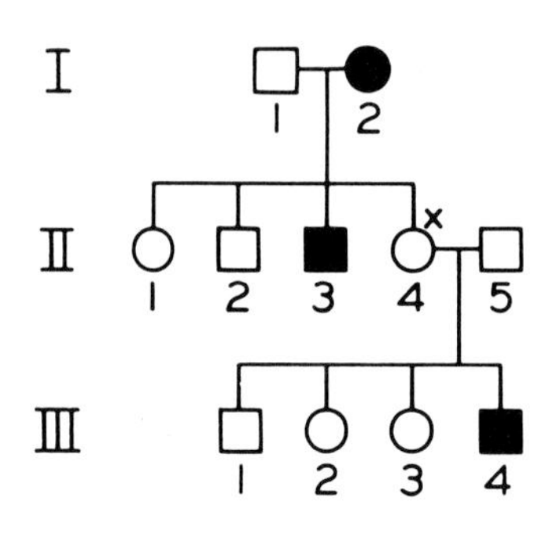

Figure 4–11 Reduced penetrance of an autosomal dominant condition. For description see text.

it, though she does not express it; it is therefore said to be **nonpenetrant.** In the individual case, penetrance is an all-or-none concept. It may be stated in mathematical terms as the percentage of genetically susceptible individuals who actually show the trait.

If, on the other hand, a trait takes somewhat different forms in different members of a kindred, it is said to exhibit **variable expressivity.** Expressivity may range from severe to mild, and members of the same kindred may express the same gene in different ways and with different severity. Expressivity may be roughly equated with clinical severity.

Reduced penetrance and variable expressivity may both appear in the pedigree of a single trait. Figure 4–12 is a pedigree of a family segregating for an unusual form of polydactyly, involving a bifid second metatarsal bone. Figure 4–13 shows the abnormality as expressed in the propositus. The trait is transmitted as an autosomal dominant, but its

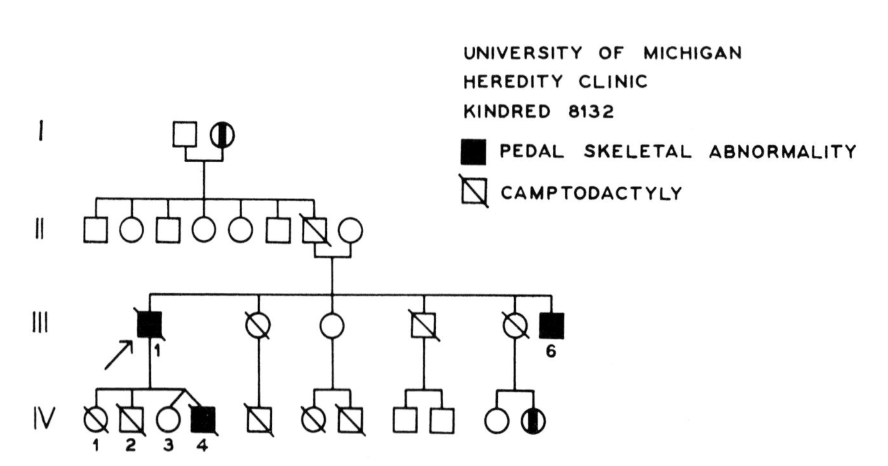

Figure 4–12 *Reduced penetrance as seen in a family with an unusual form of polydactyly which also exhibits variable expressivity. Dark symbols, full expression; half-filled symbols, partial expression. Note that the father of the propositus exhibits only camptodactyly (flexion deformity of interphalangeal joints), so the condition is virtually nonpenetrant in his case. (From Neel and Rusk 1963.* Amer. J. Hum. Genet. *15:288–291, by permission.)*

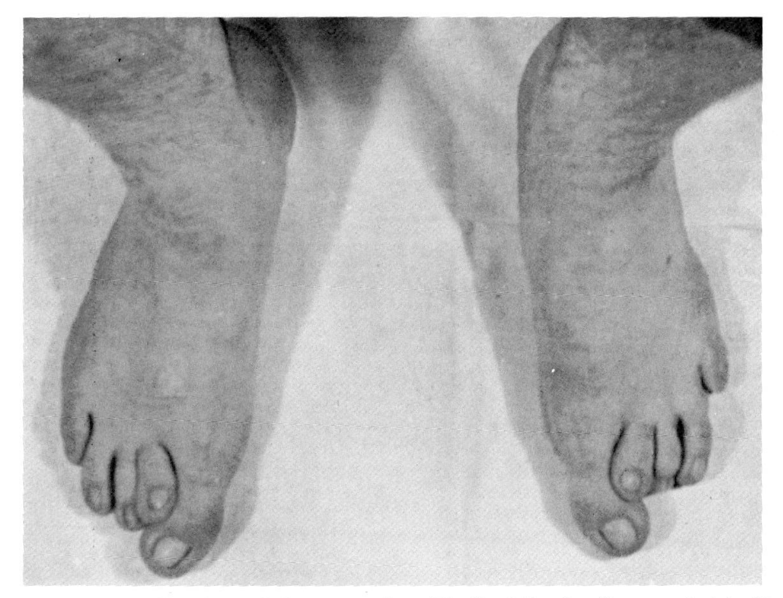

Figure 4–13 *The feet of the propositus (III–1) of the family recorded in Figure 4–12. (From Neel and Rusk 1963.* Amer. J. Hum. Genet. *15:288–291, by permission.)*

expression differs in various members of the family. The accessory spur of the second metatarsal may be large or small, polydactyly may or may not be present and a variety of other foot deformities may appear. Furthermore, in some individuals the gene is nonpenetrant; that is, those individuals transmit the gene but have no obvious deformity. In these people, penetrance might actually be revealed by radiological examination; if so, they would be classified as examples of mild expression rather than of failure of penetrance. It is not unusual for the concepts of penetrance and expressivity thus to blur into one another at the "normal end" of the spectrum of expression.

Variability of expression is usually more marked in members of different family groups than in members of a single kindred. This suggests that expression depends at least partly upon **modifying genes,** which would differ more between than within families.

Reduced penetrance and variable expressivity are more likely to be apparent in autosomal dominant than in autosomal recessive pedigrees. Perhaps this observation is not surprising when one considers that an autosomal dominant disorder is virtually always the expression of a heterozygous genotype; the individual has one normal allele as well as an abnormal one, and both alleles are active. At least two other factors are important: (1) The *primary* action of the gene may be many developmental or biochemical steps removed from the observed effect; (2) the expression of the gene or genes at that locus is modified by the "background" genes of the organism. It is well known that in inbred mice the same mutant gene may have quite different effects against different inbred "backgrounds."

Though failure of penetrance and variability of expression are more commonly observed for autosomal dominants than for autosomal recessives, many recessives are also variable in their expression; for example, albinism may be complete or partial in different members of the same sibship, and cystic fibrosis may be much more severe in one sib than in another.

Forme Fruste

The expression of a clinical syndrome may be very mild (subclinical, or of little clinical significance), and so may be difficult to distinguish from the normal range of variation. An extremely mild and clinically insignificant expression of an abnormality, disease or syndrome is called a **forme fruste.** For example, in **Marfan's syndrome,** which is inherited as an autosomal dominant, the full clinical picture includes elongated extremities, disclocation of the lens of the eye and cardiovascular abnormalities; but a pedigree of this syndrome might show a grandfather and grandson severely affected, related to one another through an apparently normal man. Stature, of course, is determined by many genes, and also influenced by environment; accordingly, the effect of the Marfan's gene upon stature, in a person whose stature would otherwise be below average, might only raise his stature to about normal. If in this person the ocular and cardiovascular effects were minimal, he might well be recorded as completely normal and the trait would appear to "skip a generation." Clinically, this is an example of a *forme fruste.*

Tuberous Sclerosis. Tuberous sclerosis is an autosomal dominant condition in which the chief clinical signs are sebaceous adenomata of the skin, a "butterfly rash" on the face, muscle tumors (rhabdomyomata) and epileptic seizures. Phakomata (retinal tumors) are present in some carriers who are otherwise normal, as well as in some patients. The majority of cases are new mutants, but there are also cases of very mild expression in heterozygotes. It is important to attempt to distinguish between new mutants with normal parents, and cases in which one parent is heterozygous but demonstrates no signs of the disease. This is because the risk that the disorder will recur in the family is zero if both parents are genetically normal, but 50 per cent if one parent is heterozygous.

PLEIOTROPY: ONE GENE, SEVERAL EFFECTS

Each gene has only one primary effect in that it directs the synthesis of a polypeptide chain. From this primary effect, however, many different consequences may arise. It is probable that most genes, even those which have a clear-cut major effect, produce quite diverse secondary effects. Multiple phenotypic effects produced by a single mutant gene or gene pair are called **pleiotropic effects.**

In any sequence of events, interference with one early step may have ramifying effects. Thus, Gruneberg (1947) speaks of a "pedigree of causes," implying that a single defect occurring early in development can lead through branching pathways to various abnormalities in fully differentiated structures. In some cases, perhaps a fan might be a more accurate simile than a family tree; a primary gene product might be envisaged as participating in a number of unrelated biosynthetic pathways, possibly at different times.

Clinical syndromes offer many examples of pleiotropy. The literal meaning of the word syndrome is "running together," and the term is used to describe the set of signs and symptoms characteristic of a specific disorder. In some syndromes, there is a clear or at least a plausible mechanism by which the primary effect of the gene could produce the diverse features of the syndrome. Phenylketonuria is a metabolic disease, inherited as an autosomal recessive, in which the enzyme phenylalanine hydroxylase is lacking. (Phenylalanine hydroxylase is essential for the first step in the metabolism of phenylalanine: the conversion of phenylalanine to tyrosine.) The primary effect in homozygotes is the specific enzyme deficiency. There are multiple secondary effects, notably severe mental retardation, excretion of phenylketones in the urine and dilution of pigmentation. Though the pathways by which these secondary effects are produced are not known in full detail, the steps by which they are related to the primary defect can at least be conjectured. Similarly, in galactosemia a lack of the enzyme galactose-1-phosphate uridyl transferase is the primary effect of homozygosity for the recessive gene concerned. The array of secondary effects, which includes cirrhosis of the liver, cataracts, galactosuria and mental retardation, can be produced in experimental animals if they are fed a diet high in galactose. Furthermore, the secondary effects can be prevented in a genetically susceptible infant if his condition is recognized and he is placed on a galactose-free diet at birth. Thus, the characteristic features of the disease are clearly secondary to the enzyme deficiency.

In other syndromes, the various clinical signs and symptoms may be less obviously related to a single underlying defect. In Marfan's syndrome, mentioned in the preceding section, the pleiotropic effects (the skeletal, ocular and cardiovascular anomalies) may have as a common basis a defect in the elastic fibers of connective tissue. By contrast, in the rare syndrome of hypogonadism, polydactyly, deafness, obesity, retinitis pigmentosa and mental retardation known as the Laurence-Moon-Bardet-Biedl syndrome, there is certainly no one obvious common basis for the assortment of abnormalities. (This syndrome, which is inherited as an autosomal recessive, is not uncommon among Canadian Eskimos.)

Association and Linkage

There are all too many examples in the medical literature of misinterpretation of the different aspects of hereditary syndromes in terms of linkage rather than pleiotropy. Let us consider one or two examples to show the error that results in this kind of misinterpretation.

An association of hereditary juvenile glaucoma and slate-blue eye color was originally described as an example of linkage of glaucoma and eye color genes. If this were so, one would expect that the two traits would remain together for two or three generations at the most; by then, in some individual, the two genes would become separated by crossing over, and two lines would be established, one with glaucoma but not the characteristic eye color, and the other with slate-blue eyes but without glaucoma. From then on, the two traits would remain independent in those branches of the family tree. If, on the other hand, a single gene is responsible for both traits, they would always appear in the same individual, or even if one trait were not expressed in a certain family member it would nevertheless reappear in his descendants. The latter explanation fits the data, and the first does not. Furthermore, it can be shown that both the eye color and the glaucoma result from a developmental defect of the mesenchyme of the anterior chamber during embryonic development. It must be concluded that these two traits are expressions of one and the same autosomal dominant gene.

Perhaps it is more difficult to distinguish between linkage and pleiotropy for a trait such as osteogenesis imperfecta, in which the three main signs (brittle bones, blue sclerae and otosclerosis) are not obviously related in their pathogenesis. This disease, which is inherited as an autosomal dominant, has highly variable expressivity; affected persons may show any one trait, any combination of two traits or the full-blown syndrome. If a parent had all three stigmata and his son had only the blue sclerae, one might suspect a trio of linked genes, which had become separated by crossing over. Here again, however, recurrence of the three associated traits in a later generation would demonstrate that the son was genetically competent to transmit the full triad of effects.

Linkage is discussed in more detail elsewhere. It is mentioned here only to emphasize that the association of two or more traits in members of a kindred does *not* usually suggest linkage, but is more likely due to pleiotropic effects of a single gene.

GENETIC HETEROGENEITY: SEVERAL GENES, ONE EFFECT

When a genetic disorder which at first glance appears to be uniform is carefully studied, it often can be shown that it is composed of a number of separate disorders which are only superficially alike. If mutations at different loci can independently produce the same trait, or traits which are difficult to distinguish clinically, that trait is said to be **genetically heterogeneous.** Genetic heterogeneity is an important aspect of the delineation of new disease entities within one category of disease.

An excellent test of heterogeneity is pedigree analysis. For example, it may happen that two persons, both with recessively inherited congenital deafness, marry and produce only hearing children (Fig. 4–14). A possible explanation is that the parents' deafness is caused by different recessive genes, and that each has normal alleles at the locus for which the other has only abnormal genes. Thus, if the two recessive genes concerned are *d* and *e*, the father could be $d/d\ E/E$ (deaf because he is homozygous for gene *d*) and the mother could be $D/D\ e/e$ (deaf because she is homozygous for gene *e*). The children of the $d/d\ E/E \times D/D\ e/e$ mating would all be $D/d\ E/e$, and since they would have a normal dominant gene at each locus they would have normal hearing.

Linkage analysis has been used to demonstrate genetic heterogeneity in elliptocytosis, a dominantly inherited abnormality of the erythrocytes. One of the earliest known examples of autosomal linkage in man was that of elliptocytosis and Rh blood groups. Not all pedigrees of elliptocytosis show this linkage, and it appears that elliptocytosis may result from an autosomal dominant gene at either of two loci, only one of which is linked to the Rh locus. Clinically, the two genetic forms are almost indistinguishable.

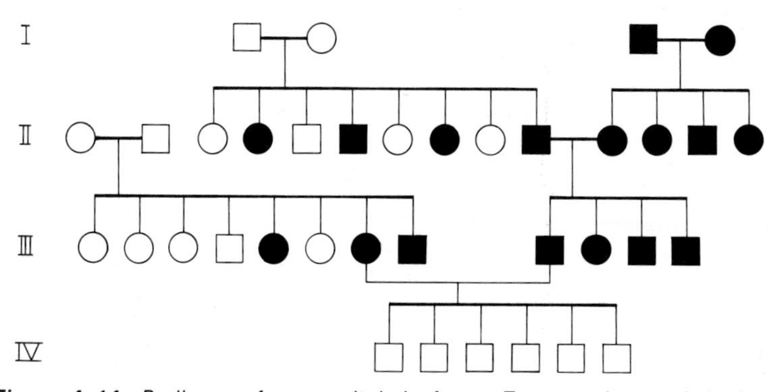

Figure 4–14 *Pedigree of congenital deafness. Two marriages of deaf parents produce only deaf children. Presumably the parents are homozygous for the same recessive gene. However, III–7 and III–9, both deaf and both apparently homozygous for a recessive form of deafness, have only normal children. Here the probable explanation is that the genes causing deafness in the parents are not at the same locus. (Adapted from Figure HD–2, page 219, "Hereditary Deaf Mutism with particular reference to Northern Ireland" by A. C. Stevenson and E. A. Cheeseman,* Annals of Human Genetics. *Used with the permission of Cambridge University Press.)*

The Muscular Dystrophies

An example of genetic heterogeneity is provided by the muscular dystrophies, a group of genetically determined primary myopathies. There are three main types, which differ from one another in their modes of inheritance, onset age and severity (Walton and Nattrass, 1954).

1. Duchenne muscular dystrophy is inherited as an X-linked recessive; only males are affected. It has its onset in very early childhood, and progresses rapidly (see page 68 and Fig. 4–9).

2. Limb-girdle muscular dystrophy is inherited as an autosomal recessive. It usually appears at about the age of puberty, and is less severe and less rapidly progressive than the Duchenne type, but usually eventually results in severe incapacity.

3. Facioscapulohumeral muscular dystrophy is inherited as an autosomal dominant. Expression is irregular, and formes frustes may occur. Onset is usually in the teens, but may be later. The disease usually progresses slowly and does not shorten life.

There are several other types of muscular dystrophy, each of which is rare. Clearly, accurate diagnosis of the specific type is of great importance both for prognosis in the individual case and for genetic counseling of the family.

SEX-LIMITED AND SEX-INFLUENCED TRAITS

If a trait is inherited by a gene on the X chromosome (or, for that matter, on the Y), the sex ratio of the affected individuals is not 1:1. An abnormal sex ratio may provide the first hint that a trait is caused by an X-linked gene, as was the case for the X-linked blood group system Xg (page 227). However, there are many traits in which the sex ratio is abnormal even though the inheritance is not X-linked. This is not surprising, because the milieu in which any gene acts is determined partly by the sexual constitution.

Sex-Limited Traits

A trait which is autosomally transmitted but expressed in only one sex is said to be **sex-limited.** An example of a sex-limited autosomal dominant trait is precocious puberty, in which heterozygous males develop secondary sexual characteristics and undergo an adolescent growth spurt at about the age of four years. Because the epiphyses of the long bones fuse at an early age, affected boys are much taller than their contemporaries but soon stop growing and end up as short men. Figure 4–15 is a portion of a large pedigree of precocious puberty, showing male-to-male transmission.

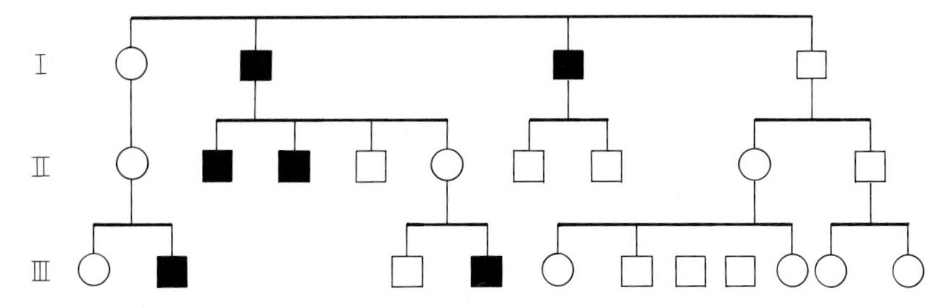

Figure 4–15 *Pedigree of autosomal dominant precocious puberty. The disorder is expressed only in males, but can be transmitted by females. X-linkage can be ruled out because male-to-male transmission occurs.*

Sex-limited expression cannot always be readily distinguished from X-linked recessive inheritance for traits in which males do not reproduce, because the most critical evidence for X-linked inheritance, namely absence of male-to-male transmission, cannot be provided. There are several good reasons for accepting the fact that the Duchenne gene is X-linked, but the case of testicular feminization has been more difficult to resolve. In this disorder, affected males are XY and have testes, but are born with female external genitalia and are raised as females. There is also a form in which affected males have either male, ambiguous or female genitalia but develop breasts and other female secondary sexual characteristics at puberty. It is believed that testicular feminization is

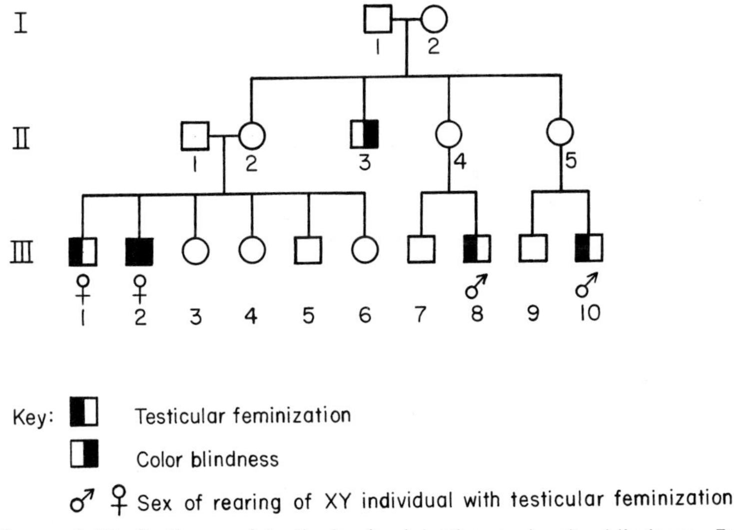

Figure 4–16 *Pedigree of testicular feminization and color blindness. For description, see text.*

TABLE 4–4 EXAMPLES OF AUTOSOMAL PHENOTYPES WITH UNEQUAL EXPRESSION IN MALES AND FEMALES*

Phenotype	Genetics	Sex Difference
Baldness	Autosomal dominant in males, recessive in females?	Great excess of males
Congenital adrenal hyperplasia	Autosomal recessive	More often recognized in females
Hemochromatosis	Autosomal dominant	Excess of males; loss of blood in menses may protect females
Hypogonadism, male	Autosomal recessive types	Limited to males
Legg-Perthés disease	Autosomal dominant type	Excess of males
Myopathy limited to females	Autosomal dominant	Limited to females
Precocious puberty	Autosomal dominant	Limited to males

*Data chiefly from McKusick, 1971.

produced not by failure of the testes to produce androgens, but by deficient response of the target tissues to androgen (possibly deficiency of the enzyme that converts testosterone to the form active within cells, dihydrotestosterone. A comparable gene in the mouse, *Tfm,* is known to be X-linked (Lyon and Hawkes, 1970). By analogy to other genes known to be X-linked in a variety of mammals, e.g., the gene for G6PD (glucose-6-phosphate dehydrogenase), it can be concluded that testicular feminization is probably X-linked. A pedigree of testicular feminization showing recombination between the testicular feminization locus and another X-linked locus (the deutan form of color blindness) is shown in Figure 4–16.

Sex-Influenced Traits

Table 4–4 lists the sex ratios in a number of different single-gene disorders. This table illustrates that many autosomal genes are expressed with quite different frequencies in the two sexes; i.e., they are sex-influenced.

ONSET AGE

Many genetic disorders are not present at birth but are manifested later in life, some at a characteristic age and others at variable ages throughout the life span. Genetic diseases are, of course, not necessarily congenital, nor are congenital diseases necessarily genetic. To speak of a disease as genetic means that genes are clearly implicated in its etiology; but to say that it is congenital means only that it is manifest at birth.

The onset of a genetic disease may occur at any age during life. Chromosomal aberrations have been found in many first-trimester abortuses, indicating that the aberrations have had a lethal effect in fetal life. Many conditions arise during differentiation of the embryo and are evinced as congenital malformations in the newborn infant (e.g., ectrodactyly or lobster claw defect). Some genetic diseases are manifested as the baby begins its independent life (e.g., phenylketonuria, galactosemia). Other defects of genetic origin appear at other characteristic ages; e.g., Tay-Sachs disease at four to six months, Duchenne muscular dystrophy when the child begins to walk and acute intermittent porphyria in the young adult.

Still other genetically caused diseases may develop at various ages (e.g., myotonic dystrophy, Huntington's chorea and diabetes mellitus). A list of single-gene disorders with characteristic onset ages is given in Table 4–5.

TABLE 4–5 CHARACTERISTIC ONSET AGES OF SOME GENETIC DISEASES

Typical Onset Age	*Condition*
Lethal during prenatal life	Some chromosomal aberrations Some gross malformations
Prior to birth	Congenital malformations Chromosomal aberrations Some forms of adrenogenital syndrome Some forms of deafness
Soon after birth	Phenylketonuria Galactosemia Cystic fibrosis
During the first year of life	Tay-Sachs disease Duchenne muscular dystrophy
Near puberty	Limb-girdle muscular dystrophy
Young adulthood	Acute intermittent porphyria Hereditary juvenile glaucoma
Variable onset age	Diabetes mellitus (0 to 80 years) Facioscapulohumeral muscular dystrophy (2 to 45 years) Huntington's chorea (15 to 65 years) Myotonic dystrophy (birth to old age)

may, in some diseases, be associated with more severe signs and symptoms, progressive worsening of a condition from generation to generation used to be regarded as the normal course for many inherited diseases. Now it is generally agreed that the phenomenon of anticipation is apparent rather than real, even in conditions like diabetes mellitus, in which there may be an onset age difference of several years between parent and child.

In myotonic dystrophy, a muscle disorder, there is still some controversy about anticipation. Since it presents a particularly wide difference between average age of onset in parent and child, it will be considered in more detail. Myotonic dystrophy is a form of muscular dystrophy inherited as an autosomal dominant with some irregularity of expression, in which affected individuals have difficulty relaxing their grip (myotonia). There are other, pleiotropic, effects such as frontal baldness (more marked in males) and cataracts. According to Penrose (1948) the average onset age in parents is 38 and in children, 15—a difference of 23 years. No other disease has such a wide difference in onset age in successive generations.

This difference is believed to represent a bias in the selection of the parents in the series. If the parent is discovered through his affected child (the propositus), the most striking parent-child pairs will be those in which parent and child have the disease at the same time. In these cases, the disease occurs simultaneously in the two generations, but parent and child differ in onset age by some 20 or 30 years.

The difficulties in studying certain combinations of onset age in a parent and child can be shown by comparing typical dates for a parent born in 1930 and a child born in 1955.

Onset Age		**Year of Onset**		**Number of Years Apart in Time**
Parent	*Child*	*Parent*	*Child*	
40	20	1970	1975	5
40	40	1970	1995	25
40	60	1970	2015	45

Clearly, the case in which onset is late in parent and early in child would be more striking and so more likely to be identified than cases which do not coincide fairly closely in time. An additional complication is that early-onset persons are, on the average, less likely to transmit a deleterious condition than are late-onset persons. Chiefly for these reasons, most workers dismiss the phenomenon of anticipation as a statistical artifact. Still, not all workers agree, and it may be that early-onset cases have the myotonic dystrophy gene and also have been exposed to the action of an unfavorable intrauterine environment during gestation in an affected mother. (When myotonic dystrophy begins in childhood, the gene has usually been inherited from the mother, not the father; Harper and Dykes, 1972.) For the present, therefore, the explanation of apparent anticipation in myotonic dystrophy must remain open.

INTERACTION OF NONALLELIC GENES

Although we have discussed genes and their effects mainly in terms of a single pair of alleles determining a single trait, no gene can be said to act independently of all other genes. In inbred mice there are many examples of mutant genes which are expressed differently on different genetic backgrounds. It is certain that the effect of background genes is important in man also, even though analysis of gene action against different backgrounds is not practical in man.

The simplest demonstration of *nonallelic interaction* is that of two gene pairs acting together to produce a phenotype. There are many good examples in classic genetics, but few are known in man. To illustrate nonallelic interaction, we will describe the ABO secretor situation.

"Secretors" are persons who secrete the antigens A, B and H in the saliva (page 209). To be a secretor of antigen A, one must have both gene *A* and gene *Se*. (This explanation overlooks additional complexities in the secretion of blood group substances.) Assume a mating of two persons, each heterozygous at both loci and therefore a secretor of A substance:

$$A/O\ Se/se \times A/O\ Se/se$$

A and *O* assort independently of *Se* and *se*, so the gametes produced by each parent could be of four different types:

A Se   *O Se*   *A se*   *O se*

The checkerboard (page 88) shows the combinations of these four gametes among the progeny.

The nonsecretors of A substance shown here are either *se/se* or *O/O* (or both). The secretor of A:nonsecretor of A ratio is 9:7. The explanation of the modification of the typical 9:3:3:1 mendelian dihybrid ratio is clear if it is recognized that two different nonallelic gene pairs are concerned, but the ratio could be wrongly interpreted if it was believed that only a single gene locus was involved. This pattern of transmission resulting in a 9:7 ratio, known as **epistasis,** is typically seen when the product of one of the two genes concerned acts as the substrate for the other.

	A Se	*O Se*	*A se*	*O se*
A Se	*A/A Se/Se* Secretor of A	*A/O Se/Se* Secretor of A	*A/A Se/se* Secretor of A	*A/O Se/se* Secretor of A
O Se	*A/O Se/Se* Secretor of A	*O/O Se/Se* Nonsecretor of A	*A/O Se/se* Secretor of A	*O/O Se/se* Nonsecretor of A
A se	*A/A Se/se* Secretor of A	*A/O Se/se* Secretor of A	*A/A se/se* Nonsecretor of A	*A/O se/se* Nonsecretor of A
O se	*A/O Se/se* Secretor of A	*O/O Se/se* Nonsecretor of A	*A/O se/se* Nonsecretor of A	*O/O se/se* Nonsecretor of A

GENERAL REFERENCES

See *General References* for Chapter 1.

PROBLEMS

1. Assume that curly hair (C) is dominant to straight hair (c).
   (a) Two curly-haired parents have a straight-haired child. What are the parental genotypes?
   (b) List all the possible genotypic and phenotypic combinations of parents that can produce a curly-haired child.

2. Brachydactyly (B) is dominant to the normal condition (b). A brachydactylous man marries a normal woman. What is the chance that:
   (a) their first child will be brachydactylous?
   (b) all their three children will be brachydactylous?
   (c) none of their three children will be brachydactylous?

3. Dentinogenesis imperfecta (D) is dominant to the normal condition. A man with brachydactyly and normal teeth marries a woman with normal fingers and dentinogenesis imperfecta.
   (a) State the genotypes of the parents with respect to both loci.
   (b) State all the possible genotypes and phenotypes of the offspring of this couple, and their expected proportions.

4. Phenylketonuria (PKU) is inherited as an autosomal recessive. Two normal parents have a child with PKU.
   (a) Give the genotypes and phenotypes of all matings that could theoretically produce a child with PKU.
   (b) To which of these mating types do the parents of this patient belong?
   (c) What is the probability that the next child will also have PKU?
   (d) What is the probability that a normal child of these parents is a carrier of PKU?

5. A normal man who is a carrier of PKU marries his first cousin.
   (a) What is the probability that their first child will have PKU?
   (b) If their first child has PKU, what is the probability that the second child will also be affected?

6. Albinism is inherited as an autosomal recessive.
   (a) If an albino man marries a carrier woman, what is the probability that their first child will be affected?
   (b) If he marries his first cousin, what is the probability that their first child will be affected?
   (c) If he marries a normal homozygote, what is the probability that their first child will be affected?

7. An albino man with normal hearing marries a woman with normal pigmentation who has a recessive form of deafmutism.
   (a) State the most probable parental genotypes.
   (b) What genotypes and phenotypes are expected in their offspring?
   (c) If a son of these parents marries a woman of the same genotype as himself, what are the possible genotypes and phenotypes of their progeny?

8. Color blindness is inherited as an X-linked recessive.
   (a) A color blind man marries a normal woman whose brother is color blind.
      (1) What is the probability that their first son will be color blind?
      (2) What is the probability that their first child will be a color blind son?
   (b) A normal man marries a normal woman whose brother is color blind. What is the probability that their first son will be color blind?
   (c) What is the one chief distinguishing feature of X-linked inheritance?

9. What proportion of the grandsons of a hemophilic male are hemophiliacs? (Assume that his marriage and his children's marriages are to normal partners.)

10. A certain trait appears in half the offspring of affected individuals.
    (a) What information does this observation convey as to its mode of inheritance?
    (b) If the trait affects half the sons and half the daughters of affected males, what is its mode of inheritance?
    (c) If the trait affects no sons but all daughters of affected males, how is it inherited?

11. A man with curly hair and albinism is married to a woman with straight hair and normal pigmentation. Their first child is a straight-haired albino.
    (a) Give the parental genotypes.
    (b) Give the genotypes and phenotypes of all possible progeny of these parents, and their expected proportions.

12. A color blind albino and a woman with normal color vision and pigmentation produce a color blind albino daughter. Give the genotypes of the parents and child.

13. A man with brachydactyly, albinism and color blindness marries a normal woman. What are the most probable genotypes and phenotypes of their children?

5

Biochemical Genetics

Many different proteins are synthesized in body cells. These proteins, either enzymes or structural components of cells and tissues, are responsible for all the developmental and metabolic processes of the organism.

The fundamental relationship between genes and proteins is that the sequence of base pairs in the DNA of a given gene codes for the sequence of amino acids in the corresponding polypeptide chain. Alteration of the base pair sequence in the gene results in the synthesis of a variant polypeptide with a correspondingly altered amino acid sequence. Proteins are composed of one or more polypeptide chains. Hence, a gene mutation results in the formation of a variant protein, which may have changed properties and function as a consequence of its altered structure.

The phenotypic changes produced by gene mutations are numerous and varied. If the mutation results in an amino acid substitution in a so-called structural protein, such as hemoglobin, the phenotypic effect will depend upon how the alteration in amino acid sequence affects such properties of the hemoglobin molecule as its affinity for oxygen or tendency to sickle. If the amino acid sequence of an enzymatic polypeptide is affected, the variant enzyme synthesized by the mutant gene may be altered quantitatively or qualitatively. Most of the variant enzymes known have less activity than the normal enzyme and may even be completely inactive. Rarely, the mutation may produce excessively high activity. Occasionally the mutation leads to the formation of an unstable polypeptide chain that is rapidly destroyed *in vivo*. In addition to determining the amino acid sequence of proteins, genes regulate the rate at which certain proteins are formed. The thalassemias (see later) may provide an example of a genetically altered regulatory mechanism.

Not all genetic changes lead to clinically abnormal phenotypes. On the contrary, many proteins are known to occur in two or more relatively common, structurally different, genetically distinct "normal" forms. Such a situation is known as **polymorphism** (page 237). However, in other cases the altered protein may lead to a variety of clinical and pathological consequences.

The subject matter of this chapter falls into three main areas. The first of these is the hemoglobins, normal and abnormal, which have done more than any other type of protein to clarify the relationship between genes, proteins and disease. Next we will examine the variety of inborn errors of metabolism determined by gene mutations. Finally, we will describe briefly some examples of genetic variations in response to drugs (pharmacogenetics).

THE HEMOGLOBINS

The genetic control of protein synthesis has chiefly been elucidated by the study of the hemoglobinopathies, blood disorders in which the primary defect is an alteration in the genetically determined molecular structure of hemoglobin. The first step in understanding the genetics of the hemoglobinopathies was taken by Neel (1949), who showed that patients with a blood disorder known as sickle cell disease were homozygous for a gene which produced a similar but much milder abnormality, sickle cell trait, in both parents, who were heterozygous. Shortly afterwards Pauling et al. (1949) designated sickle cell disease as the prototype of the "molecular diseases," in which an abnormal hemoglobin molecule was the basic defect (see later). Ingram (1956) then showed that the abnormality in the hemoglobin present in sickle cell disease consisted of a change in only one of the 300 amino acids present in the hemoglobin half-molecule.

The Structure of Hemoglobin. Hemoglobin is the oxygen-carrying pigment of the red blood cells. A hemoglobin molecule is composed of four subunits arranged to form two identical half-molecules. Each of the four subunits is made up of two parts: a polypeptide chain, globin, and an iron-containing constituent, heme. The heme portion is alike in all genetically different forms of hemoglobin, and genetic studies are concerned with the structure of the globin portion only.

Two different kinds of polypeptide chains are combined to form a normal hemoglobin molecule. Normal adult hemoglobin (Hb A) is made up of two identical α chains and two identical β chains. The four polypeptide chains are folded and fitted together to form a roughly globular molecule with a molecular weight of approximately 68,000 (Fig. 5–1). Its structural "formula" is written $\alpha_2\beta_2$. The two kinds of chains are roughly equal in length, the α chain containing 141 amino acids and the β chain containing 146 amino acids. The α and β chains resemble one another markedly in primary structure (amino acid sequence) and

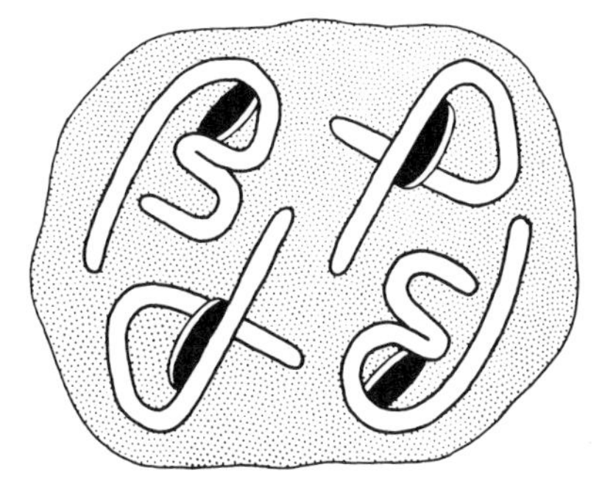

Figure 5–1 *Schematic representation of a molecule of normal adult hemoglobin. The molecule is composed of two α and two β chains, each of which is associated with a heme moiety (black disc). The peptide chains have been fancifully represented as the corresponding Greek letters. (After Ingram 1959.* Nature *183:1795–1798.)*

tertiary structure (three-dimensional configuration). They also resemble myoglobin, the oxygen-carrying protein of muscle, though less closely; the myoglobin molecule has only a single polypeptide chain, but similarities in amino acid sequence and tertiary structure suggest that hemoglobin and myoglobin molecules evolved from a common ancestral polypeptide.

Five distinct structural gene loci, designated α, β, γ, δ, and ϵ, direct the synthesis of the globin part of hemoglobin. Each locus is responsible for the structure of one type of polypeptide chain. Besides Hb A, at least four other normal forms of hemoglobin occur. Fetal hemoglobin (Hb F) is found in the developing fetus and disappears after birth; it is composed of two α and two γ chains and is designated $\alpha_2\gamma_2$. (It is now known that there are two different fetal hemoglobins, differing at only a single site, one having glycine and the other alanine at position 136.) About 2 per cent of the hemoglobin present in the blood of the adult is a form designated Hb A_2, and consists of two α chains and two δ chains ($\alpha_2\delta_2$).

Two types of embryonic hemoglobin, known as Gower I and Gower II, are present during the first 90 days of prenatal life. Gower I is believed to be composed of four ϵ chains (ϵ_4), and Gower II of two α and two ϵ chains ($\alpha_2\epsilon_2$).

The normal and abnormal hemoglobins referred to in this section are listed in Table 5–1.

ABNORMAL HEMOGLOBINS

Most of the abnormal hemoglobins are produced by mutations in the structural genes which determine the amino acid sequence of the globin portion of the hemoglobin molecule. More than 100 abnormal hemoglobins have been described. The first to be detected electrophoretically and still the most important clinically is sickle cell hemoglobin (Hb S).

TABLE 5–1 EXAMPLES OF NORMAL AND ABNORMAL HEMOGLOBINS*

Normal Hemoglobins	*Formulas*
Hb A	$\alpha_2\beta_2$
Hb A_2	$\alpha_2\delta_2$
Hb F	$\alpha_2\gamma_2{}^{136\ gly}$ $\alpha_2\gamma_2{}^{136\ ala}$
Gower I	ϵ_4
Gower II	$\alpha_2\epsilon_2$
Variant Hemoglobins	***Formulas***
Hb S	$\alpha_2\beta_2{}^S$ $(\alpha_2\beta_2{}^{6\ glu\rightarrow val})$
Hb C	$\alpha_2\beta_2{}^C$ $(\alpha_2\beta_2{}^{6\ glu\rightarrow lys})$
Hb Lepore	$\alpha_2\delta\text{-}\beta_2$
Hb Hopkins-2	$\alpha_2{}^{Ho-2}\ \beta_2\ (\alpha_2{}^{112\ his\rightarrow asp}\ \beta_2)$
Hb H	β_4
Hb Bart's	γ_4
Hb C Georgetown	$\alpha_2\beta_2{}^{6\ glu\rightarrow val\ +}$
Hb C Harlem	$\alpha_2\beta_2{}^{6\ glu\rightarrow val\ +\ 73\ asp\rightarrow asn}$
Hb Gun Hill	$\alpha_2\beta_2{}^{93\text{-}97\ deletion}$
Hb Freiburg	$\alpha_2\beta_2{}^{23\ deletion}$
Hb Constant Spring	$\alpha_2{}^{CS}\ \beta_2\ (\alpha_2{}^{+142-171}\ \beta_2)$

*In most of these formulas, the abnormal chain is identified and the nature and location of the alteration in amino acid sequence, if known, are indicated.

Sickle Cell Disease

Sickle cell disease is a severe hemolytic disease characterized by a tendency of the red cells to become grossly abnormal in shape ("sickled cells") under conditions of low oxygen tension (Fig. 5–2). The clinical manifestations include anemia, jaundice and "sickle cell crises" marked by impaction of sickled cells, vascular obstruction and painful infarcts in various tissues such as the bones, spleen and lungs. The disease has a characteristic geographic distribution, occurring most frequently in Equatorial Africa, less commonly in the Mediterranean area and India, and in countries to which people from these regions have migrated. About 0.25 per cent of American Negroes are born with this disease, which is often fatal in early childhood.

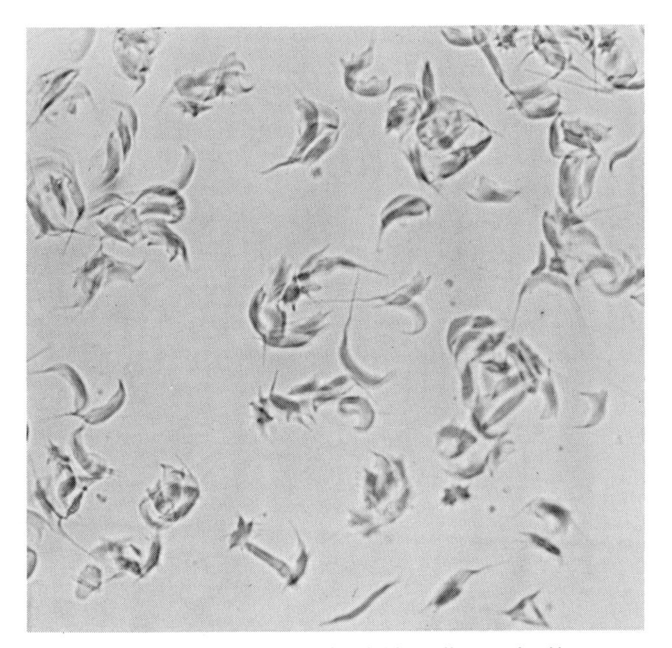

Figure 5–2 *The red cell phenotype in sickle cell anemia. Homozygotes for the sickle cell hemoglobin gene form sickle cell hemoglobin, but no normal adult hemoglobin. Under conditions of reduced oxygen tension the abnormal hemoglobin aggregates to form long projections which distort the cells. (Photomicrograph courtesy of J. H. Crookston.)*

Parents of affected children, though usually clinically normal, have red cells which sickle when subjected to very low oxygen pressure *in vitro* (that is, they show a positive sickling test). This heterozygous state, known as sickle cell trait, is present in approximately 8 per cent of American Negroes. Persons with sickle cell anemia are homozygous for the mutant gene (Neel, 1949).

The physicochemical abnormality of sickle cell hemoglobin was demonstrated in 1949 by Pauling, Itano, Singer and Wells, who studied Hb A and Hb S by electrophoresis. They found that Hb A and Hb S were readily distinguishable from one another by their mobility in an electrical field, and concluded that their globin molecules were different. They also found that the hemoglobin of persons with sickle cell trait behaved like a mixture of normal and sickle cell hemoglobin. The structure of sickle cell hemoglobin was, in their words, "a clear case of a change produced in a protein by an allelic change in a single gene involved in synthesis."

The precise nature of the change in the protein molecule predicted by these workers was identified by Ingram, who applied the technique of "fingerprinting" (not to be confused with fingerprinting for dermatoglyphic analysis, which is discussed in Chapter 14). Fingerprinting is a technique developed by Sanger for determining the structure of proteins by breaking them down with trypsin and then separating the resulting small peptides by electrophoresis in one direction and by chromatography at right angles to it (Fig. 5–3). Ingram demonstrated that the difference between normal and sickle cell hemoglobin lay in the β chain, and involved only *one* of the 146 amino acids in the chain, the amino acid sixth in position from the N-terminal end of the β chain. At this position, the amino acid valine has replaced the glutamic acid of normal hemoglobin. The sequences in the corresponding sections of the two hemoglobins are as follows:

Hb A: val-his-leu-thr-pro-*glu*-glu-lys

Hb S: val-his-leu-thr-pro-*val*-glu-lys

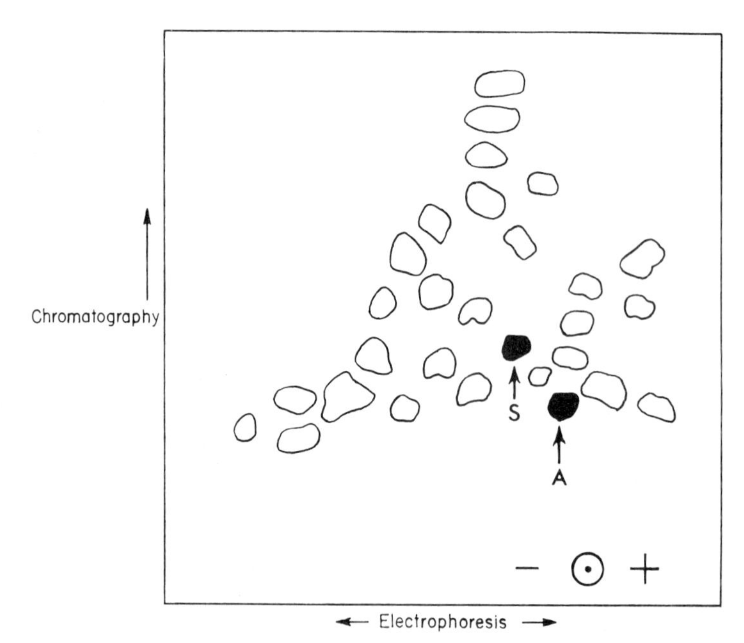

Figure 5–3 *Diagram to compare the "fingerprints" of normal adult hemoglobin and sickle cell hemoglobin. In the fingerprinting technique, the polypeptide chains of globin are first split into smaller peptide fragments (tryptic peptides) by digestion with the proteolytic enzyme trypsin, which cleaves the chain wherever there is a lysine or an arginine residue. The tryptic peptides are then separated in two dimensions: horizontally by electrophoresis and vertically by chromatography. After staining with ninhydrin, each peptide appears as a dark spot, in a characteristic position in the "fingerprint."*

The fingerprints of normal adult hemoglobin (Hb A) and sickle cell hemoglobin (Hb S) differ in only a single peptide. Spot A is present in Hb A but absent in Hb S; conversely, spot S is present in Hb S but absent in Hb A.

The substitution of a valine for a glutamic acid in each of the two β chains of the hemoglobin molecule is the only biochemical difference between the two types of hemoglobin. This difference, which results in an altered electrical charge, explains the difference in electrophoretic mobility of the two hemoglobins. All the clinical manifestations of sickle cell hemoglobin are the consequence of this relatively minor change. The physical basis of the sickling phenomenon is the tendency of the abnormal hemoglobin molecules to aggregate, forming rod-like masses which distort the red cells into sickle shapes under conditions of low oxygen tension.

Because the abnormality of Hb S is localized in the β chain, the formula for sickle cell hemoglobin is written $\alpha_2{}^A\beta_2{}^S$, or simply $\alpha_2\beta_2{}^S$, since it is not essential to show the superscript for a normal chain. This symbol indicates that the α chain is normal, but that the β chain is of the sickle cell hemoglobin type. Note that the abnormality affects both β chains of the molecule. For a reason as yet unknown, hybrid molecules such as $\alpha_2{}^A\beta^A\beta^S$ do not occur. A heterozygote has a mixture of the two types of hemoblobin, A and S, not a compound type. The relationships among clinical status, hemoglobin types and genes can be summarized as follows:

Clinical Status	***Hemoglobin Type***	***Hemoglobin Formula***	***Genotype***
Normal	Hb A	$\alpha_2{}^A\beta_2{}^A$	$\alpha/\alpha\ \beta/\beta$
Sickle cell trait	Hb A + Hb S	$\alpha_2{}^A\beta_2{}^A + \alpha_2{}^A\beta_2{}^S$	$\alpha/\alpha\ \beta/\beta^S$
Sickle cell disease	Hb S	$\alpha_2{}^A\beta_2{}^S$	$\alpha/\alpha\ \beta^S/\beta^S$

Hemoglobin C

Hemoglobin C (Hb C) is another relatively common variant hemoglobin, occurring mainly in Equatorial Africa. The anemia produced by homozygous hemoglobin C is milder than that in sickle cell disease, and sickling does not occur. Biochemical analysis has shown that the abnormality in Hb C is in the β chain, and involves precisely the same position in the amino acid sequence as is involved in sickle cell hemoglobin. In Hb C, at the sixth amino acid from the N-terminal, lysine replaces the glutamic acid of Hb A. The sequence is as follows:

Hb C: val-his-leu-thr-pro-*lys*-glu-lys

The mutation producing Hb C is allelic to that which produces Hb S, since both affect the synthesis of the same polypeptide. Allelism was first demonstrated by family studies. Not only is the same polypeptide involved, but the same amino acid of the polypeptide; thus, since the sequence of amino acids in a polypeptide chain is believed to be determined by the sequence of base triplets (codons) in the DNA molecule, the same codon must be affected by both mutations.

Now that the genetic code for the amino acids is known, we can visualize the process of mutation in the codon which, in Hb A, specifies that a certain position be occupied by a glutamic acid (see Table 3–1).

Amino Acid	*mRNA Code*
Glutamic acid (glu)	GAA or GAG
Valine (val)	GUU, GUC, GUA or GUG
Lysine (lys)	AAA or AAG

Note that a single change (A to U in the second base of the triplet) in the codon for glutamic acid will alter it to code for valine, and that substitution of A for G in the first base of the same glutamic acid codon will change it to code for lysine. Thus, a mutation from Hb A to Hb S involves only the alteration of one base in the complete β locus, and the mutation from Hb A to Hb C involves a different change in the same codon.

Other Abnormal Hemoglobins

Many other abnormal hemoglobins which have been analyzed in detail have been shown to have amino acid substitutions in either the β chain or, less commonly, the α chain. A few rare hemoglobins have other types of abnormalities.

Beta Chain Abnormalities. Besides Hb S and Hb C, more than 70 other hemoglobin variants have amino acid substitutions in the β chain. Substitutions are known at many positions other than position 6, which is substituted in both Hb S and Hb C. Though the mutations which produce all the variant hemoblobins are allelic in that they affect the amino acid sequence of the same polypeptide chain, the mutation is at a different site within the gene; in other words, a different codon has been affected. The term **heteroallele** has been coined for allelic mutations in which different codons are involved, to distinguish them from **eualleles,** in which the same codon is affected.

Hemoglobin Lepore. Hemoglobin Lepore is a special type of abnormal hemoglobin in which the α chain is normal but the non-α chain has been shown to comprise portions of the N-terminal portion of a normal δ, and the C-terminal portion of a normal β chain. It has presumably arisen by unequal crossing over between the two closely linked and very similar genes which code for these two polypeptides. There are only 10 differences in amino acid sequence between the δ and β chains, so that aberrant pairing could occur at that region in meiosis. If crossing over occurs between the two mismatched genes, a Lepore gene is produced. (Fig. 5–4). The clinical and hematological consequences resemble β thalassemia (see later). "Anti-Lepore" hemoglobin, in which the first part of the gene resembles β and the second part resembles δ, is also known.

Alpha Chain Abnormalities. Normal hemoglobin molecules, whether Hb A, Hb A_2, or Hb F, have two α chains in each molecule. (One of the two embryonic hemoglobins has four ε chains only, but embryonic hemoglobin is not a normal constituent of blood in postnatal life and the embryonic hemoglobins are therefore omitted from further discussion here.) A mutation in the gene coding for the α chain would affect all three kinds of hemoglobin, not just one as is the case if a β, γ or δ chain is abnormal. The first six α chain variants to be analyzed in detail each involved a different amino acid of the chain. Independent segregation of an α chain abnormality (Hb Hopkins-2) and a β chain abnormality (Hb S) provided evidence that the α and β genes are at separate loci. In addition, there are some hemoglobin variants in which the α chain is completely missing

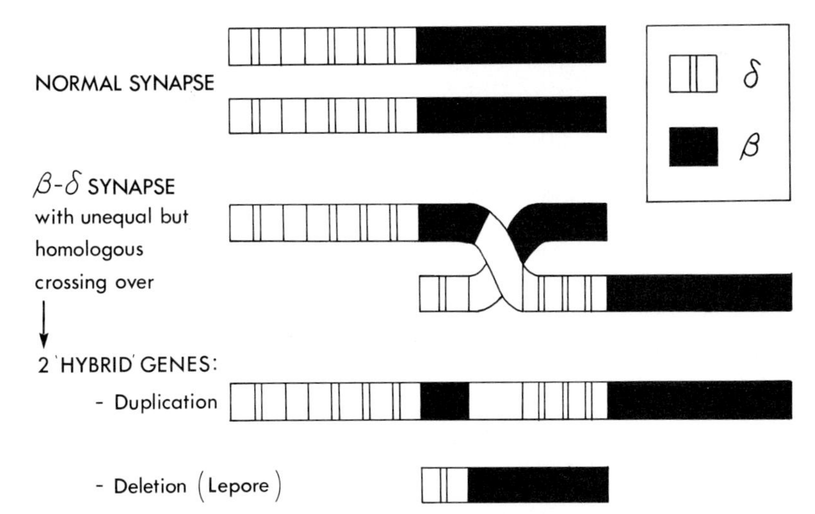

Figure 5–4 *The formation of a Lepore gene by unequal crossing over. The genes for the δ and β chains of hemoglobin are very closely linked and their sequences differ at only 10 sites (shown on the δ gene as vertical lines). If a δ gene accidentally lines up with a β gene when homologous chromosomes pair at prophase of meiosis, crossing over can occur and two hybrid genes can be formed, one with a deletion (the Lepore gene) and one with a duplication (not as yet identified). (Adapted from Giblett 1969.* Genetic Markers in Human Blood. *Blackwell Scientific Publications, Oxford, p. 564, by permission.)*

and the molecule is a tetramer of four identical chains. One of these, hemoglobin H (Hb H) is composed of four β chains (β_4). It is found in patients with one variety of thalassemia and may also occur normally. Hb Bart's (so called because it was first identified at St. Bartholomew's Hospital, London) is made up of four γ chains (γ_4). In persons with these tetrameric hemoglobins, it appears that α chain production is suppressed, but β (or γ) chains are formed and conjugate to form tetramers.

Delta Chain Abnormalities. A mutation in the gene coding for the δ chain leads to production of an abnormal Hb A_2. A few persons with both an abnormal Hb A_2 and sickle cell trait have been identified. By marriage to a normal homozygote, these double heterozygotes have produced children with neither abnormality and children with both, but none with only one or the other, thus providing genetic evidence that the β and δ genes are closely linked.

Double Mutations. At least two abnormal hemoglobins possess two separate amino acid substitutions in the same polypeptide chain, Hb C Georgetown and Hb C Harlem. These two hemoglobins are identified as Hb C because they migrate electrophoretically in the same position as Hb C, not because they have the same amino acid substitution. On the contrary, each has the same mutation as Hb S, plus a second substitution at a different site. The mechanism may be either a second muta-

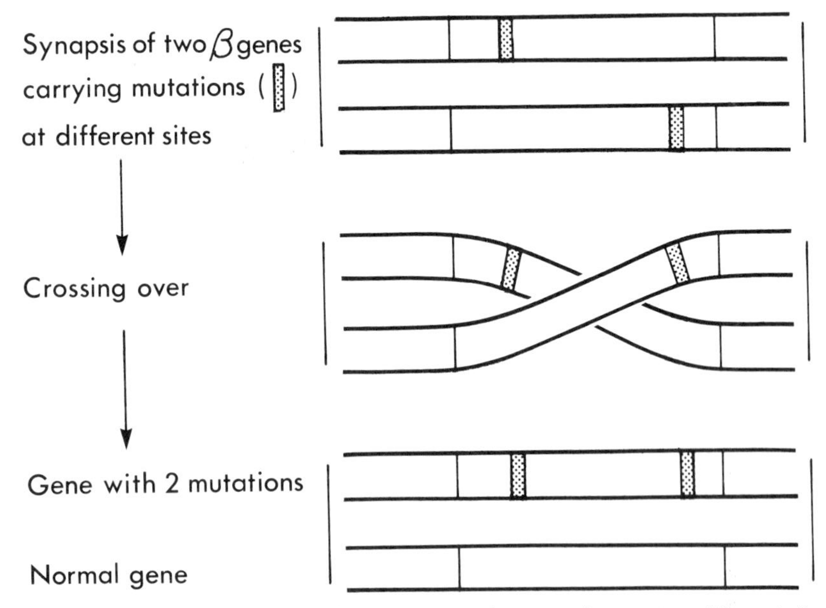

Figure 5–5 *Possible origin of a β gene carrying mutations at two different sites, by intragenic crossing over.*

tion in a β^S allele, or intragenic crossing over during meiosis between two β genes carrying mutations at different sites (Fig. 5–5).

Deletions. As previously mentioned, the α hemoglobin chain is five amino acids shorter than the β chain. From comparison of amino acid sequences, it seems likely that amino acids 54 to 58 of the β chain are missing in the α chain. The variant hemoglobin Gun Hill ($\alpha_2\beta_2^{93\text{-}97\ \text{deletion}}$) appears to have lost five amino acid residues, and Hb Freiburg ($\alpha_2\beta_2^{23\ \text{deletion}}$) has lost one. All of these deletions could have arisen as a result of abnormal pairing and intragenic crossing over during meiosis (Fig. 5–6).

Chain-Termination Mutations. Two types of chain-termination mutations can occur—mutation to a termination codon (UAA or UAG) at some position within the gene, or mutation of the termination codon itself. The first of these types has not yet been definitely identified in any human protein. It would presumably lead to the production of a short, unstable polypeptide. Perhaps more intriguing are mutations of the codons which normally terminate the chain, so that the growing polypeptide continues to add amino acids until the next termination codon is reached. In Hb Constant Spring, which may be the result of such a chain-termination mutation, the α chain is 172 amino acids long instead of the normal 141. Among the interesting features of this mutation is the fact that the additional 29 amino acid residues do not resemble the amino acid sequence of

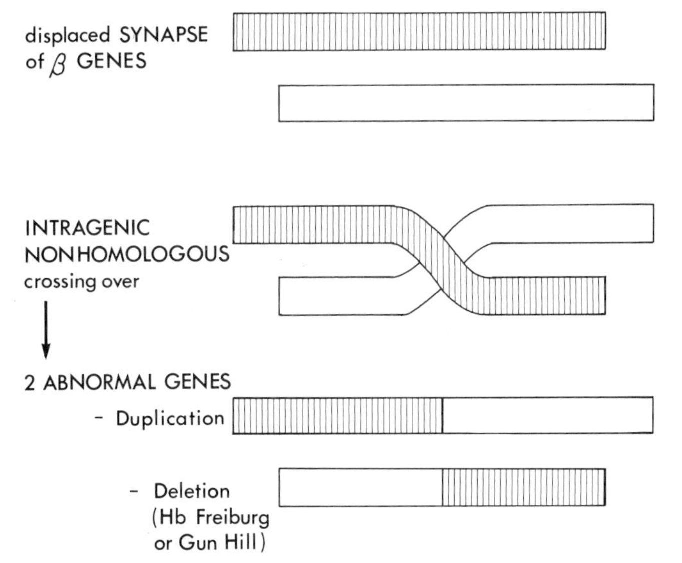

Figure 5–6 *Possible origin of a β gene carrying a deletion, by intragenic nonhomologous crossing over. For discussion, see text. (Adapted from Giblett 1969.* Genetic Markers in Human Blood, *Blackwell Scientific Publications, Oxford, p. 562, by permission.)*

any known protein; hence, Hb Constant Spring may help elucidate the composition of the chromosomal DNA *between* the structural genes.

Hereditary Persistence of High Fetal Hemoglobin. This rare genetic trait results in the presence of large amounts of fetal hemoglobin in adult life. The homozygote has no normal adult hemoglobin, neither Hb A nor Hb A_2. No clinical abnormality is associated with the presence of high fetal hemoglobin. The genetic importance of this disorder is the light it casts on the mechanism which normally switches from fetal to adult hemoglobin production; in affected persons the β and δ loci fail to be switched on and the γ locus remains active in adult life. Family studies indicate that the gene for "high F" is either very closely linked or allelic to the β gene.

The Thalassemias. The thalassemias (Cooley's anemia, Mediterranean anemia) are a group of hereditary anemias in which the erythrocytes typically have a target cell appearance. They have recently been reviewed by Bannerman (1972). Unlike the hemoglobinopathies already described, in the thalassemias there is no associated abnormality in the primary amino acid structure of hemoglobin, but there is decreased synthesis of a particular globin chain. There are two forms: a severe form, thalassemia major (homozygous), and a mild form, thalassemia minor or thalassemia trait (heterozygous).

Thalassemia occurs in a characteristic geographical distribution, in a band around the Old World—the Mediterranean, the Middle East, India, and the Orient. The name is derived from the Greek word for sea, *thalassa,* and signifies that the disease was first discovered in persons of Mediterranean stock. Thalassemia minor is sufficiently common in Canada and the United States to pose an important problem of differential diagnosis from iron deficiency anemia.

The defect in rate of globin chain synthesis may involve β chains, δ chains, both β and δ chains, or α chains; hence, we speak of β thalassemia, δ thalassemia and so forth. The genetic mechanism is not at all clear. In β thalassemia, family studies indicate that the thalassemia gene is either allelic to the β chain gene or very closely linked (perhaps adjacent) to it; in the latter case, perhaps the β thalassemia mutation affects an "operator gene," as in the model of genetic regulation proposed by Jacob and Monod (page 44). The control of hemoglobin synthesis is discussed further in Chapter 10.

THE EVOLUTION OF HEMOGLOBIN

Because of the technical feasibility of determining the amino acid sequences of polypeptides and the development of the concept that the amino acid sequence of a given polypeptide is encoded in a specific gene, it is possible to begin to reconstruct evolutionary events at the molecular level.

The amino acid sequences of the different globin chains of man have extensive homologies. The α chain is composed of 141 amino acids; the β, γ and δ chains each are composed of 146 . After allowances are made for the difference in length of the chains, the α chain differs from the γ chain in 83 amino acids. The β chain differs from the α chain in 77 amino acids, and from the γ chain in 38. The β and δ differ in only 10 amino acids. Forty-one sites are invariant in all four chains.

The muscle protein myoglobin has a single polypeptide chain rather than a four-subunit molecule, but in size, amino acid sequence and three-dimensional configuration it resembles the hemoglobin chains.

According to an evolutionary model proposed by Ingram (1963), and by Epstein and Motulsky (1965), the α gene arose by duplication of a primitive gene for a myoglobin-like molecule. Mutation gradually brought about divergence of the two primitive genes. Further gene duplication led to the formation of "new" genes which in turn diverged from one another (Fig. 5–7) to result eventually in the amino acid sequences of the globin chains of present-day man.

There is a limit to the kind of amino acid substitution that is acceptable in terms of natural selection. Only those substitutions that do not impair function can be maintained, and it is of interest that the sites of heme attachment are invariant in the different globin chains (and in those of other vertebrate species that have been examined; Perutz, 1965). Another constraint is that those amino acids which are classified as polar

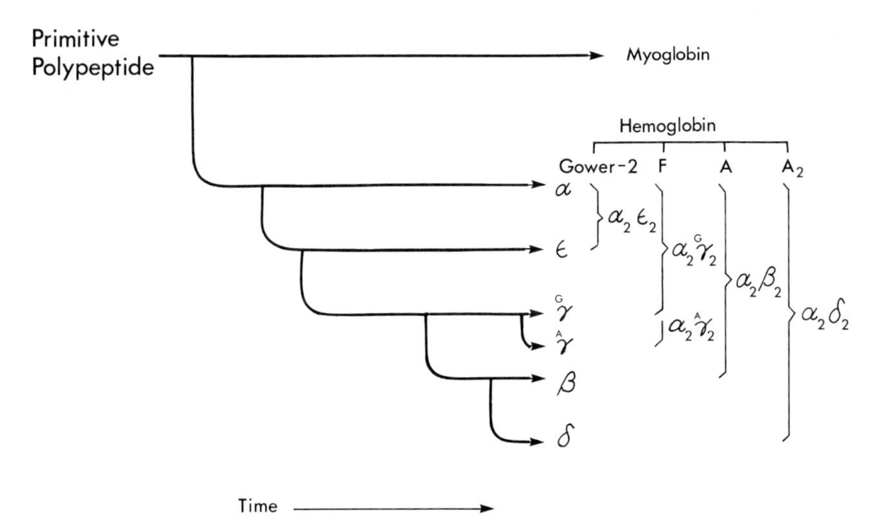

Figure 5–7 *Evolution of hemoglobin. A primitive gene gave rise by duplication to genes for myoglobin and the α chain of hemoglobin, which then diverged by accumulation of mutations. Successive gene duplications gave rise to the ϵ, γ, β and δ genes. A recent duplication accounts for the single amino acid difference between $^G\gamma$ and $^A\gamma$. (Adapted from Epstein and Motulsky 1965.* Progr. Med. Genet. *4:85–127 (Grune & Stratton, Inc.); and from Giblett 1969.* Genetic Markers in Human Blood. *Blackwell Scientific Publications, Oxford, p. 402, by permission.)*

(hydrophilic) tend to occupy the surface of the hemoglobin molecule, where they are exposed to water, whereas those which are nonpolar (hydrophobic) are in the interior, where they play a role in maintaining the structure of the molecule. Consequently, substitutions are not likely to be preserved if they do not have the right type of polarity. The fact that 41 sites are alike in all four chains indicates that selection places rigid constraints on some types of changes in specific proteins.

As a generalization, many proteins have "sensitive areas," in which mutations cannot occur without affecting function, and "insensitive areas," where mutations are tolerated more freely.

A NOTE ON HEMOGLOBIN NOMENCLATURE

New hemoglobins are being discovered too rapidly for their nomenclature to follow any firm rules. The normal hemoglobins, Hb A, and Hb A_2 and Hb F, and the first abnormal hemoglobin to be discovered, Hb S, are obviously named for their associations with adult, fetal, and sickle cell forms of hemoglobin. Other hemoglobins were at first named alphabetically in order of their discovery, but this system soon broke down in practice. At present, newly identified hemoglobins are usually named from the place where they were discovered (e.g., Hb Bart's, Hb Hopkins-2), or by their resemblance to some previously discovered hemoglobin (e.g., Hb C Georgetown, Hb C Harlem). From time to time, international conferences are held at which the nomenclature is brought up to date in the light of recent findings.

INBORN ERRORS OF METABOLISM

An inborn error of metabolism is a genetically determined biochemical disorder in which a specific enzyme defect produces a metabolic block that may have pathological consequences. The concept of inborn errors of metabolism as causes of disease was first proposed by the eminent physician Garrod in 1902, shortly after the rediscovery of Mendel's laws. Garrod noted that in alcaptonuria and several other disorders the patients appeared to lack the ability to perform one specific metabolic step. (In the case of alcaptonuria, this step was the breakdown of the benzene ring of the amino acid tyrosine. Failure to perform this breakdown led to the excretion of an abnormal constituent of urine, homogentisic acid or alcapton.) Garrod also noted that these disorders showed a striking familial pattern of distribution, in that two or more sibs might be affected though the parents and other relatives were usually normal, and a higher than normal incidence of consanguineous marriage was found among the parents. Through discussions with Bateson, a leading biologist and pioneer geneticist, Garrod came to interpret this pattern as just what would be expected on the basis of mendelian recessive inheritance, and thus described in alcaptonuria the first example of autosomal recessive inheritance in man.

Garrod's concept of an inborn error of metabolism as a genetically determined enzyme defect leading to interruption of a metabolic pathway at a specific point was far in advance of its time. The first actual demonstration of a specific enzyme defect in an inborn error was not provided until a half century later (1952), when Gerty Cori showed that von Gierke's disease (one of the glycogen storage diseases) is due to loss of activity of the enzyme glucose-6-phosphatase. The specific deficiency of the enzyme homogentisic oxidase, which is the cause of alcaptonuria, was not directly demonstrated until 1958 (by La Du et al.).

The majority of inborn errors of metabolism are inherited as autosomal recessives. Several, including some of great clinical or genetic interest, are X-linked; for example, glucose-6-phosphate dehydrogenase (G6PD) variants, hypoxanthine-guanine phosphoribosyl transferase (HGPRT) deficiency (Lesch-Nyhan syndrome), the type of glycogen storage disease associated with phosphorylase kinase deficiency and pseudohypoparathyroidism (an X-linked dominant). Not many true inborn errors of metabolism are autosomal dominants. One type of cystinuria is incompletely dominant (or, if one prefers, incompletely recessive). Acute intermittent porphyria and congenital spherocytosis are autosomal dominants which can also be classified as inborn errors of metabolism.

A large number of biochemical defects which produce inborn errors of metabolism are now known. (Table 5–2 lists those mentioned in this chapter.) In the following sections we will illustrate some of the mechanisms involved, and the variety of clinical and genetic problems related to some specific enzyme deficiencies.

PATHOLOGICAL CONSEQUENCES OF ENZYME DEFECTS

Metabolism is performed as a stepwise series of reactions, each step being catalyzed by a specific enzyme. At any step the pathway may be blocked either completely or partially by impaired activity of the required enzyme resulting from a mutation in the gene which codes for the normal enzyme. Most of these changes are probably substitutions of single amino acids, though by analogy with the hemoglobins we assume that such mechanisms as double substitution or deletion also take place.

A metabolic block may have a variety of consequences, which are not necessarily pathological but which may lead to a metabolic disturb-

(*Text continued on page 110*)

TABLE 5–2 SUMMARY OF BIOCHEMICAL DISORDERS MENTIONED IN CHAPTER 5

Disorder	*Chief Clinical Manifestations*	*Specific Defective Protein*	*Inheritance*	*Heterozygote Detection*	*Prenatal Diagnosis in Cultured Amniotic Fluid Cells*
Acatalasemia	Oral gangrene; may be entirely normal	Catalase	Autosomal recessive	Low activity of catalase in some but not all types	–
Albinism	Absence of pigmentation in skin and eyes, visual disorders	Tyrosinase (in one common type)	Heterogeneous; usually autosomal recessive, but autosomal dominant and X-linked types exist	–	–
Cystinuria	Aminoaciduria, renal lithiasis	Not known; affects transport of cys, lys, arg, orn	Type I autosomal recessive; Types II III incompletely recessive	In Types II and III, increased urinary excretion of cys, lys, arg and orn	–
Galactosemia	Hepatosplenomegaly, cirrhosis of liver, cataracts, mental retardation	Galactose-1-phosphate uridyl transferase	Autosomal recessive; milder Duarte variant is allelic	Reduced enzyme activity	Absent enzyme activity
Gaucher's disease	Hepatosplenomegaly, neurological problems, anemia, thrombocytopenia	Glucocerebrosidase	Autosomal recessive; two or more major types	Reduced activity of beta glucosidase in cultured fibroblasts	Absent enzyme activity
Glucose-6-phosphate dehydrogenase variants	Hemolytic anemia in response to certain foods and drugs	Glucose-6-phosphate dehydrogenase	X-linked; numerous variants	Electrophoretic variants; variants with altered enzyme activity	Theoretically possible

Table 5–2 *Continued on opposite page.*

Glycogen storage disease Type I (von Gierke's disease)	Hepatomegaly, enlarged kidneys, growth retardation, hypoglycemia, xanthomata, acidosis	Glucose-6-phosphatase	Autosomal recessive	Doubtful	–
Glycogen storage disease Type VIII (of McKusick, 1970)	Hepatomegaly, mild acidosis, hypoglycemia	Phosphorylase kinase	X-linked	–	–
GM_2 gangliosidosis (Tay-Sachs disease)	Degenerative neurological changes, cherry-red spot on macula, severe physical and mental retardation; onset 4 to 6 months of age, death 2 to 4 years	Hexosaminidase A	Autosomal recessive	Reduced level of hexosaminidase A	Absence of hexosaminidase A
Hartnup disease	Skin rash, neurological changes	Not known; affects tryptophane transport	Autosomal recessive	–	–
Hemoglobinopathies	Anemia; secondary manifestations depend upon nature of hemoglobin abnormality	Constituent polypeptide of hemoglobin α or β chain	Autosomal clinical manifestations usually recessive; numerous variants	Usually by electrophoresis	–
Homocystinuria	Arachnodactyly, dislocated lenses, mental retardation	Cystathionine synthetase	Autosomal recessive	Not well established	Absent enzyme activity with accumulation of methionine and homocystine
Isoniazid inactivation, slow	Neurological problems	Acetyltransferase	Autosomal recessive	–	–
Lesch-Nyhan syndrome	Uric aciduria, cerebral palsy, self-mutilation, mental retardation	Hypoxanthine-guanine phosphoribosyl transferase (HGPRT)	X-linked	Autoradiographic technique to measure HGPRT activity in cultured fibroblasts	Autoradiographic or biochemical method.

Table 5–2 *Continued on following page.*

TABLE 5–2 SUMMARY OF BIOCHEMICAL DISORDERS MENTIONED IN CHAPTER 5 (*Continued*)

Disorder	*Chief Clinical Manifestations*	*Specific Defective Protein*	*Inheritance*	*Heterozygote Detection*	*Prenatal Diagnosis in Cultured Amniotic Fluid Cells*
Mucopolysaccharidoses Type I (Hurler's syndrome)	"Gargoyle" facies, mental retardation, hepatosplenomegaly, hearing defect, corneal clouding, cardiovascular problems, dwarfism	Not known; produces mucopolysaccharide accumulation	Autosomal recessive	Excessive storage of certain mucopolysaccharides in cultured fibroblasts	Excessive storage of certain mucopolysaccharides
Type II (Hunter's syndrome)	Similar to Hurler's though corneal clouding is rare	Not known; produces mucopolysaccharide accumulation	X-linked	Excessive storage of certain mucopolysaccharides in cultured fibroblasts	Excessive storage of certain mucopolysaccharides
Niemann-Pick disease	Hepatomegaly, splenomegaly, severe central nervous system damage, cherry-red spot on macula	Sphingomyelinase	Autosomal recessive	Potentially, by assay of sphingomyelinase in cultured fibroblasts	Absence of sphingomyelinase
Orotic aciduria	Megaloblastic anemia, urinary excretion of orotic acid	Two enzymes: orotidylic pyrophosphorylase, orotidylic decarboxylase	Autosomal recessive	Urinary excretion of orotic acid; reduced enzyme activity in cultured fibroblasts	–
Pentosuria	Excretion of a reducing substance in urine, but no symptoms of diabetes mellitus	Xylitol dehydrogenase	Autosomal recessive	Intermediate level of xylitol dehydrogenase; abnormal response to loading test	–

Table 5–2 *Continued on opposite page.*

Phenylketonuria	Mental retardation, microcephaly, diluted pigmentation	Phenylalanine hydroxylase	Autosomal recessive	Plasma phe and phe/tyr ratio above normal	–
Porphyria, acute intermittent	Episodes of abdominal pain, neurological problems, excessive urinary excretion of δ-amino levulinic acid (ALA)	Excessive production of hepatic ALA synthetase	Autosomal dominant	Patients are usually heterozygous; latent cases may be detected by the presence of porphobilinogen in urine	–
Pseudohypoparathyroidism	Skeletal and developmental defects	Not known; produces deficient end-organ response to parathyroid hormone	Probably X-linked dominant	Patients are probably heterozygous females or hemizygous males	–
Spherocytosis, congenital	Episodes of hemolytic anemia and jaundice	Not known; produces defect in red cell membrane	Autosomal dominant	Patients are usually heterozygous	–
Suxamethonium (succinylcholine) sensitivity	Prolonged apnea in response to suxamethonium	Serum cholinesterase	Clinically significant in homozygotes or compounds for unusual types, i.e., autosomal recessive	Reduced enzyme activity	–
Tyrosinemia	Acute liver disease, usually fatal in first year of life	Parahydroxyphenylpyruvate oxidase; (in a second type, tyrosine transaminase)	Autosomal recessive	Probably by impairment of tyrosine oxidation	–
Wilson's disease	Cirrhosis of liver, Kayser-Fleisher ring in cornea, neurological problems	Not known; affects Cu metabolism	Autosomal recessive	Decreased serum ceruloplasmin and serum Cu in some but not all heterozygotes	–

ance and culminate in a clinical disorder. The consequences are of two main types:

1. Accumulation of a precursor prior to the block.
   a. The accumulated precursor may itself be harmful.
   b. Alternate minor pathways may open, with overproduction of toxic metabolites.
2. Deficiency of product.
   a. The product itself may be a substrate for a subsequent reaction, which is then unable to proceed normally.
   b. The "feedback inhibition" type of control mechanism may be impaired.

Consider the hypothetical sequence of biochemical reactions shown in Figure 5–8 (Rosenberg, 1973). This sequence includes a membrane transport enzyme T_A, and three intracellular enzymes, E_{AB}, E_{BC} and E_{CD}, each catalyzing a specific reaction in the sequence $A \rightarrow B \rightarrow C \rightarrow D$. A minor pathway $A \rightarrow F \rightarrow G$ exists. The final product of the sequence, D, exerts feedback control over the enzyme E_{AB}.

A defect of any of these enzymes will have consequences that can be classified either as precursor accumulation or as product deficiency. These are illustrated below with examples of specific diseases.

1. The membrane transport enzyme T_A is responsible for active transport of precursor A into the cell. In Hartnup disease, there is defective transport of the amino acid tryptophan, which is a precursor of

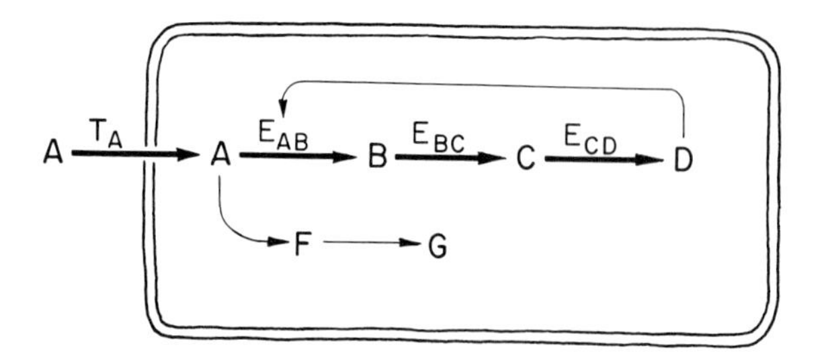

A, B, C, D - Substrate and Products of Major Pathway
F, G - Products of Minor Pathway
T_A - Transport System for A
E_{AB}, E_{BC}, E_{CD} - Enzymes Catalyzing Conversion of A to B, B to C, and C to D
|| - Cell Membrane

Figure 5–8 *A hypothetical sequence of biochemical reactions. For discussion, see text. (Reprinted from Bondy, P. K., and Rosenberg, L. E. (eds.) 1974.* Duncan's Diseases of Metabolism, *7th ed. W. B. Saunders Company, Philadelphia.)*

nicotinamide, across the intestinal mucosa. Unless the diet is supplemented with niacin, symptoms of nicotinamide deficiency develop.

2. Deficiency of any of the intracellular enzymes E_{AB}, E_{BC} or E_{CD} could cause accumulation of the immediate precursor. In galactosemia, deficiency of the enzyme galactose-1-phosphate uridyl transferase causes accumulation of galactose, which is the direct cause of the severe clinical disease. More remote precursors can also accumulate. Thus, in homocystinuria, deficiency of the enzyme cystathionine synthetase causes accumulation of both homocystine, the immediate precursor, and methionine, a precursor of an earlier reaction.

3. Opening of the minor pathway $A \rightarrow F \rightarrow G$ because of a defect in E_{AB} preventing conversion of A to B could lead to difficulty if the products of the alternate pathway are toxic. It is believed that in phenylketonuria the severe mental retardation results from toxic effects on the developing brain due to abnormal metabolites of an alternate pathway of phenylalanine metabolism.

4. Deficiency of product because of blockage of an earlier synthetic step could have a variety of pathological consequences, depending upon the particular physiological role of the product. There are many examples. One form of albinism results from a failure of melanin formation that is due to deficiency of the enzyme tyrosinase.

Deficiency of product D could also remove feedback inhibition of E_{AB}, causing overproduction of product B. This may be the cause of the great overproduction of uric acid in Lesch-Nyhan syndrome.

The inborn errors of metabolism are clinically very diverse, representing as they do the secondary pathophysiological consequences of a wide assortment of enzyme defects. They range in severity from those which are relatively harmless to those which cause death in early infancy. (Note that all the known genetic defects of enzymes are compatible with life at least to the time of birth; consequently, they must be milder in their effects than the unknown mutations which are lethal in the prenatal period.)

DISORDERS OF AMINO ACID METABOLISM

Phenylketonuria

Phenylketonuria (PKU) is a disorder of metabolism of the amino acid phenylalanine, due to a defect in the enzyme phenylalanine hydroxylase, which normally converts phenylalanine to tyrosine as the first step in its metabolism. (Note that this statement implies that the structure of phenylalanine hydroxylase is determined by the normal allele at the PKU locus.) The discovery of PKU by Følling in 1934 marked the first demonstration of an enzyme defect as a cause of mental retardation. Because phenylalanine is not converted to tyrosine in individuals with PKU, phenylalanine in the diet or formed by normal tissue breakdown is degraded through an alternate pathway, which produces phenylpyruvic acid and other abnormal metabolites. These metabolites are excreted in the urine. The disorder is believed to have an incidence of about one in 10,000 births in North America, so the carrier incidence is about one in 50.

The characteristic feature of PKU is severe mental retardation; many untreated patients have I.Q.'s below 20, although very rarely the intelligence is within the normal range. Phenylketonurics represent 1 to 2 per cent of all institutionalized mental defectives. Other neurological features are not marked; there is some accentuation of reflexes, and convulsive seizures may occur. Affected children may have fairer hair and skin than their normal sibs, but the effect of PKU on pigmentation is variable and depends to some extent on the background genes for pigmentation (Fig. 5–9).

Because children with PKU are normal at birth and become retarded only if they are fed phenylalanine, PKU is the prototype of those inborn errors of metabolism for which screening of the newborn population is valuable. A standard screening test for PKU has been the ferric chloride test. A few drops of ferric chloride are placed on a wet diaper. If the urine contains phenylpyruvic acid, a bright green color develops. The color fades to brown within minutes, so the test must be read promptly, and even then false positives or false negatives may occur. The phenylalanine level in the serum of PKU patients is elevated to about 30 times normal, and a more accurate screening test, the Guthrie "inhibition assay," makes use of the high serum phenylalanine and the principle that the growth of a certain strain of bacteria is inhibited by the presence of phenylalanine in the growth medium. An advantage of the blood test is that it can be used very soon after birth, whereas the ferric chloride test is usually not suitable until the baby is a few weeks old.

The consequences of the enzymatic block may be partly circumvented by reducing the amount of precursor, i.e., by giving a diet low in phenylalanine. Phenylketonuric children are normal at birth because the maternal enzyme protects them during prenatal life. The results of treatment are best when the diagnosis is made soon after birth and treatment is begun promptly. If the child is fed phenylalanine for some time, irreversible mental retardation occurs, apparently because of toxic accumulations of abnormal metabolites of phenylalanine in the brain.

Genetics. PKU is an autosomal recessive disorder, but recently a mild variant has been discovered which is probably caused by a different mutation of the same gene locus responsible for "classic" PKU. There are also a transient type, present in early infancy but not later, and a benign but persistent type. The genetic relationships of these types to classic PKU have not been well established.

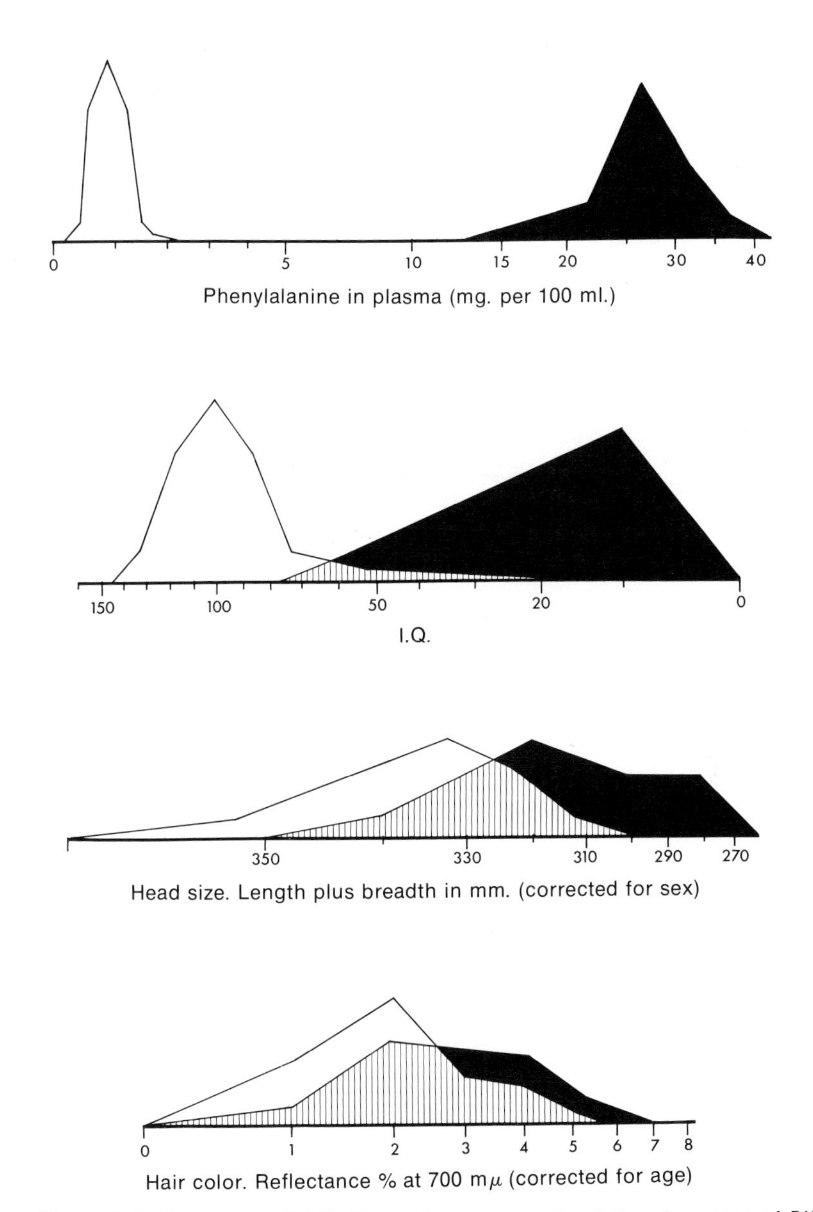

Figure 5–9 *Frequency distributions of some aspects of the phenotype of PKU patients (black) as compared with controls (white). Hair color, head size and I.Q. show varying degrees of overlap between patients and controls. Only the primary effect of the gene—plasma phenylalanine level—completely distinguishes PKU patients from controls. (Redrawn from Penrose 1951.* Ann. Eugen. *16:134.)*

The occurrence of a number of different types of PKU and hyperphenylalaninemia causes difficulty in management. Phenylalanine is an essential amino acid. It is extremely difficult to maintain the concentration in serum at a level high enough to provide for normal growth and development, but low enough to prevent brain damage in homozygotes for classic or mild PKU. Fortunately, after the age of about three years the brain has sufficiently matured so that it is unnecessary to stick rigidly to the low-phenylalanine diet.

Maternal Phenylketonuria. PKU can now be treated with enough success that affected individuals have almost normal prospects for marriage and parenthood. Because phenylketonurics who have been effectively managed at birth can function adequately in many ways, it has been disconcerting to learn that almost all the children of female phenylketonurics are mentally retarded. Most of these children are heterozygotes (a few are homozygotes, having heterozygous fathers). Their retardation is apparently produced not by their own genetic constitution but by their intrauterine development in mothers who have unduly high phenylalanine levels. About 25 per cent of these children (vs. about 3 per cent of controls) also have congenital malformations of various types.

Disorders of Tyrosine Metabolism, Alcaptonuria and Albinism

In the pathway of phenylalanine metabolism, at least three other enzymatic blocks are known to be associated with clinical problems. The clinical disorders are tyrosinosis or tyrosinemia (several types), alcaptonuria and albinism (several types). The relationship of these to one another and to PKU is shown schematically in Figure 5–10. Block 1 produces phenylketonuria; 2, a type of tyrosinemia; 3, alcaptonuria; and 4, albinism.

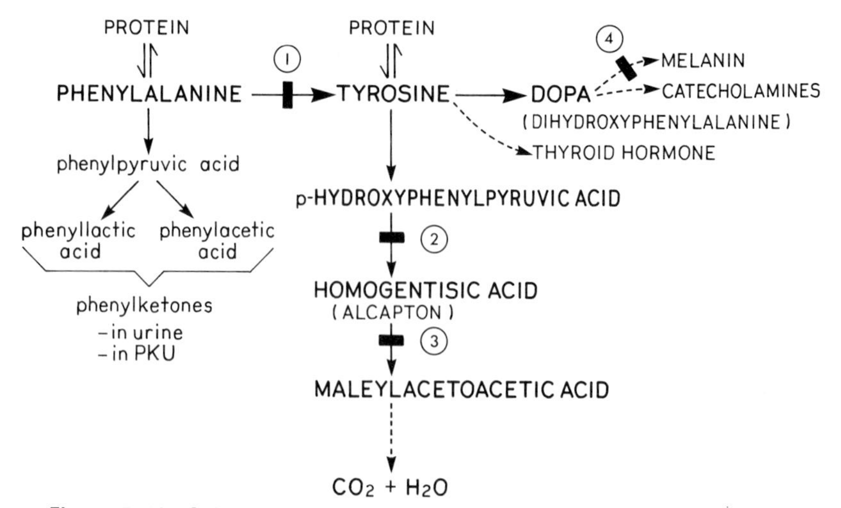

Figure 5–10 *Scheme of metabolic pathways of phenylalanine and tyrosine. (Based on Rosenberg and Scriver. In: Bondy, P. K. (ed.) 1969.* Duncan's Diseases of Metabolism, *6th ed. W. B. Saunders Company, Philadelphia.)*

DISORDERS OF CARBOHYDRATE METABOLISM

The inborn errors of carbohydrate metabolism include galactosemia, the nine different glycogen storage diseases, the relatively harmless trait pentosuria and several other disorders of intermediary metabolism or transport. Possibly diabetes mellitus should be included in the list, but it is omitted from discussion here because it is not a single-gene disorder with a block in a specific metabolic step. Galactosemia will be used to illustrate this group of inborn errors.

Galactosemia, as the name implies, is the result of inability to metabolize galactose, a monosaccharide which is a component of lactose (milk sugar). The inheritance is autosomal recessive. Affected infants completely lack the enzyme galactose-1-phosphate uridyl transferase (gal-1-PUT), which normally catalyzes the following reaction:

$$\text{Galactose-1-phosphate} + \text{UDP-glucose}^{*} \xrightarrow{\text{gal-1-PUT}} \text{UDP-galactose}^{\dagger} + \text{glucose-1-phosphate}$$

Infants with galactosemia are usually normal at birth but begin to develop gastrointestinal problems, cirrhosis of the liver and cataracts as soon as they are given milk. If untreated, galactosemia is usually fatal, though in older children an alternative pathway for galactose metabolism eventually develops. Complete removal of milk from the diet can protect against the harmful consequences of the enzyme deficiency.

A different mutation at the galactosemia locus, the Duarte variant, leads to the production of an enzyme with only about half the normal activity, but without associated clinical problems (Beutler et al., 1965).

The frequency of galactosemia is estimated to be about one in 40,000 births, but about one person in 200 is homozygous for the Duarte variant. The corresponding frequencies of the two types of heterozygotes are one in 100 and one in seven, respectively.

Heterozygotes for galactosemia can be identified by assaying for the enzyme in red cells or cultured skin fibroblasts. Typically, heterozygotes for the galactosemia allele have about 50 per cent and Duarte heterozygotes have about 75 per cent of normal activity. Galactosemia can also be identified in cultured amniotic fluid cells.

DISORDERS OF LIPID METABOLISM

Tay-Sachs Disease

As an example of the various disorders of lipid metabolism, we will consider Tay-Sachs disease (GM_2 gangliosidosis, infantile amaurotic

---

*UDP-glucose is an abbreviation of uridine diphosphoglucose.
†UDP-galactose is an abbreviation of uridine diphosphogalactose.

familial idiocy). There are several well-known related disorders, e.g., GM_1 gangliosidosis (a generalized disorder of ganglioside and mucopolysaccharide storage), Niemann-Pick disease (a sphingomyelin lipidosis) and Gaucher's disease (a glucosyl ceramide lipidosis). All are inherited as autosomal recessives.

Tay-Sachs disease (TS) is one of several different forms of amaurotic familial idiocy, not all of which are gangliosidoses. The term *amaurotic* is derived from *amauros* (dark or obscure) and referred originally to the obscure cause of the progressive blindness which is one of the characteristic features of the disorder. Affected children are normal at birth, but at four to six months begin to show retardation in physical and mental development. A cherry-red spot develops on the macula. The disease is fatal by the age of two to four years.

TS results from a known enzyme lesion, namely, marked deficiency in activity of hexosaminidase A in a wide variety of tissues. It is much more common in one subpopulation (Ashkenazi Jews) than in other population groups. It can readily be identified in heterozygotes by screening blood samples for hexosaminidase activity in serum, and can be diagnosed by biochemical study of cultured amniotic fluid cells from affected fetuses, thus making selective abortion possible (page 346).

Because of these characteristics, TS has become the prototype of those inborn errors in which it is possible to screen whole population groups to find marriages of two carriers, and then to monitor their pregnancies in order to identify and abort homozygous affected fetuses. In other words, TS has become an almost entirely preventable disease. Though it does occur in other population groups, the frequency is at least 100 times as high in Ashkenazi Jews (about one affected child in 3600 births) as in the general population (one in 360,000). Carriers have a frequency of one in 30 among Ashkenazi Jews, one in 300 elsewhere. (In some centers the carrier frequently is even higher, e.g., about one in 15 in the Jewish population of Toronto.)

Other Disorders of Lipid Metabolism

Niemann-Pick Disease. There are several genetically different forms of Niemann-Pick disease, varying in onset age and severity. All are characterized by accumulation of sphingomyelin, which is a phospholipid found in the red cell membrane and in many other cell types; thus, it appears that the disorder results from a genetically determined defect of sphingomyelinase.

Niemann-Pick disease is characterized by neurological degeneration which begins soon after birth and progresses to death by the age of five years. The cherry-red spot on the macula, characteristic of amaurotic familial idiocy, is seen in about half of all patients with Niemann-Pick disease. Splenomegaly and hepatomegaly are present. "Foam cells" full of lipid, especially sphingomyelin, occur in many tissues. The neurological degeneration and pathological changes are very similar to those seen in amaurotic familial idiocy.

Gaucher's Disease. This disorder is characterized by an enlarged spleen and by the presence in the spleen and marrow of "Gaucher cells" containing cerebroside; Gaucher cells resemble the foam cells seen in Niemann-Pick disease. Gaucher's disease usually affects adults, and may be compatible with normal life, especially after splenectomy. An infantile form also exists. The lipid which is accumulated is glucocerebroside, a substance which is normally hydrolyzed to glucose and ceramide by a specific enzyme, glucocerebrosidase. In other words, Gaucher's disease results from deficiency (or impaired activity) of glucocerebrosidase.

MUCOPOLYSACCHARIDOSES

The chief components of the ground substance of connective tissues are acid mucopolysaccharides. In at least six different disorders known as mucopolysaccharidoses (MPS), there is excessive tissue deposition and urinary excretion of these substances. Clinically, the patients have characteristic skeletal malformations and mental retardation. In some forms hepatosplenomegaly, cardiac and pulmonary changes, corneal clouding and progressive deafness are present. In different types of MPS the chief mucopolysaccharide excreted in the urine may be dermatan sulfate, heparan sulfate or keratan sulfate.

Hurler's syndrome, the most common MPS, has been known since 1919. The coarse facial features of affected children account for its older name of "gargoylism." The inheritance is autosomal recessive. Dermatan sulfate and heparan sulfate are excreted in the urine.

Hunter's syndrome was the first mucopolysaccharidosis to be described (in 1917 by a Winnipeg physician, Dr. Charles Hunter). It resembles Hurler's syndrome closely but is considerably less common and has X-linked inheritance. Because the affected males do not live to reproduce, a large proportion (two-thirds) of the patients are new mutants with no affected relatives. (See also page 342.)

Genetic Considerations

The problem of distinguishing between these two defects can produce difficulty in genetic counseling. Consider the pedigree shown in Figure 5–11. Two apparently normal parents have two affected sons. This pattern is consistent with either of the following possibilities:

1. Both parents might be carriers of an autosomal recessive gene. If so, the risk that any of their future children will be affected is one in

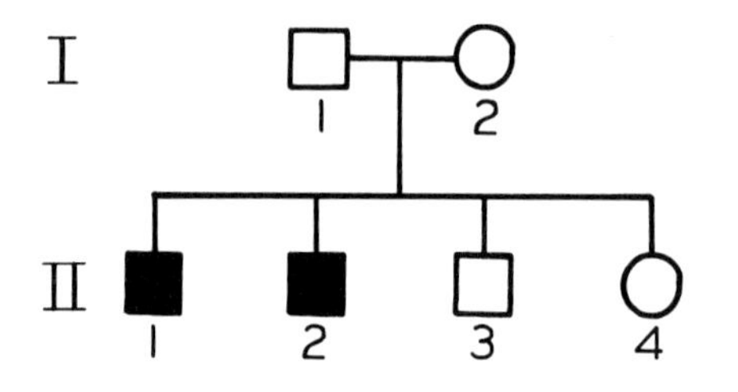

Figure 5–11 *A hypothetical pedigree of a mucopolysaccharidosis. For discussion, see text.*

four. However, the disorder will appear in the next generation only if the normal children, II–3 and II–4, *are* carriers and *marry* carriers.

The risk that either II–3 or II–4 is a carrier is 2/3.

The risk of marrying a carrier of the same gene (assuming random mating) is probably about 1/160. (Hurler's syndrome has a frequency of about one in 100,000, and a carrier frequency of about one in 160.)

Consequently, the risk that II–3 or II–4 will have an affected first-born child is only about $2/3 \times 1/160 \times 1/4 = 1/1000$, which is negligible.

2. Alternatively, I–2 could be a carrier of the X-linked gene for Hunter's syndrome. In this case, II–3 is normal; he does not have the gene and cannot pass it on. However, the risk that his sister is a carrier is one in two, and if she is a carrier, her sons have a one in two chance of being affected. Thus, the overall chance that one of her sons will be affected is $1/2 \times 1/2 = 1/4$ and the chance that a daughter will be a carrier is also one-fourth.

These calculations emphasize the importance of distinguishing between autosomal and X-linked inheritance. If the pedigree is not sufficiently informative, two other possibilities exist:

1. Clinically, it may be possible to sort out the two phenotypes. The chief difference is that corneal clouding is present in Hurler's syndrome but rare in Hunter's syndrome.

2. Carrier detection. In both disorders, cultured fibroblasts of heterozygotes as well as of affected individuals contain metachromatic granules, and produce mucopolysaccharides in measurable amounts. If the father and mother are both heterozygotes, the disorder is autosomal recessive; if the mother (and/or a sister) is heterozygous, and the father and phenotypically normal brothers are normal, probably the disorder is X-linked. (However, note that when an X-linked gene is as severely deleterious as the Hunter's gene, an affected male may be a new mutant with *no* affected or heterozygous relatives.)

All the other mucopolysaccharidoses are autosomal recessive. It is not definitely known whether they represent different mutations at the Hurler locus, or mutations at different loci, but it is probable that some types are allelic and others not allelic to the Hurler gene. It is important to note that if two different allelic mutations exist in a population, "compound" genotypes (having one of each of the two mutant alleles) may be quite frequent (McKusick et al., 1972) and may produce a phenotype clinically different from that of either homozygote. It is not always possible to distinguish compounds from true homozygotes. A patient whose parents are consanguineous, or are members of the same genetic isolate, has probably inherited the same allele from both parents and is therefore probably a homozygote rather than a compound.

DISORDERS OF PURINE AND PYRIMIDINE METABOLISM

Lesch-Nyhan Syndrome

Lesch and Nyhan (1964) were the first to describe this rare X-linked disorder, characterized by uric aciduria, cerebral palsy, mental retardation and a compulsive behavior of self-mutilation by gnawing the lips and fingers. The affected boys completely lack the enzyme hypoxanthine-guanine phosphoribosyl transferase (HGPRT), which plays a role in the regulation of purine synthesis (Fig. 5–12), converting guanine and hypox-

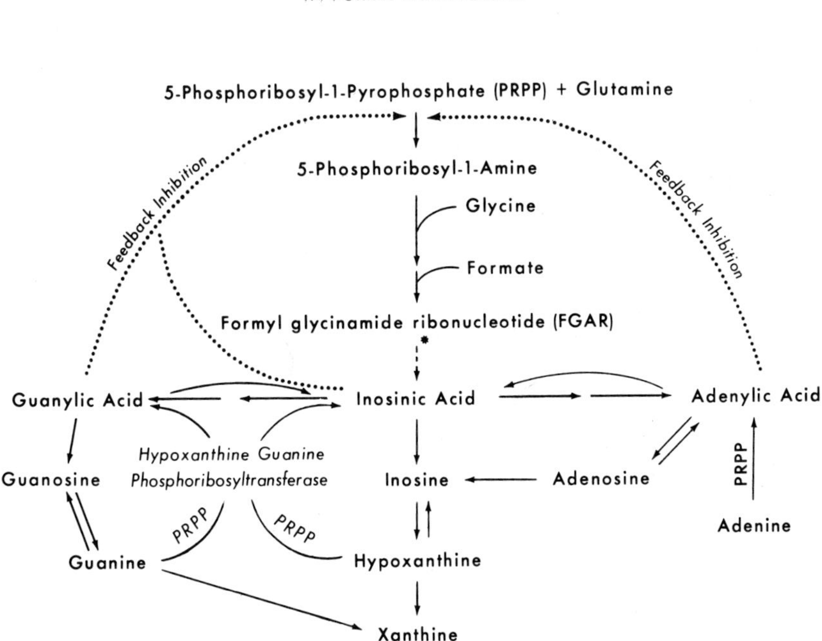

Figure 5–12 *Feedback control mechanism for regulation of purine synthesis. (From Seegmiller, J. E., et al. 1967.* Science *155:682. Copyright 1967 by the American Association for the Advancement of Science.)*

anthine to their respective nucleotides. Deficiency of HGPRT leads to overproduction of purines and consequent excessive excretion of uric acid. Not all forms of HGPRT deficiency completely lack the enzyme, and a much less severe disorder results if even a very small amount of enzymatic activity remains.

Many patients who survive for several years with HGPRT deficiency develop gouty arthritis. Gout is characterized by recurrent and disabling inflammation, usually involving the first metatarsophalangeal joint. The underlying biochemical finding is hyperuricemia; not all hyperuricemics develop clinical gout, but perhaps 25 per cent of them deposit urate crystals in and around the joints, in the cartilage of the outer ear (tophi) and sometimes in the kidney as kidney stones. The hyperuricemia of gout has a heterogeneous group of causes, but many cases appear to be due to genetic defects of HGPRT.

Apart from its clinical significance, HGPRT deficiency has proved to be a useful mutation for research in somatic cell genetics. There are several reasons for its research importance. First, it can be recognized phenotypically in several kinds of cultured cells. (Not all genes are expressed in the types of cells commonly used in somatic cell genetics, such as skin fibroblasts, lymphoblasts or amniotic fluid cells.) Secondly, it is possible to select either for or against HGPRT-deficient cells simply by changing the medium in which the cells are growing. Mutants lacking HGPRT will grow in media containing analogs of hypoxanthine or guanine which are toxic to normal cells, such as 8-azaguanine or 6-mercaptopurine. On the other hand, they will not grow in "HAT" medium – medium containing *hypoxanthine* as the sole exogenous source of purine, *aminopterin* to block endogenous purine synthesis, and *thymidine,* which must be provided when aminopterin is present. (Mutants of this type, having special nutritional requirements, are known as auxotrophs.)

Because the HGPRT-deficient phenotype can be identified, heterozygous carriers can be detected. The defect can also be identified in early prenatal life by assaying for the enzyme in cultured amniotic fluid cells.

Orotic Aciduria

This disorder is extremely rare, only about six families having been described so far. It is an autosomal recessive defect in which a child who is normal at birth soon develops a severe megaloblastic anemia and becomes physically and mentally retarded. Orotic acid, a precursor of pyrimidine nucleotides, is excreted in the urine. The enzymatic defect involves *two* enzymes of pyrimidine metabolism, orotidine-5′-phosphate pyrophosphorylase and orotidine-5′-phosphate decarboxylase. Heterozygotes can be identified because they excrete excessive quantities of orotic acid in the urine.

WILSON'S DISEASE: A DISORDER OF COPPER METABOLISM

Much of the copper ingested is excreted and only a small amount is retained, most of which is specifically bound to a serum protein, ceruloplasmin. In Wilson's disease (hepatolenticular degeneration) there is excessive storage of copper in liver, brain, cornea and other tissues, associated with a decreased level of serum ceruloplasmin. There are serious neurological consequences which are progressive and ultimately fatal. The disease is inherited as an autosomal recessive with onset in childhood or youth. It is very rare, with an incidence of about one in 200,000 births. There is some possibility that clinically distinguishable forms of the disease are caused by different mutations. Many heterozygotes can be recognized by low serum copper and ceruloplasmin levels.

There are two main reasons for including Wilson's disease in this brief discussion of metabolic diseases:

1. Study of the disorder, which is the best known genetic abnormality of copper metabolism, can elucidate details of normal human copper metabolism.

2. From the standpoint of therapy, Wilson's disease provides an example of treatment of a genetic disorder by depletion of the stored substance. Penicillamine treatment has the effect of removing the stored copper, and will usually stabilize or even reverse the course of the disease.

GENETIC DEFECTS OF AMINO ACID TRANSPORT: CYSTINURIA

"Active" transport, as opposed to passive transport, means transport against a gradient and requires energy. Active transport mechanisms are involved in the transport of amino acids and many other substances from the extracellular fluid to the interior of the cell, or across a cell layer such as the gut mucosa or the lining of the proximal tubules of the kidney.

Though membrane transport is by no means fully understood, genetic disorders of transport systems are helping to clarify the process. In amino acid transport, genetically determined proteins in or near the cell surface are thought to bind amino acids, assisting their entry into the cell. These proteins are known as "carriers," "permeases" or simply "reactive sites."

Cystinuria was originally regarded by Garrod as a prototype inborn error of metabolism. Cystinurics form renal calculi composed of cystine and excrete large amounts of cystine in the urine. Although for many years it was believed that the defect in cystinuria was a block in cystine catabolism, with accumulation of cystine prior to the block and its eventual urinary excretion, we now know that the defect is in the reabsorption of cystine by the proximal portion of the renal tubule.

Additional features of cystinuria are that the defect in absorption involves the dibasic amino acids, lysine, arginine and ornithine, as well as cystine, and that the transport defect involves uptake from the gut as well as renal tubular reabsorption.

Genetics

Though originally cystinuria was considered to be a typical autosomal recessive disorder, three subtypes are now recognized. Homozygotes for the three types are clinically indistinguishable, but heterozygotes differ. Type I, by far the most common type, is "completely recessive" in that heterozygotes have normal urinary amino acid excretion and cannot be distinguished biochemically from normal individuals. Types II and III are "incompletely recessive," i.e., heterozygotes excrete elevated amounts of cystine and the dibasic amino acids, and may even form cystine stones. Types II and III are distinguished from one another chiefly by the degree of clinical severity in heterozygotes, which is greater in Type II than in Type III.

The genes for the three types of cystinuria may be allelic, and compound genotypes (i.e., I/II, I/III and II/III) have been described (Rosenberg, 1967). If the different forms are allelic, the same carrier protein must be abnormal in the different types, though the specific alteration varies. However, it is not yet definite that the various mutations are all at the same locus. One alternative possibility is that they affect different subunits of a protein containing two or more different polypeptide chains, a situation analogous to mutations of the α and β hemoglobin chains.

PHARMACOGENETICS

Pharmacogenetics is the special area of biochemical genetics that deals with drug responses and their genetic modification. Broadly speaking, pharmacogenetics can be said to encompass any genetically determined variation in response to drugs; for instance, the effect of barbiturates in precipitating attacks of porphyria in genetically susceptible persons, and the effect of cortisone in pregnant mice upon the incidence of cleft palate in the progeny. In a narrower sense, pharmacogenetics can be restricted to those genetic variations that are revealed *only* by response to drugs. The classic examples of pharmacogenetic variations include atypical serum cholinesterase, slow isoniazid inactivation, primaquine sensitivity, inability to taste phenylthiocarbamide and acatalasemia. Each of these is briefly discussed in this chapter.

The origin of polymorphisms for drug response and the mechanisms by which they are maintained pose a problem. Obviously they have not developed in response to drugs, since they antedate the drugs concerned. The biotransformation of drugs requires many specific biochemical reactions, and the enzymatic sequences involved may be used in the metabolism of ordinary food substances as well as drugs. For example, an extract of the potato, solanine, is an inhibitor of serum cholinesterase. It is interesting to speculate that possibly, in some ancient age, an atypical serum cholinesterase phenotype might have conferred a selective advantage. Although there is no definite proof that any one genetic polymorphism in drug response has become established in response to a food, it is a reasonable assumption. At least one drug response, primaquine sensitivity, like a number of other abnormalities of the red blood cells (e.g., sickle cell anemia and thalassemia) may confer protection against malaria.

Pharmacologists recognize that there is normal variation in response to drugs by defining the "potency" of a drug as that dose which produces a given effect in 50 per cent of a population. For genetic traits, continuous variation is usually best explained on the basis of polygenic inheritance (inheritance by many genes at different loci, with small additive effects—see page 270) or by a combination of genetic and environmental factors. But response to drugs can also show discontinuous variation, with sharp distinctions between different degrees of response. Discontinuous variation lends itself more readily to genetic analysis than does continuous variation, and the examples used here are all of the discontinuous type.

SERUM CHOLINESTERASE AND SUCCINYLCHOLINE SENSITIVITY

Serum cholinesterase is an enzyme of human plasma. It has the property of hydrolyzing choline esters, such as acetylcholine, to form free choline and the corresponding organic acid. It is frequently called pseudocholinesterase because its hydrolytic action on acetylcholine is slow, compared to the speed with which "true" cholinesterase, a red cell enzyme, destroys acetylcholine at the neuromuscular junction (endplate).

The function of serum cholinesterase is obscure. It may be a "protective enzyme" because it can hydrolyze some choline esters which in high concentrations inhibit acetylcholinesterase. A low level of serum cholinesterase, or even its complete absence, is fully compatible with normal development and health, so it cannot play a major physiological role.

Succinylcholine (suxamethonium) is a drug widely used as a muscle relaxant in anesthesia and in connection with electroconvulsive therapy. Chemically it is made up of two molecules of acetylcholine, so it is rapidly hydrolyzed by serum cholinesterase. The rapidity of its hydrolysis effectively reduces the amount of succinylcholine that reaches the motor endplates, and this hydrolysis is allowed for in the dosage given to the average patient. Occasionally, however, a patient will respond to the administration of succinylcholine by developing prolonged apnea lasting from one to several hours.

The abnormal response is not always genetic. Some 50 per cent of patients who respond abnormally have an abnormal genotype, but in the remaining 50 per cent the underlying problem is a nongenetic, pathological change, or a technical problem in administration of the anesthetic.

Genetics of Serum Cholinesterase Variants

The activity of cholinesterase in the plasma is determined by two codominant alleles, known as E_1^u and E_1^a. (E_1 signifies the first esterase locus to be described; the superscripts u and a denote genes responsible, respectively, for the *u*sual and *a*typical forms of the enzyme.) Cholinesterase alteration occurs in persons who are homozygous for the mutant allele E_1^a; the enzyme produced by E_1^a/E_1^a homozygotes is qualitatively altered and has lower activity than the usual type.

Serum cholinesterase phenotypes cannot be determined with certainty on the basis of cholinesterase levels in serum, because the values thus obtained show considerable overlap. Kalow and his collaborators distinguish between normal and abnormal phenotypes by using an inhibitor of cholinesterase, dibucaine (Nupercaine), which is a well-known local anesthetic. The "dibucaine number" (DN) of a serum sample expresses its per cent inhibition by dibucaine. The following relationships exist:

DN	*Phenotype*	*Genotype*	*Approximate Frequency (Canadian Data)*
About 80	Usual	E_1^u/E_1^u	0.9625
About 60	Intermediate	E_1^u/E_1^a	0.0370
About 20	Atypical	E_1^a/E_1^a	0.0005

The Silent Allele. In about 20 per cent of atypical individuals, the cholinesterase phenotypes observed cannot be explained on a two-allele hypothesis. Very rarely, individuals have been observed whose serum completely lacks cholinesterase activity. To explain these rare instances, a third allele E_1^s (*s*ilent) has been postulated. Either E_1^a/E_1^a or E_1^a/E_1^s can produce the atypical phenotype, and E_1^s/E_1^s produces no activity at all.

The "Fluoride-Resistant" Allele. A fourth allele at the E_1 locus, E_1^f, determines the structure of a type of cholinesterase in which the enzyme is unusually resistant to inhibition by sodium fluoride. Most serum samples are classified identically regardless of whether dibucaine or fluoride is used as the inhibitor, but there are a few rare sera which are resistant to inhibition by fluoride but not by dibucaine. The enzyme determined by the E_1^f allele has lower activity than the usual type, and some degree of suxamethonium sensitivity is present in individuals who are E_1^a/E_1^f, E_1^s/E_1^f, or E_1^f/E_1^f.

The E_2 Locus. A second locus determining cholinesterase activity, E_2, accounts for a serum cholinesterase isozyme which is detectable only by electrophoresis. About 10 per cent of Caucasians have a variant allele E_2^+, which produces an extra isozyme band, C_5 (the four usual isozymes are C_{1-4}), and 25 per cent higher activity than the more common E_2^- type.

Linkage. The E_1 locus is linked to the locus for the serum protein transferrin (Tf) (Robson et al., 1966). This pair of autosomal loci has not yet been assigned to any one chromosome. The two loci are about 14 map units apart.

ISONIAZID METABOLISM

Isoniazid is a drug used in the treatment of tuberculosis. The rate of inactivation of isoniazid shows polymorphism, as measured, after a test dose, either by the free isoniazid excreted in the urine during the following 24 hours or by the plasma concentration after six hours. "Slow inactivators" are homozygous for a recessive gene which is believed to lead to lack of the hepatic enzyme acetyltransferase, the enzyme which normally acetylates isoniazid as one step in the metabolism of this drug. "Rapid inactivators" are normal homozygotes or heterozygotes. The same enzyme acetylates the drugs sulfamethazine, hydralazine and sulfisoxazole, but not all sulfa drugs; sulfanilamide, for example, has been shown to be acetylated in the red blood cells, not in the liver.

The gene for slow inactivation shows marked differences in geographic distribution (Table 5–3). It is very low in Eskimos, for example; but up to 80 per cent of African Negroes and of some European populations are either heterozygous or homozygous for the slow-inactivation gene.

Surprisingly, the isoniazid inactivation phenotype appears to have no influence upon the response of patients with tuberculosis to isoniazid treatment. However, it does seem to have a bearing on the development of side effects; slow inactivators are more likely than rapid inactivators

TABLE 5–3 RACIAL DIFFERENCES IN SPEED OF INACTIVATION OF ISONIAZID*

Racial Origin	Per Cent Rapid Inactivators	Per Cent Slow Inactivators
American and Canadian White	45.0	55.0
American Negro	47.5	52.5
Eskimo	95.0	5.0
Latin American	67.0	33.0

*After Kalow, 1962.

to develop polyneuritis as a complication of isoniazid treatment. Slow inactivators develop toxic reactions at the dosage necessary to maintain adequate blood levels in fast inactivators, and this should be considered when isoniazid is used in the treatment of tuberculosis. On the other hand, there is a danger that rapid inactivators (a group that includes most of the Canadian native populations) may not always be able to maintain adequate blood levels of the drug unless they receive their medication on a carefully regulated schedule (Jeanes et al., 1972).

GLUCOSE-6-PHOSPHATE DEHYDROGENASE VARIANTS

Primaquine is a drug used in the treatment of malaria. From the time it was first introduced, it was known to be capable of inducing hemolytic anemia in some patients, especially in Negro males. Further investigation of this phenomenon showed that the red cells of primaquine-sensitive subjects are deficient in glucose-6-phosphate dehydrogenase (G6PD), an ubiquitous enzyme involved in glucose metabolism. The trait is inherited as an X-linked recessive.

Favism, a severe hemolytic anemia in response to ingesting the broad bean *Vicia faba,* has been known since ancient times in parts of Italy, and its hereditary nature has been recognized. The basis of favism, like that of primaquine sensitivity, is G6PD deficiency.

Primaquine sensitivity and favism are only two of some 50 genetic variants of G6PD, many of which are associated with some degree of deficiency of activity of the enzyme. About 100 million people throughout the world have G6PD deficiency. For a deleterious gene to reach such a high population frequency, a selective advantage must be postulated, and it is believed that many G6PD variants, like sickle cell hemoglobin, may confer some protection against malaria (see also page 301).

The specific biochemical role of G6PD is in the hexosemonophosphate shunt pathway, which is the minor pathway for red cell glycolysis (Fig. 5–13). Ninety per cent of red cell glycolysis takes place through the

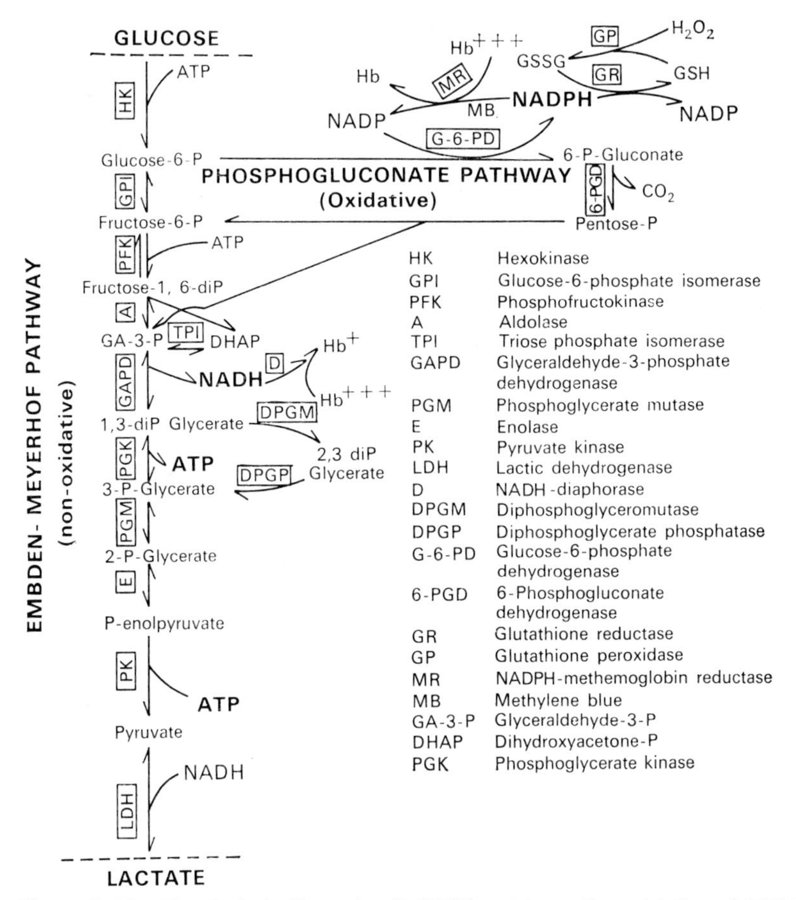

Figure 5–13 *Glycolysis in the red cell. G6PD catalyzes the oxidation of G6P to 6PG in the hexosemonophosphate shunt pathway. Enzymes are encircled. [From Giblett 1969.* Genetic Markers in Human Blood, *p. 445 (originally in Simon, Giblett and Finch 1966.* Red Cell Manual*), Blackwell Scientific Publications, Oxford, by permission.*]

anaerobic pathway, in which no enzyme polymorphisms are known. It appears that as a general rule, polymorphisms are biologically tolerable in minor pathways, but less so in major ones.

The normal and common abnormal variants of G6PD are listed in Table 5–4. The normal is known as type B, which is the conventional label given to normal electrophoretic variants of a number of genetic markers. About 20 per cent of American Negro males of African descent have the faster, nearly normal A variant. The two deficient types migrate electrophoretically at the same rate as A and B but have much lower activity, and so are A– and B– respectively. Because the Gd locus is on the X chromosome, males have only a single *Gd* gene, but heterozygous females (e.g., AB heterozygotes) show two bands. The electrophoretic patterns are shown in Figure 5–14.

TABLE 5–4 COMMON G6PD PHENOTYPES

Gd Type	Gene Symbol	Electrophoretic Mobility	Enzyme Activity (% normal, approx.)	Approximate Population Distribution
B	Gd^{B}	Normal	100	Normal
A	Gd^{A}	Fast	90	20% of American Negro males
A–	Gd^{A-}	Fast	15	10% of American Negro males
B–	Gd^{B-}	Normal	4	Common in Mediterranean areas

Lyonization

The Lyon hypothesis of gene action on the X chromosome is discussed on page 163, but must be briefly noted here. In females, an X chromosome is randomly inactivated early in embryonic life, and this

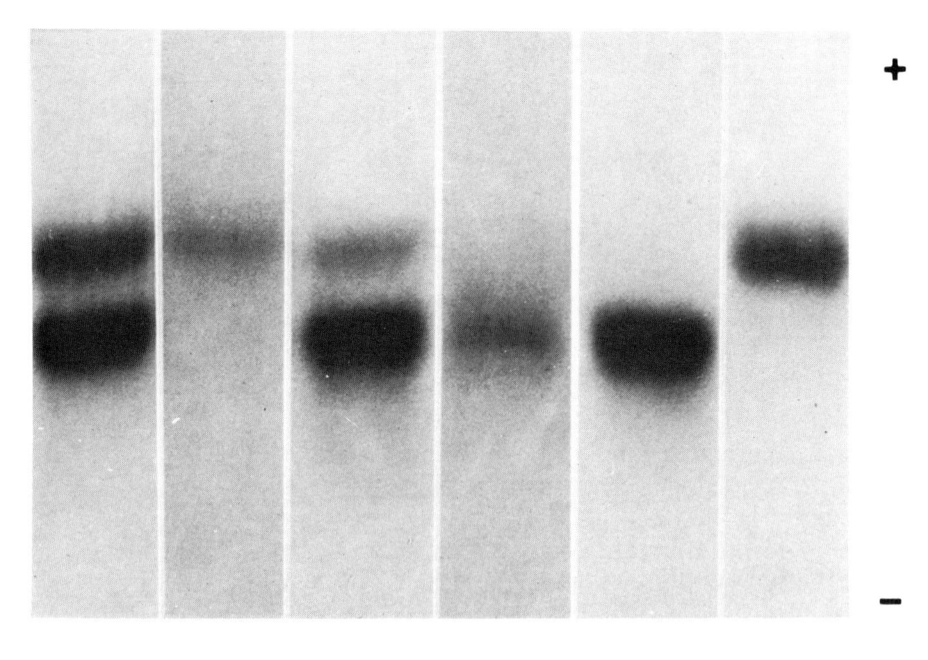

Figure 5–14 *Starch gel electrophoretic patterns of six different G6PD phenotypes in red cell hemolysates. (From Giblett 1969.* Genetic Markers in Human Blood. *Blackwell Scientific Publications, Oxford, facing page 452, by permission.)*

event leads to mosaicism is heterozygous females as well as provides a mechanism for "dosage compensation," i.e., for equalizing the amount of product of X-linked genes in males (with a single X) and females (with two). Because G6PD is X-linked and its variants so common, it has been very useful in testing the Lyon hypothesis.

Linkage Studies

The Gd locus is very close to the locus for hemophilia A, and to the protan and deutan loci for red-green color blindness, but not close to the Xg blood group locus. Because variants at the locus are so frequent, it is one of the most useful markers on the X.

Clinical Aspects

Many of the G6PD variants are associated with a high risk of hemolytic crises on exposure to certain drugs which stress the shunt pathway. These drugs include sulfanilamide, trinitrotoluene, quinidine, primaquine, naphthalene (moth balls) and many others. When an infection or some other precipitating factor is present, an even longer list of drugs can cause hemolysis—for example, acetylsalicylic acid. It is noteworthy that the Negro type of deficiency (A–) is much less susceptible to hemolysis than the non-Negro type (B–); for example, favism is common in the Mediterranean region but almost unknown in Negroes. Deficient infants of the Mediterranean type can develop hemolytic crises if dressed in clothes that have been in moth balls and not thoroughly cleaned.

Congenital nonspherocytic hemolytic anemia associated with very low enzyme activity is a consequence of some of the rare variants, especially in Caucasians.

Though G6PD deficiency is far more common in males, it is not impossible for females to inherit two abnormal alleles and consequently be affected. About 2 per cent of American Negro females are genetically Gd^{A-}/Gd^{A-} and clinically susceptible to drug-induced hemolysis.

TASTE SENSITIVITY TO PTC

The human polymorphism for taste sensitivity in response to the clinical phenylthiourea (also known as phenylthiocarbamide and, commonly, as PTC) has already been described (page 51) as a classic example of a difference determined by a single pair of alleles, designated as T and t, t being recessive to T. Tasters are T/T or T/t, and nontasters are t/t. About 70 per cent of Caucasians are tasters and 30 per cent are nontasters.

When taste sensitivity to PTC was analyzed more precisely, it became apparent that discrimination between the two phenotypes is not absolute. Harris and Kalmus have therefore designed a more elaborate test, in which the ability to taste PTC at different dilutions is measured. By this test it is possible to show that the threshold of taste sensitivity is distributed over a rather wide range of variation, but is bimodal (Fig. 5–15). About two-thirds of subjects fall in the group able to taste very dilute concentrations of PTC, and one-third can taste it only in very high concentrations. Some overlap exists between the two groups, so a few people cannot be classified accurately. The distribution of taste levels within each genotype probably depends on modifying genes. There is also a slight sex difference in taste sensitivity, with a slightly higher frequency of tasters among females than among males, and a slight loss in sensitivity with increasing age.

A number of other drugs (e.g., thiouracil and its derivatives) evoke a similar bimodal taste response. The chemical grouping NC=S is found in all these drugs, including PTC, and so is believed to be responsible for the taste response.

The distribution of PTC sensitivity varies strikingly in different populations. Nontasters are almost unknown among American Indians and Eskimos, and rare among Negroes. Some typical figures are given in

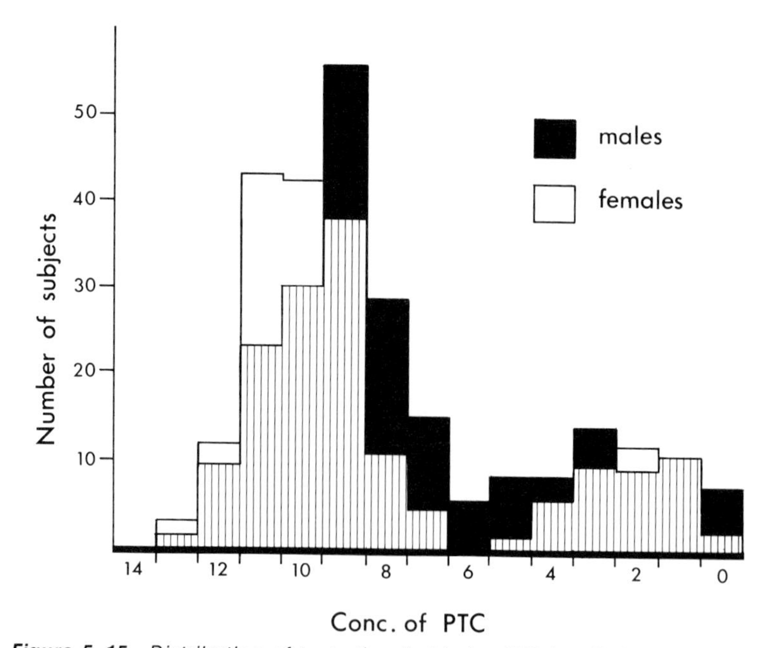

Figure 5–15 *Distribution of taste thresholds for PTC in a Belgian population of 225 males and 200 females. The concentration of PTC in solution 1 was 0.13 per cent, and each successive solution had half the strength of the preceding one. (Redrawn from Kalow 1962.* Pharmacogenetics: Heredity and Response to Drugs. *W. B. Saunders Company, Philadelphia, p. 121. Data from Leguebe, 1960.)*

TABLE 5–5 RACIAL VARIATION IN THE ABILITY TO TASTE PTC

Race	*Per Cent Tasters*
American Indian	90–98
American Negro	91
American White	65–75
African Negro	91–97
Chinese	89–94

Table 5–5. An intrepid group of investigators went so far as to test PTC sensitivity in anthropoid apes and reported that of 27 chimpanzees, 20 were tasters. By chance or otherwise, this is close to the relative frequency of the two genotypes in Caucasians.

PTC occurs in nature in some plants, and may be found in the milk of cattle pastured on these plants. PTC is therefore a normal constituent of food, but its concentration is below the taste threshold even for tasters.

The substances most similar to PTC are goitrogenic (i.e., antithyroid) drugs, and PTC itself has been shown to have a goitrogenic effect in mice. This observation has led to research into the PTC sensitivity of patients with various types of thyroid disease, and several curious relationships have been revealed. Nontasters are more frequent than expected among patients with hypothyroid types of disorder (athyrotic cretinism and nodular goiter) (Table 5–6). These observations suggest that the PTC-tasting polymorphism may have some relation to protection against thyroid disease. The mechanism of this protection is still unknown.

Parenthetically, it should be noted that PTC and related chemicals are not the only substances for which variation in taste thresholds is known. For example, quinine shows wide variation in taste threshold, but the distribution over the range is continuous, not bimodal. The taste threshold for salt and its relation to heart disease have been studied by Fallis et al. (1962), who reported that hypertensives showed a highly significant diminution in their ability to taste salt, as compared with

TABLE 5–6 ASSOCIATION OF ABILITY TO TASTE PTC AND THYROID DISEASE

Condition	*Per Cent Tasters*
Normal	70
Nodular goiter	60
Athyrotic cretinism	15

controls; consequently, these authors believe hypertensives probably take in more salt than do controls.

ACATALASEMIA

Acatalasemia (or acatalasia) is characterized, as the name implies, by the absence or greatly reduced activity of catalase in the blood and other tissues. The physiological role of catalase is obscure, but clinically its absence can lead to severe infections of the mouth. The original patient was seen by a Japanese otorhinolaryngologist, Takahara, for treatment for oral gangrene. Takahara noticed that when hydrogen peroxide was applied as a disinfectant, no bubbles were formed and the blood that came in contact with the hydrogen peroxide turned brownish-black. Absence of catalase, which in normal persons promptly degrades hydrogen peroxide, explains these observations.

Acatalasemia has since been reported from Switzerland and other western European countries, and has proved to be not a single disorder but a group of related mutations. Most Japanese heterozygotes have enzyme activity in the 50 per cent range, but in one family heterozygotes have quite normal activity. In the Swiss variant, homozygotes for the abnormal allele are asymptomatic and activity in heterozygotes is 60 to 85 per cent.

Acatalasemic mice have been bred (Feinstein et al., 1964), and attempts have been made to treat their genetic defect by infusing pellets containing catalase into their blood stream (Chang and Poznansky, 1968). This is a possible line of research for the eventual treatment of some genetic disorders.

GENERAL REFERENCES

Aebi, H., and Suter, H. 1971. Acatalasemia. In: *Advances in Human Genetics* 2:143–199. (Harris, H. and Hirschhorn, K., eds.) Plenum Publishing Corporation, New York.

Bondy, P. K., and Rosenberg, L. E. (eds.) 1974. *Duncan's Diseases of Metabolism.* 7th ed. W. B. Saunders Company, Philadelphia.

Brady, R. O., and Kolodny, E. H. 1972. Disorders of ganglioside metabolism. *Progr. Med. Genet.* 8:225–241.

Gardner, L. I. 1969. *Endocrine and Genetic Diseases of Childhood.* W. B. Saunders Company, Philadelphia.

Giblett, E. R. 1969. *Genetic Markers in Human Blood.* Blackwell Scientific Publications Ltd., Oxford.

Goodman, R. M., ed. 1970. *Genetic Disorders of Man.* Little, Brown and Company, Boston.

Harris, H. 1970. *The Principles of Human Biochemical Genetics.* North-Holland Publishing Company, Amsterdam and London; American Elsevier Publishing Co., Inc., New York.

Hsia, D. Y. Y. 1970. Phenylketonuria and its variants. *Progr. Med. Genet.* 7:29–68.

Kalow, W. 1962. *Pharmacogenetics: Heredity and the Response to Drugs.* W. B. Saunders Company, Philadelphia.

Kirkman, H. N. 1971. Glucose-6-phosphate dehydrogenase. In: *Advances in Human Genetics* 2:1–60. (Harris, H., and Hirschhorn, K., eds.) Plenum Publishing Corporation, New York.
Kirkman, H. N. 1972. Enzyme defects. *Progr. Med. Genet.* 8:125–168.
Lehmann, H., and Carrell, R. W. 1969. Variations in the structure of human haemoglobin: with particular reference to the unstable haemoglobins. *Brit. Med. Bull.* 25:14–23.
McKusick, V. A. 1970. Human genetics. In: *Annual Review of Genetics* 4:1–46. (Roman, H. L., ed.) Annual Reviews, Inc., Palo Alto, California.
Stanbury, J. B., Wyngaarden, J. B., and Fredrickson, D. S., eds. 1972. *The Metabolic Basis of Inherited Disease.* 3rd ed. McGraw-Hill Book Company, New York.
Witkop, C. J. 1971. Albinism. In: *Advances in Human Genetics* 2:61–142. (Harris, H. and Hirschhorn, K., eds.) Plenum Publishing Company, New York.

PROBLEMS

1. A woman who has atypical cholinesterase and is a carrier of Wilson's disease marries a man who has intermediate cholinesterase and Wilson's disease.
   (a) What are the possible genotypes and phenotypes of any children they might have?
   (b) Give the expected phenotypic ratios.

2. A normal woman whose father was G6PD-deficient marries a normal man.
   (a) What proportion of their sons is expected to be G6PD deficient?
   (b) If the husband were himself G6PD-deficient, what effect would this have upon the answer to part (a)?

3. In view of the lethal effect of sickle cell anemia, what are the phenotypes of the parents of affected children?

4. If the brother of a child who dies of sickle cell anemia marries the sister of another such child, what is the probability that their first-born child will be affected?

5. A woman who has atypical cholinesterase and normal hemoglobin is married to a man with usual cholinesterase and sickle cell trait. What possible combinations of types of cholinesterase and hemoglobin could appear in their children? State genotypes and phenotypes and give the expected ratios.

6. A man has sickle cell trait and is heterozygous for the Hopkins-2 type of hemoglobin. His wife is normal. Give the genotypes of both, and the possible genotypes and phenotypes of their children.

7. What are the possible combinations of genes in the gametes formed by an individual who is of the intermediate phenotype for cholinesterase, is heterozygous for dentinogenesis imperfecta and has sickle cell trait?

8. A man has sickle cell trait and his wife is heterozygous for hemoglobin C. What proportion of his children will have normal hemoglobin?

6

Chromosomal Aberrations

A new era in medical genetics opened in 1959 with the demonstration by Lejeune and Turpin that "enfants mongoliens" (children with mongolism or, as it is now more commonly called, Down's syndrome) have 47 chromosomes in each body cell instead of the normal 46. Cytogenetic abnormalities in clinical medicine are much more frequent and varied than was originally anticipated. Chromosomal aberrations are a significant cause of birth defects and fetal loss, occurring in an estimated 0.7 per cent of live births and one-third of spontaneous first-trimester abortions. There are many excellent reviews of medical cytogenetics, some of which are listed at the end of the chapter. A complete and authoritative reference work covering the field to 1970 is Hamerton's *Human Cytogenetics* (1971).

To be recognizable under the microscope with ordinary staining techniques, chromosomal aberrations must be abnormalities of number or gross abnormalities of structure. Any structural alteration involving less than about one-tenth of a chromosome arm would be undetectable even under ideal conditions of preparation and microscopy. Development of new "banding" techniques has made it possible to identify and define many previously unrecognized or unclassified structural changes, but even with the use of banding, many structural alterations may be beyond the limits of resolution. Examination of human cells in meiosis is still inadequate, even though recently improved by a technique for staining the centromeres (see Fig. 2–13).

Patients with chromosomal aberrations usually have characteristic phenotypes, closely resembling those of other patients with the same abnormality. The phenotypic anomalies have their basis in developmental confusion resulting from imbalance of the genetic material. Balanced translocations (see later), in which all the genetic material is present but in an abnormal arrangement, are not necessarily associated with abnormal phenotypes.

Trisomy (the presence of three representatives of a particular chromosome instead of the usual pair) is the general term used to describe the usual kind of chromosomal aberration found in Down's syndrome. Trisomy has been demonstrated in a large number of species of plants. In the Jimson weed or thorn apple, Datura, which has 12 pairs of chromosomes, there are 12 different trisomic varieties, each with an extra member of a different chromosome pair. Each of the 12 trisomic lines has its own unique phenotype. This parallels the condition in man, where there is a characteristic abnormal phenotype for each autosomal trisomy and a somewhat less distinctive phenotype for sex chromosomal trisomy. Although in any two members of the same species the majority of the genes on any chromosome are probably identical, each of us carries an assortment of variant genes, which are either genes for common polymorphisms or rare mutations. Therefore, the general phenotypic similarity of subjects with a specific chromosomal aberration, such as Down's syndrome, is modified by individual differences related to the genotype. If the chromosomal aberration is a deletion of part of a chromosome (or loss of a whole chromosome), the subject is hemizygous for all the gene loci that are deleted, and any recessive mutant genes present on the chromosome or chromosome segment homologous to the deletion are expressed.

CLASSIFICATION OF CHROMOSOMAL ABERRATIONS

Abnormalities of the chromosomes may be either numerical or structural, and may affect either autosomes, sex chromosomes (gonosomes) or, rarely, both in the same karyotype. In this section we will define the more common types of chromosomal aberrations and will introduce the terms used to describe them.

NUMERICAL ABERRATIONS

Numerical changes arise chiefly through the process of **nondisjunction** (failure of paired chromosomes or sister chromatids to disjoin at anaphase, either in a mitotic division or in the first or second meiotic division) (Fig. 6–1). Anaphase lag, when the members of a chromosome pair fail to synapse and therefore do not move apart correctly on the spindle, is a type of nondisjunction which can result in one member of the pair failing to be included in either daughter cell.

Any species has a characteristic chromosome number. In man the characteristic **diploid** number (in somatic cells) is 46 and the characteristic **haploid** number (in gametes) is 23. Any number which is an exact

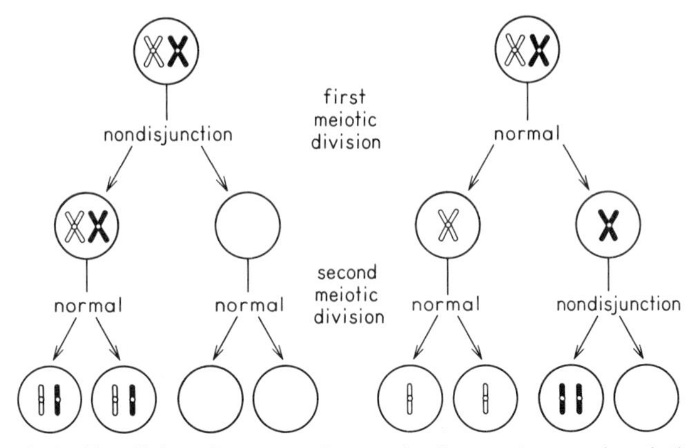

Figure 6–1 *Nondisjunction occurring at the first and second meiotic divisions. Nondisjunction at meiosis I produces gametes containing both members of the pair of homologous chromosomes concerned, or neither member. Nondisjunction at meiosis II produces gametes containing (or lacking) two identical chromosomes both derived from the same member of the homologous pair. In experimental organisms and probably also in man, nondisjunction occurs much more frequently at meiosis I than at meiosis II. Gametes lacking a chromosome (other than the X) are usually apparently unable to form a viable zygote.*

multiple of the haploid number is **euploid.** Euploid numbers need not be normal; thus $3n$ (triploid) and $4n$ (tetraploid) chromosome numbers are very unusual in man, only a few triploids having been born alive and tetraploids having been seen only in early abortuses. Chromosome numbers such as $3n$ and $4n$, which are exact multiples of n but greater than $2n$, are said to be **polyploid.** Polyploidy can arise by a variety of mechanisms, but triploidy probably results from failure of one of the maturation divisions, either in ovum or sperm, and tetraploidy results from failure of completion of the first cleavage division of the zygote.

Any number which is not an exact multiple of n is **aneuploid.** Some types of aneuploids are **trisomics,** with $2n+1$ chromosomes and three members of one particular chromosome, as in Down's syndrome; **monosomics,** with $2n-1$ chromosomes and only one member of some chromosome; double trisomics ($2n+1+1$), with an extra member for each of two chromosomes) and so forth.

Any chromosome number which deviates from the characteristic n and $2n$ is **heteroploid,** whether it is euploid or aneuploid.

How Aneuploidy Arises

The chief cause of aneuploidy is a mistake in a meiotic division, leading to unequal distribution of one pair of homologous chromosomes to the daughter cells so that one daughter cell has both and the other has neither chromosome of a pair. This kind of error is called **nondisjunction** on the assumption that it is due to failure of a mated pair to disjoin. However, since it probably usually arises by failure of pairing of two homologous chromosomes, followed by their random assortment rather than segregation, in most cases the term nondisjunction may not be technically correct.

If nondisjunction of chromosome 21 occurs at meiosis, the gametes formed have either an extra chromosome 21 ($n + 1 = 24$ chromosomes in all) or one too few ($n - 1 = 22$). Fertilization of the 24-chromosome gamete by a normal gamete would produce a 47-chromosome zygote, trisomic for chromosome 21.

Nondisjunction can occur at either the first or the second meiotic division. The consequences are rather different (Fig. 6–1). If nondisjunction occurs at meiosis I, the gamete with $n + 1$ chromosomes will contain both the paternal and the maternal representatives of that chromosome; if it involves the two chromatids of one chromosome at meiosis II, the gamete with $n + 1$ chromosomes will contain a double complement of *either* the paternal *or* the maternal chromosome. Nondisjunction can occasionally occur at successive meiotic divisions, or in both male and female gametes, so that zygotes with bizarre chromosome numbers may be formed, although these "multisomics" have been described only with respect to the X chromosome. Double aneuploidy (trisomy for two different chromosomes at once) has been observed.

Nondisjunction can also occur at *mitosis* after formation of the zygote, in which case the nondisjoining objects are halves of a single chromosome (chromatids), as in meiosis II. If this happens at an early cleavage division, a trisomic and a monosomic cell line are established; the trisomic one might persist, but the monosomic one probably would not. Again the X chromosomes are an exception, because XO lines are viable.

ABERRATIONS OF CHROMOSOME STRUCTURE

Much of our knowledge of human structural aberrations is based on work with other organisms, especially the fruit fly and certain plants. Structural rearrangements result from chromosome breakage, followed by reconstitution in an abnormal combination. Chromosome breaks occur normally at a low frequency, but may also be induced by a wide variety of breaking agents (clastogens) such as ionizing radiation, some virus infections and many chemicals.

The changes in chromosome structure resulting from breakage may be either stable (i.e., capable of passing through cell division unaltered) or unstable. The stable types of aberration are **deletions, duplications, inversions, translocations, insertions** and **isochromosomes.** The unstable types, which fail to undergo regular cell division, are **dicentrics, acentrics** and **rings.** These aberrations are shown diagrammatically in Figure 6–2 and elsewhere in this chapter.

Deletion

Deletion is loss of a portion of a chromosome, either terminally following a single chromosome break, or perhaps more often interstitially

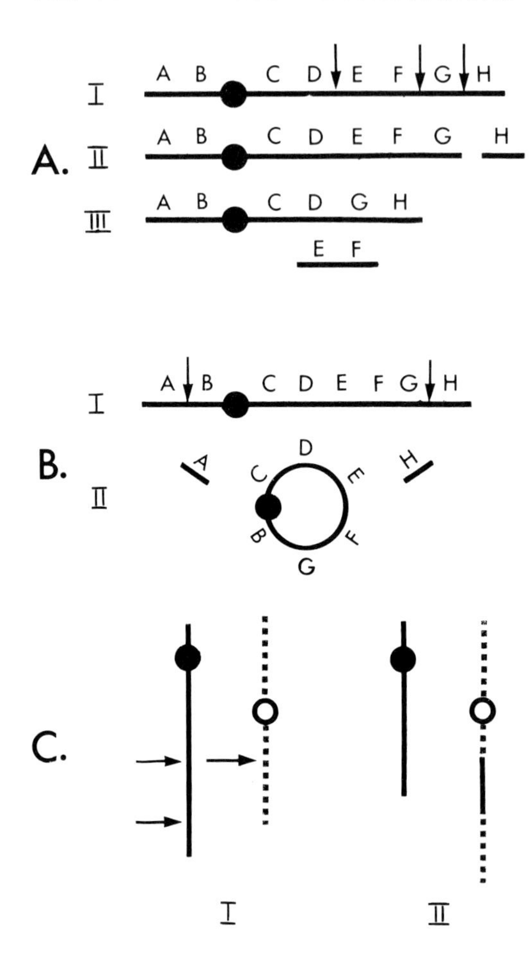

Figure 6–2 *Structural rearrangements of chromosomes.*

A, Deletion. I. Chromosome with possible positions of breaks indicated.

II. Single break between G and H producing terminal fragment, which, because it lacks a centromere, is usually lost in a subsequent division.

III. Two breaks, between D and E and between F and G, with reconstitution of the deleted chromosome without the interstitial fragment EF. The fragment EF, which lacks a centromere, is usually lost.

B, Formation of ring chromosome. Terminal fragments A and H have been deleted, and the broken ends of the chromosome have united to form a ring.

C, Insertion. This type of structural change requires three breaks (shown by arrows, I), followed by insertion of the segment between the breaks of one chromosome into the break in the nonhomologous chromosome (II).

between two breaks. The deleted portion, if it lacks a centromere, is an acentric fragment, which will fail to move on the spindle because it has no centromere and will eventually be lost at a subsequent cell division. The affected chromosome lacks whatever genetic information was present in the lost fragment. The common example of deletion in humans is the *cri du chat* syndrome, in which part of the short arm of chromosome 5 is deleted (see later).

Figure 6–2A shows a chromosome in which the order of the genes is given as ABCDEFGH (I). A terminal deletion might produce a chromosome lacking H (II). A deletion produced by two breaks, between D and E and between F and G with loss of the fragment between them, might produce a chromosome ABCDGH (III). If the deleted portion does not involve the centromere, the chromosome will replicate and divide in the normal way at subsequent divisions, but the acentric fragment will be unable to orient itself on the spindle, will fail to move at anaphase and will probably be lost.

A ring chromosome is a type of deletion chromosome in which both ends have been lost and the two broken ends have reunited to form a ring (Fig. 6–2B). If it has a centromere, a ring chromosome can replicate, but it may undergo alteration in structure.

Duplication

Duplication is the presence of an extra piece of chromosome, which usually has originated by unequal crossing over (see Fig. 5–6). The reciprocal product is a deletion. Duplications are more common and much less harmful than deletions. In fact, small duplications ("repeats") may be an evolutionary mechanism for the acquisition of new genes, which may then evolve into genes with quite different functions from the genes from which they originated. Duplications of whole genes or, much less frequently, of parts of genes are considered to be important factors in evolution (see hemoglobin loci, Lepore hemoglobin, haptoglobin). A classic example of the effect of a duplication is the variant *Bar eye* in the fruit fly *Drosophila*.

Duplication of parts of chromosomes may occur as a consequence of various structural rearrangements. For example, if a parent is a translocation heterozygote, unbalanced gametes may be formed which have, in effect, a duplication attached to some nonhomologous chromosome (see later). Partial duplications also result from crossing over in inversion heterozygotes, or from isochromosome formation (see later).

Inversion

Inversion involves fragmentation of a chromosome by two breaks, followed by reconstitution with inversion of the section of the chromosome between the breaks (ABCDEFGH might become ABCFEDGH). If the inversion is in a single chromosome arm it is **paracentric** (beside the centromere), but if it involves the centromere region it is **pericentric** (around the centromere).

Because inversions interfere with pairing between homologous chromosomes in inversion heterozygotes, crossing over may be suppressed within them. This can lead to the retention by a species of groups of genes which can then evolve as units. Inversions are therefore of evolutionary significance.

Usually a change in gene order produced by an inversion does not lead to an abnormal phenotype. The medical significance of inversions is for the subsequent generation. It arises from the consequences of crossing over between a normal chromosome and one with an inversion (Fig. 6–3).

For the homologous chromosomes to pair, one of them must form a loop in the region of the inversion. If the inversion is paracentric, the centromere lies outside the loop. When a crossover occurs within the loop, a dicentric chromatid and an acentric fragment are formed, as well

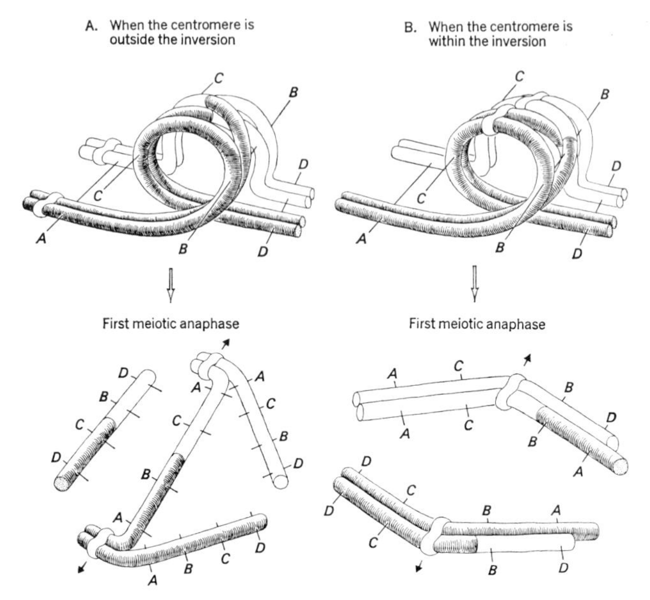

Figure 6–3 *Crossing over within the inversion loop of an inversion heterozygote results in aberrant chromatids with duplications or deficiencies. (From* General Genetics, *Second Edition, by Adrian M. Srb, Ray D. Owen and Robert S. Edgar. W. H. Freeman and Company. Copyright © 1965.)*

as a normal and an inverted one. Both aberrations are unstable. If the inversion is pericentric, the centromere lies within the loop. If a crossover now takes place, each of the two chromatids involved in the crossover has both a duplication and a deletion. If gametes are formed with these abnormal chromosomes, the resulting progeny will be monosomic for part of the chromosome and trisomic for another part.

With standard staining techniques it is difficult to recognize an inversion, unless it is a pericentric inversion large enough to make the position of the centromere clearly abnormal.

Translocation

Translocation is the transfer of part of one chromosome to a nonhomologous chromosome. The process requires breakage of both chromosomes, with repair in an abnormal arrangement. Translocations are often, but not always, reciprocal. A translocation does not necessarily lead to an abnormal phenotype, but, like inversions, translocations can lead to the formation of unbalanced gametes and therefore carry a high risk of abnormal offspring.

Robertsonian translocations are a special type in which the breaks occur at the centromeres and whole chromosome arms are exchanged. This process is also called centric fusion. When it occurs in man it usually involves two acrocentric chromosomes, e.g., 21/22 or 14/21. The relation of translocation to Down's syndrome is described later.

Insertion. An **insertion** is a type of translocation in which a broken part of a chromosome is inserted into a nonhomologous chromosome (Fig. 6–2C). This process requires three breaks, and has not been identified with certainty in man, though it has been postulated to explain some unusual situations.

Gametogenesis in Heterozygotes for Reciprocal Translocations. Translocations interfere with normal chromosome pairing and segregation at meiosis I, and can lead to unbalanced gametes and unbalanced offspring. The consequences of gametogenesis in an individual carrying a reciprocal translocation are shown in Figure 6–4. The two normal and the two translocated chromosomes synapse as a cross-shaped figure, which may open up into a ring or chain unless the arms of the chromosomes are held together by chiasmata.

The most frequent types of gametes formed include a normal combination, an abnormal but balanced combination and two abnormal unbalanced combinations. The first two types can lead to normal progeny, but the last two can produce unbalanced progeny with a duplication or a deletion of part of a chromosome.

Gametogenesis in Carriers of Robertsonian Translocations. A model of synapsis and disjunction in carriers of Robertsonian translocations is shown in Figure 6–5. The clinical significance of this phenomenon

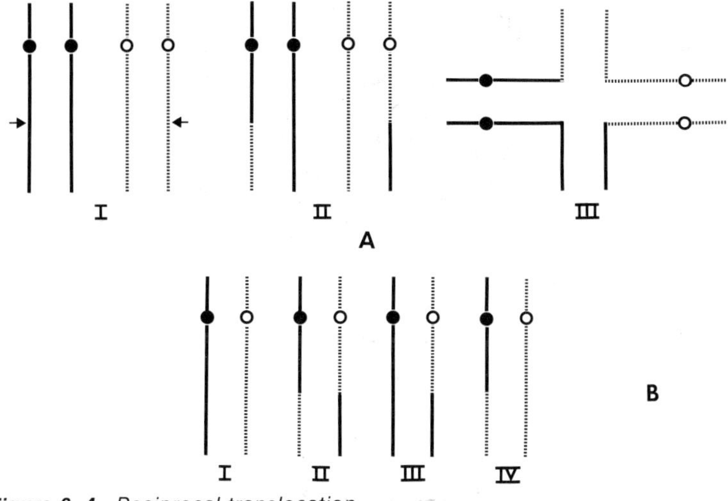

Figure 6–4 *Reciprocal translocation.*

A, *Origin of a reciprocal translocation by breakage of two nonhomologous chromosomes (I) followed by restitution with the broken ends interchanged (II). At meiosis I, the two normal chromosomes and their reciprocally translocated partners may form a cross-shaped figure (III).*

B, *Gametes typically formed by a translocation heterozygote. I, normal; II, balanced; III and IV, unbalanced. When fertilized, I forms a normal zygote and II a balanced heterozygote (like the parent); III and IV form unbalanced zygotes, each partially monosomic and partially trisomic.*

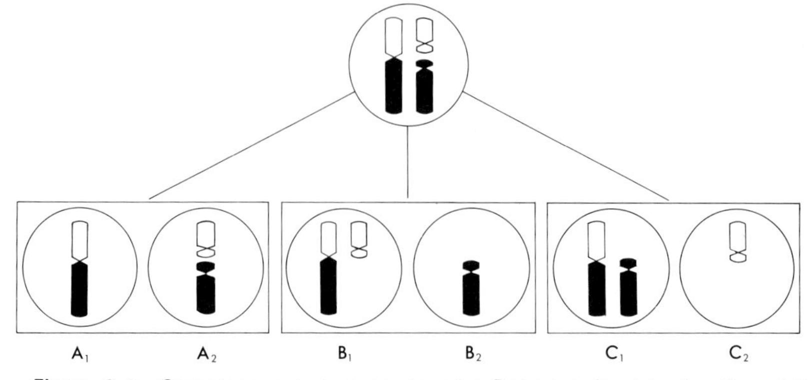

Figure 6–5 *Gametogenesis in a carrier of a Robertsonian translocation of the long arms of chromosome 14 (black) and chromosome 21 (outline). For details, see text.*

is that carriers of D/G (14/21 or occasionally 15/21) or G/G (21/22 or rarely 21/21) translocations have a high risk of producing offspring with Down's syndrome. The theoretical risk that one-third of the offspring of a translocation carrier should have Down's syndrome is much higher than the observed risks of 11 per cent for female carriers and 5 per cent for males (Hamerton, 1971).

Consider gametogenesis in a carrier of a translocation involving the long arms of chromosomes 14 and 21. The carrier has 45 chromosomes and is phenotypically normal, though the short satellited arms of both chromosomes are missing. (If the translocation was reciprocal, the reciprocal product has been lost in a subsequent cell division.) Theoretically, he forms six types of gametes in equal proportions:

A_2 Normal, 23 chromosomes
A_1 Balanced, 22 chromosomes including 14/21
B_1 Abnormal, 23 chromosomes including 14/21 and 21
B_2 Abnormal, 22 chromosomes lacking 21
C_1 Abnormal, 23 chromosomes including 14/21 and 14
C_2 Abnormal, 22 chromosomes lacking 14

Two of these types (B_2 and C_2) lack rather large amounts of chromosomal material. Assuming fertilization by a normal gamete, they would produce zygotes monosomic for chromosome 21 and chromosome 14, respectively. These are inviable (except for rare instances of children monosomic for chromosome 21). C_1 would produce a child with trisomy 14. This karyotype also appears to be inviable.

The three remaining types of gametes can produce viable offspring. Type A_2 is entirely normal. Type A_1 produces a phenotypically normal translocation heterozygote like the parent. Type B_1 produces a child with "translocation Down's syndrome," i.e., with 46 chromosomes, one of which is a 14/21 translocation. The karyotype could be formally described as 46,D–,t(DqGq)+, to indicate that there are 46 chromosomes in all, but that a D is missing and a translocation involving the long arms (q) of a D and a G is present.

Isochromosomes

During cell division the centromere of a chromosome may divide perpendicularly to the long axis of the chromosome instead of parallel to it (Fig. 6–6). If this mistake in division occurs in a submedian centromere (e.g., in the centromere of an X chromosome), the chromosome, instead of dividing into two identical halves, divides into a long and a short chromosome, both with metacentric centromeres; these are isochromosomes. Isochromosomes have been described for the X and presumptively also for D group chromosomes and chromosome 21. (A 21/21 chromosome might originate either as an isochromosome or by centric fusion of a pair of 21's.) Isochromosomes are partly duplications and partly deletions of the normal chromosomes. A patient with a normal X and an isochromosome of the long arm of the X (XXqi) is monosomic for genes on the short arm of the X and trisomic for genes on the long arm.

MIXOPLOIDS (MOSAICS)

If nondisjunction occurs during gametogenesis and the gamete with the heteroploid chromosome number is fertilized and begins to develop, every cell of the embryo will have the same chromosome number. However, if the nondisjunctional event happens during an early cleavage division of the zygote, the baby may have two or more lines of cells with different chromosome numbers. Such individuals are termed mixoploids or mosaics. There is a great variety of mixoploids, depending on the original karyotype of the zygote and the stage of development in which the error occurs.

A chromosomal mosaic has at least two cell lines, with different karyotypes, derived from a single zygote. The alterations in the karyotype may be either numerical or structural. Many different mosaics have been

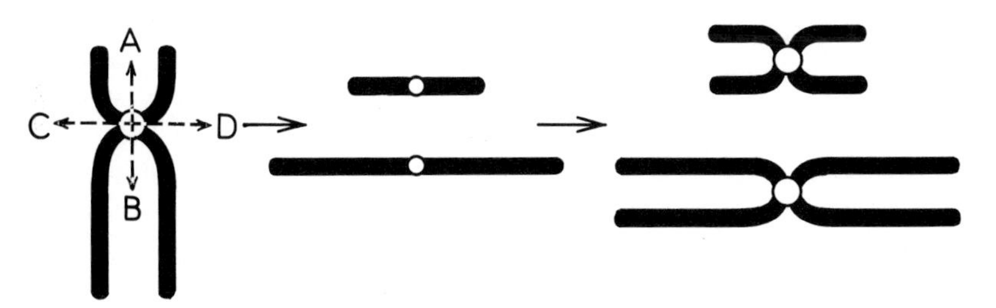

Figure 6–6 *Formation of isochromosomes. The submedian centromere divides in the plane C-D instead of the normal plane A-B; consequently, two abnormal metacentric chromosomes are formed, each containing duplication of the chromosomal material of one arm and no chromosomal material of the other arm of the original chromosome.*

described, most of which have cell lines with different numbers of sex chromosomes. A few children with Down's syndrome have a mixture of 46-chromosome and 47-chromosome tissues.

The proportions of normal and abnormal cells vary from tissue to tissue within the same patient, as well as from patient to patient. At present it is not possible to correlate the degree of mosaicism with clinical severity, although on the average mosaics are less severely affected than patients in whom all of the cells have an extra chromosome.

There are practical difficulties in the investigation of mosaicism. The relative proportions of the two or more cell lines are usually different in different tissues. The only tissues that are studied frequently are peripheral blood, marrow and skin, and there is a limit to the number of samples of these tissues that can be obtained from any one patient, so even if mosaicism cannot be demonstrated, it cannot definitely be ruled out. Because normal and abnormal cells may survive and multiply at different rates in culture, the proportions observed in a chromosome preparation may not mirror closely the proportions in the living patient. The proportions in the patient at the time of study may in any case be quite different from the proportions during the critical period of differentiation and development, the first trimester of prenatal life. These factors complicate attempts to demonstrate and interpret mosaicism.

CAUSES OF CHROMOSOMAL ABERRATIONS

Although by now we understand in a general way the mechanisms that produce aberrations of chromosome number and structure, we do not know much about the genetic and environmental factors that predispose to the formation of these aberrations. We are also ignorant about the developmental steps by which the abnormal karyotype produces its characteristic phenotype.

Late maternal age is a major factor in the production of Down's syndrome but is of less significance in other trisomies. The reason for the correlation of late maternal age and the nondisjunctional event that leads to Down's syndrome is unknown. Because paternal age probably has no effect upon nondisjunction, attempts have been made to explain the marked frequency of nondisjunction in oogenesis in older mothers on the basis of the difference between oogenesis and spermatogenesis. Developing ova are in the first meiotic prophase before the time of birth, and remain suspended at this stage of development until ovulation, which for a particular ovum may not be until 40 or more years later. Spermatozoa, on the other hand, undergo a developmental period of no more than 48 days after meiosis begins.

In vitro studies of maturation of human oocytes have cast some light on the mechanism of nondisjunction in causing aneuploidy. Chiasmata play a role in holding homologous chromosomes together at meiosis.

The number of chiasmata in the meiotic chromosomes of oocytes declines with increasing maternal age. This could cause homologues to fall apart prematurely, with consequent formation of aneuploid gametes and, eventually, the birth of aneuploid offspring. It has been postulated that oocytes are formed sequentially in the fetal ovary and released from the mature ovary in the same "production line" sequence. If so, the errors in chromosome disjunction that make older mothers more likely than younger ones to have children with abnormal chromosome numbers would relate to the position of the oocyte in the fetal ovary rather than directly to the increased time between germ cell differentiation and ovulation.

Genes predisposing to nondisjunction probably exist in humans, since such genes are known in other organisms. There are case reports of patients trisomic for two different chromosomes at once (double aneuploids), and pedigrees with clusters of aneuploids of the same or different kinds. Such reports suggest a predisposing genetic mechanism.

Autoimmune disease seems to have some role in the pathogenesis of nondisjunction, in view of an observed correlation between high thyroid autoantibody levels and chromosomal anomalies in families.

Radiation has been postulated as a cause of nondisjunction, and it is true that many parents (especially mothers) of aneuploids have unusual radiation histories, but adequate data for the assessment of this factor are not yet at hand. Radiation is known to be a cause of chromosome damage such as deletions and translocations (Bloom and Tjio, 1964), but its possible role in nondisjunction is less clear.

Ten studies of parental exposure to irradiation in relation to the risk of having a baby with Down's syndrome have now been made, but it is still not certain that such an association exists. In the most recent report (Alberman et al., 1972), mothers of Down's children had both more X-rays and more total X-ray dose than controls, but the difference was not significant. The cautious conclusion was that the risk of a Down's child is increased when the mother has had radiation damage by small doses, added to aging. The "doubling dose" (the X-ray dosage that doubles the risk of Down's syndrome) is estimated as 2r, which is equivalent to 10 abdominal X-rays or 360 chest X-rays.

Viruses have been shown to cause fragmentation of chromosomes. For example, Nichols et al. (1963) have demonstrated that the measles virus produces visible fragmentation. The possible effect of this process on the genetic material requires further investigation.

Chromosomal abnormalities themselves lead to abnormal segregation of chromosomes. The most obvious example is the transmission of a translocation, as just described, but there are other aspects. It has been noted that the chromosomes most commonly involved in rearrangements are members of the acrocentric groups, which are satellited. In preparations of chromosomes in mitotic metaphase, these satellited chromosomes can frequently be seen in "satellite association," as though the satellited ends were sticky. Satellite association could interfere with normal chromosomal segregation.

CLINICAL ASPECTS OF ABERRATIONS OF THE AUTOSOMES

There are several well-recognized clinical conditions in man which involve either the autosomes or the sex chromosomes. Anomalies of the sex chromosomes are discussed in Chapter 7. The four best known autosomal anomalies are described in this section: Down's syndrome, trisomy 18, D trisomy (trisomy 13) and cri du chat syndrome.

DOWN'S SYNDROME (TRISOMY 21, MONGOLISM)

Down's syndrome is the most common and best known of the chromosomal anomalies (Fig. 6–7). The condition was first described by Langdon Down in 1866, but its cause remained a deep mystery for nearly

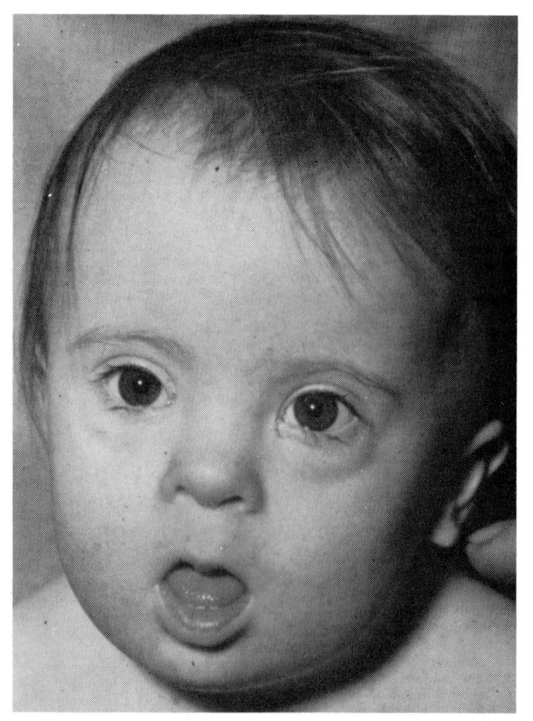

Figure 6–7 *A child with Down's syndrome. (From Smith 1970.* Recognizable Patterns of Human Malformation. *W. B. Saunders Company, Philadelphia.)*

a century. There are two noteworthy features of the distribution of Down's syndrome in the population: late maternal age, and a peculiar distribution in twins and families—concordance in all monozygotic twins but almost complete discordance in dizygotic twins and other relatives. It was suggested by Waardenburg in 1932 that a chromosomal anomaly could explain these observations, but the chromosomal anomaly was not demonstrated until 1959, when Lejeune and Turpin showed that children with Down's syndrome had 47 chromosomes, the extra one being a small acrocentric, now conventionally regarded as chromosome 21.

The old name of mongolism refers to the somewhat Oriental cast of countenance produced by the epicanthal folds, which give the eyes a slanting appearance. It is not an appropriate name and is gradually being replaced by the term Down's syndrome.

Diagnostic Criteria

Down's syndrome can usually be identified at birth or shortly thereafter by the presence of a combination of some of the following characteristics:

Hypotonia: Often the first sign of abnormality, noticed in the hospital nursery.

Mental retardation: I.Q. usually in the 25 to 50 range.

Head: Brachycephalic, with flat occiput.

Eyes: Epicanthal folds, Brushfield spots (speckled iris).

Nose: Low bridge.

Tongue: Usually protruding, furrowed, lacking a central fissure.

Hands: Short and broad, with single palmar crease (simian crease) and clinodactyly (incurving) of the fifth finger. Characteristic dermal patterns, with a distal axial triradius and other peculiarities.

Feet: Wide gap between first and second toes, with furrow extending backward along plantar surface. Characteristic dermal patterns in the hallucal area.

Heart: Cardiac anomaly in about 35 per cent of cases.

Acetabular and iliac angles: Decreased (a useful radiological criterion).

Stature: Below average.

Often the diagnosis presents no particular difficulty, but karyotyping is usually indicated to determine whether the child has the typical trisomy 21 karyotype (95 per cent of cases), is a mosaic (1 per cent of cases) or has a translocation (4 per cent of cases).

The Chromosomal Basis of Down's Syndrome

There is no doubt that Down's syndrome involves trisomy for chromosome 21, but about 4 per cent of the patients have this chromosomal material arranged not as a separate chromosome but as a translocation of the long arm of chromosome 21 onto either a D chromosome (14 or sometimes 15) or another G (21 or 22). A standard trisomy 21 karyotype is shown in Figure 6–8, and the karyotype of t(DqGq) translocation Down's syndrome is shown in Figure 6–9.

A patient with a Dq Gq translocation (which may be abbreviated to D/G) has 46 chromosomes in all, including the translocation chromosome. This karyotype is effectively trisomic for chromosome 21 and its phenotypic consequence is indistinguishable from standard trisomy 21.

By definition, a translocation is an exchange of material between nonhomologous chromosomes, so one would expect a translocation involving chromosome 21 and another G group chromosome to be a 21/22

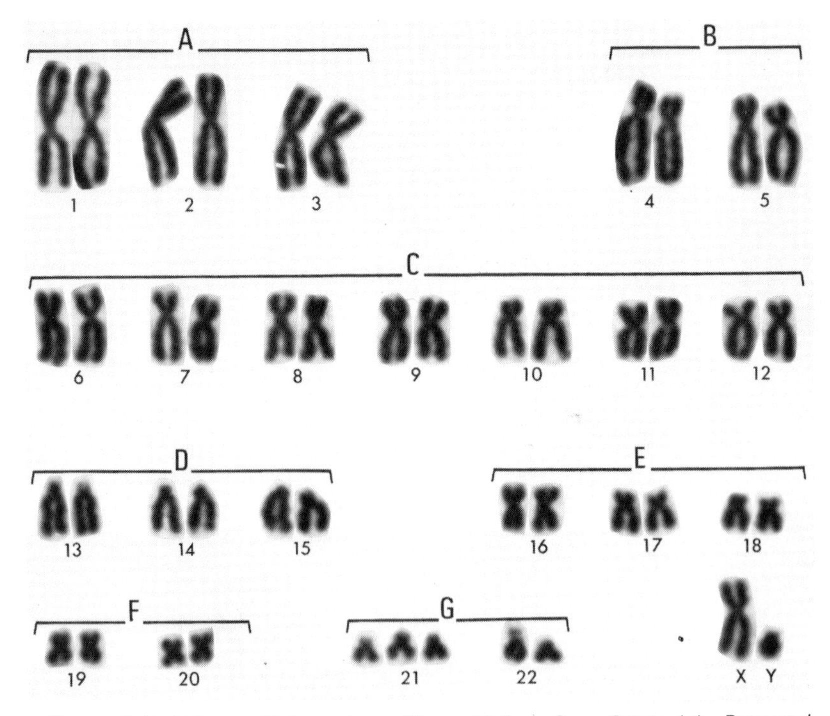

Figure 6–8 *Trisomy 21 karyotype. (From Walker, Carr, Sergovich, Barr and Soltan 1963.* J. Ment. Defic. Res. *7:150–163.)*

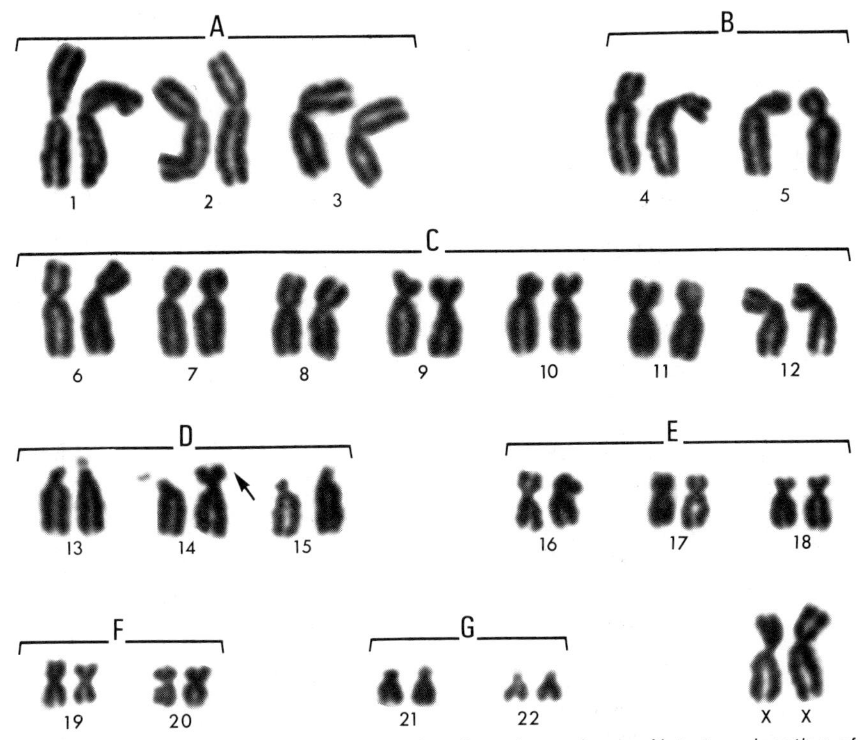

Figure 6–9 *Karyotype of translocation Down's syndrome. Note translocation of the long arm of a chromosome 21 to the long arm of a chromosome 14. (Modified from Walker, Carr, Sergovich, Barr and Soltan 1963.* J. Ment. Defic. Res. *7:150–163.)*

translocation. However, there are a few pedigrees in which it is clear that there has been duplication of the twenty-first chromosome, though these may be isochromosomes of the long arm of the chromosome 21 rather than translocations. Whether the aberrant chromosome is actually an isochromosome for the long arm of chromosome 21 (21 qi) or a 21 q 21 q translocation, a normal parent who carries this chromosome can have *only* children with Down's syndrome (or, very rarely, children with monosomy 21. This is because the parental gamete must have two representatives of 21q or none.)

About 1 per cent of Down's syndrome patients are mosaics, usually with a mixture of 46-chromosome and 47-chromosome cells. Such patients usually have relatively mild stigmata, and are more intelligent than the usual trisomics. However, their risk of having Down's children may be high if some or all of their reproductive cells have the extra chromosome.

Recurrence Risk of Down's Syndrome

A frequent problem in genetic counseling is the risk that a woman who has had a child with Down's syndrome will have another at a later pregnancy. The risk varies with the woman's age, the karyotypes of her child, her husband and herself, and the family history with respect to Down's syndrome.

Estimates of the incidence of Down's syndrome average about one in 700 live births. Carter and MacCarthy (1951) found the incidence to be one in 666 in a London survey. The risk varies with the mother's age and it is probable that in recent years the decline in average maternal age and curtailment of the reproductive period as a consequence of oral contraceptives have lowered the birth incidence to perhaps one in 1000 births. The childhood death rate has fallen sharply since antibiotics have become available to treat the respiratory infections that at one time led to the death of many young patients. Consequently, the prevalence of living cases of Down's syndrome has increased almost fourfold in the last generation. (See also page 349.)

The average maternal age at the birth of a child with Down's syndrome is about 34 years (as compared with the current overall mean maternal age of about 26 years). Though the *risk* is lower at early maternal ages, there are so many more births to younger mothers that the absolute number of Down's patients born to younger mothers is quite high. The age distribution of mothers of translocation Down's patients is the same as the usual maternal age distribution.

The probability that the Down's syndrome child of a woman under 30 has a translocation is about 9 per cent, whereas for mothers over 30 it is 1.5 per cent, i.e., only one-sixth as high. Few of these babies have inherited the translocation from a parent. In a summary of 34 series of translocation Down's patients (Hamerton, 1971), about 40 per cent of the translocations were inherited and the remainder were sporadic. (The true proportion may not be so high, because it is difficult to be sure that at least some of the patients were not selected for study because of a family history of Down's syndrome, which would increase the proportion of patients with a hereditary translocation.)

The risk of Down's syndrome is high if the mother is a translocation carrier (page 140). The data are summarized in Table 6–1.

TABLE 6–1 PROGENY OF TRANSLOCATION CARRIERS

	Normal	*Carrier*	*Affected*
Theoretical (if parent a translocation carrier)	0.33	0.33	0.33
Observed			
Mother Dq Gq carrier	0.49	0.40	0.11
Father Dq Gq carrier	0.39	0.59	0.02
Mother Gq Gq carrier	0.46	0.52	0.01
Father Gq Gq carrier	0.34	0.66	–

These figures show that the risk of having an affected child is far below the theoretical one-third, and the risk of a translocation carrier child is somewhat higher. Carrier males are less likely than carrier females to have affected children.

There is a slight additional risk of Down's syndrome in later children even when the affected child is a 21 trisomic. Accurate risk figures should eventually come from amniocentesis studies in pregnant women who are in older age ranges or who have previously had a child with Down's syndrome. Table 6–2 gives tentative estimates of the risk of Down's syndrome by maternal age, for women in general and for women who have produced an affected child.

Reproduction in Down's Syndrome

Little is known about the transmission of most chromosomal anomalies, and indeed many of them produce such severe abnormalities that transmission to the next generation is impossible. Males with Down's syndrome have not been known to reproduce, but a few affected females have had children. About half the offspring have been affected and half normal. This is precisely what would be expected if a 47-chromosome mother formed 23-chromosome and 24-chromosome gametes in equal proportions. Reproduction has not yet been described in women with translocation Down's syndrome.

TRISOMY 18 (E SYNDROME)

Trisomy 18 is a syndrome of multiple congenital malformations associated with trisomy for an E-group chromosome, no. 18 (Fig. 6–10).

TABLE 6–2 RISK OF DOWN'S SYNDROME

	Risk of Down's Syndrome in Child	
Age of Mother	*At Any Pregnancy**	*After the Birth of a Baby with Down's Syndrome***
- -29	1 in 1500	1 %–2 %
30–34	1 in 700	Normal
35–39	1 in 300	risk
40–44	1 in 100	for
45–49	1 in 50	age
All mothers	1 in 910	

*Based on the data of Collmann and Stoller, 1963.

**Estimate of Smith and Wilson (1973) for subsequent sibs of trisomic Down's patients or Down's patients whose chromosomes have not been studied.

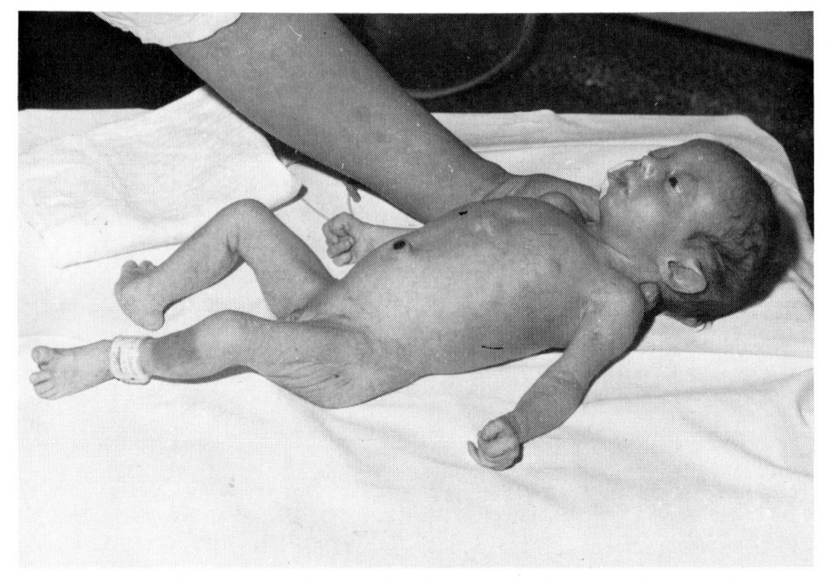

Figure 6–10 *Infant with trisomy of a chromosome in group E, probably chromosome 18. Note the overlapping fingers, rocker-bottom feet, simplified patterns of the ear and dorsiflexion of the big toe. (Courtesy of D. H. Carr.)*

The syndrome was first described by Edwards et al. in 1960. It is more severe than Down's syndrome, most of the affected infants dying by the age of six months.

Some of the characteristic stigmata of trisomy 18, other than the chromosomal anomaly, are listed below. Numerous review papers and case reports have appeared and may be consulted for additional details. There are many malformations seen in this syndrome, but those listed include some of the most useful criteria for discriminating between trisomy 18 and other malformation syndromes.

Mental retardation.

Failure to thrive.

Excess of females (78 per cent).

Ears: Low set and malformed.

Hands: Fists clenched, with second digit overlapping third, and fifth overlapping fourth. Single palmar crease (simian crease). Arch patterns on six or more fingers.

Feet: "Rocker bottom" deformity.

Heart: Characteristic type of malformation.

D TRISOMY (TRISOMY 13)

Trisomy for a chromosome in the D group is less common than trisomy 18, perhaps because the anomalies associated with trisomy for this chromosome are more severe. A typical patient is shown in Figure 6–11. Central nervous system defects are present; and although variable, they are severe and associated with mental retardation. The more obvious external anomalies include cleft lip and palate, polydactyly and characteristic dermal pattern anomalies. Abnormalities of the heart, viscera and genitalia are also present.

CRI DU CHAT SYNDROME

Partial monosomy (deletion of a portion of the short arm) of chromosome 5 (5p–) produces a characteristic spectrum of anomalies termed the *cri du chat* syndrome because of a fancied resemblance of the infant's cry to the mewing of a cat. The affected children are mentally retarded, are microcephalic, have a characteristic facial appearance with marked hypertelorism (Fig. 6–12) and have characteristic dermal patterns.

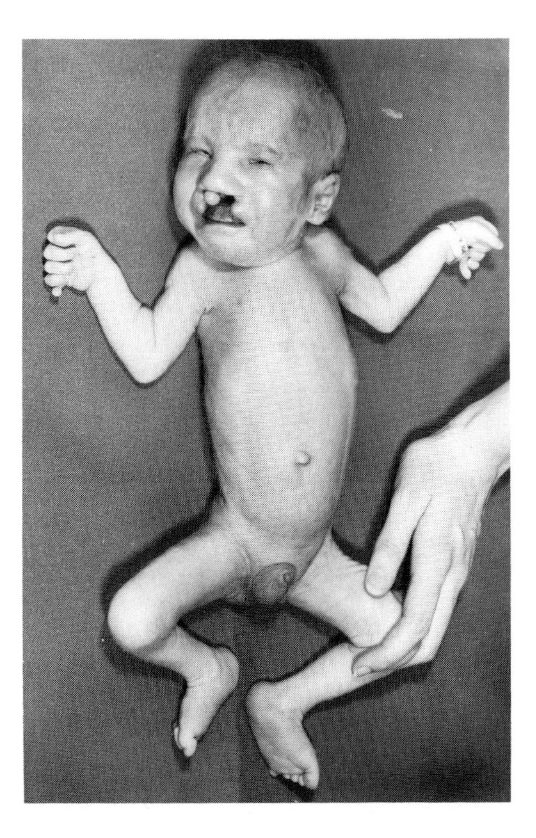

Figure 6–11 *Infant with D trisomy. Note particularly the bilateral cleft lip and polydactyly. (Courtesy of P. E. Conen.)*

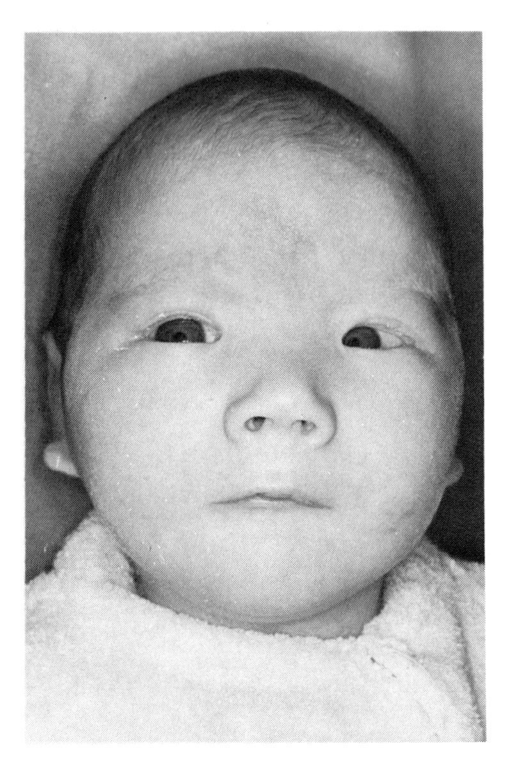

Figure 6–12 *Infant with cri du chat syndrome, resulting from a partial deletion of the short arm of chromosome 5 (5p–). Note characteristic facies with hypertelorism, epicanthus and retrognathia.*

Cri du chat syndrome can occur in the child of a translocation carrier who has formed an unbalanced gamete with a deficiency. In the same family, patients with a corresponding duplication may be found.

Deletion of the short arm of the other B chromosome, no. 4 (4p–) is rarer than 5p–. It produces a different and much more severely abnormal phenotype, with low birth weight, failure to thrive, coloboma of the iris, abnormal facial appearance and many other defects. Children with this anomaly do not have the mewing cry of *cri du chat* syndrome.

"ANTISYNDROMES"

Autosomal abnormalities which are in a sense the reverse of the more common ones include 21 monosomy or deletion (21q–), and deletion of the short or long arm of chromosome 18 (18p– or 18q–). Ring chromosomes are special types of deletions, theoretically missing portions of both chromosome arms.

In some respects "antisyndromes" have phenotypes that differ from the normal in the opposite direction from their better-known partners, though the comparison cannot be pushed too far. For example, 18q–

patients have a total fingerprint ridge count that is higher than average, rather than lower as in trisomy 18; but they are mentally and developmentally retarded.

In any deletion syndrome, the patient is hemizygous for any genes that have their loci on the deleted portion. Consequently, any recessive gene on the homologous part of the normal chromosome can be expressed. This may account in part for the clinical severity of some deletions.

CHROMOSOMES AND ABORTION

About 15 per cent of all recognizable pregnancies end in abortion, and about one-third of all spontaneous first-trimester abortions are heteroploid (abnormal in chromosome number). The incidence is higher in earlier abortions and lower in later ones. Typical figures (Carr, 1971) are as follows:

Up to 90 days (measured from last menstrual period)	40 per cent
91 to 120 days	25 per cent
Over 120 days	3.5 per cent

About 20 per cent of all chromosomally abnormal abortuses are 45,X, as compared with an incidence in live births of one in 2500. About 45 per cent are trisomic and 25 per cent polyploid (either triploid or tetraploid). The remaining 10 per cent are monosomic for some autosome, or have a structural anomaly. Many of the numerical anomalies seen in abortuses have never been seen in liveborn infants.

In families with habitual abortion, a translocation in one or the other parent is somewhat more frequent than in the general population. This is not surprising, because the presence of a translocation can lead to "unbalanced" gametes and therefore to chromosomally abnormal offspring (Figs. 6–3 and 6–4).

CYTOGENETIC ABNORMALITIES AND LEUKEMIA

The Philadelphia Chromosome

The Philadelphia chromosome (Ph[1]) is a characteristic anomaly found in the leukocytes of patients with chronic granulocytic leukemia, It is a deleted chromosome 22, and is probably a specific marker for the leukemic cells of persons with this type of leukemia.

Other Associations of Chromosomal Anomalies and Leukemia

The enhanced susceptibility of patients with Down's syndrome to leukemia was recognized before the cytogenetic basis of Down's syndrome was known. Down's patients are about six times as susceptible to leukemia as controls in the same age range.

Pedigrees in which leukemia and Down's syndrome appear in different members of the same kindred also suggest an association between abnormal chromosome complements and leukemia. For example, Miller et al. (1961) described a family in which Down's syndrome, Klinefelter's syndrome and leukemia had all appeared.

GENERAL REFERENCES

Carr, D. H. 1971. Chromosomes and abortion. *Advances in Human Genetics* 2:201–258 (Harris, H., and Hirschhorn, K., eds.) Plenum Publishing Corporation, New York.

Hamerton, J. L. 1971. *Human Cytogenetics*. Vol. I. *General Cytogenetics*. Vol. II. *Clinical Cytogenetics*. Academic Press, Inc., New York.

Penrose, L. S., and Smith, G. F. 1966. *Down's Anomaly*. J. and A. Churchill, London.

Smith, D. W., and Wilson, A. C. 1973. *The Child with Down's Syndrome (Mongolism)*. W. B. Saunders Company, Philadelphia.

PROBLEMS

1. A woman of blood group AB, married to a man of blood group O, has a child with trisomy 21. If the ABO blood group locus is on chromosome 21, what blood group genes will the Down's child have:
   (a) If nondisjunction of the twenty-first pair of chromosomes occurred in the maternal gamete at the first meiotic division?
   (b) If nondisjunction occurred at the second meiotic division?
   (In your answer, ignore the probability that crossing over might have occurred between the gene locus concerned and the centromere.)

2. A female carrier of a Dq Gq translocation has a normal phenotype.
   (a) What types of gametes would she theoretically produce?
   (b) What is the theoretical risk of mongolism in her liveborn offspring?

3. In a female Down's syndrome patient with a Dq Gq translocation:
   (a) What types of gametes would be produced?
   (b) Assuming fertilization by a normal sperm, what would be the karyotypes and phenotypes of her offspring?

7

The Sex Chromosomes

The importance to medical genetics of the human sex chromosomes and their abnormalities entitles them to a chapter of their own. Because sex chromosomes are crucial in normal and abnormal sexual development, it is appropriate to begin by reviewing briefly the development of the human reproductive system.

EMBRYOLOGY OF THE REPRODUCTIVE SYSTEM

The embryonic germ cells are wandering cells derived from the gut endoderm. By about the eighth week of intrauterine life, they settle down in the cortex and medulla of the primitive, bipotential gonad. Normally, if a Y chromosome is present the gonad differentiates in the male direction, the medullary tissue forming typical testes with seminiferous tubules and Leydig cells which become capable of androgen secretion. If two X chromosomes are present, or even if there is one X and no Y, the gonad differentiates in the female direction, the cortex developing and the medulla regressing. If two X's are present, oogonia develop in the cortex, and before the time of birth the first meiotic division has begun in the sex cells.

By about the fourth week of intrauterine life thickenings in the genital ridges indicate the positions of the wolffian (male) and müllerian (female) duct systems. In the presence of androgen-secreting Leydig cells the wolffian duct normally differentiates to form the typical male duct system. At the same time the Leydig cells produce a second male organizer which causes regression of the müllerian duct system. If an ovary is present, or if the gonad remains undifferentiated, the müllerian system normally develops toward a typical female duct system and the wolffian system regresses.

In the early embryo the external genitalia consist of a genital tubercle, two labioscrotal swellings and two urethral folds. The genital tubercle can form either penis or clitoris. In the male, under the influence of androgens, the labioscrotal folds fuse to form the scrotum and the urethral folds fuse to form the urethra including its penile portion, while the genital tubercle takes on a male (penile) configuration. In the absence of a differentiated gonad, or in the presence of an ovary, the labioscrotal folds remain separate to form the labia majora, and the urethral folds remain unfused to form the labia minora. The genital tubercle assumes the female (clitoral) configuration.

Abnormalities of genes, chromosomes or environment can affect any of these stages. In either 46,XX or 46,XY individuals, if no differentiated gonad forms, the phenotype remains female. If the androgenic male organizer is deficient, the wolffian system does not develop, but if the second male organizer is present the müllerian system also fails to develop and the internal genitalia are rudimentary, typical of neither sex. Similarly, refractoriness of the tissues to androgen may produce female external genitalia in an XY individual. In an XX fetus, androgen production (as from an abnormal suprarenal gland) or exogenous hormones crossing the placenta from the mother may lead to varying degrees of masculinization of the external genitalia.

THE SEX CHROMOSOMES AND SEX CHROMATIN

THE X CHROMOSOME

The X chromosome is a submetacentric chromosome which is a member of the C group. Genes on the X are said to be X-linked (or sex-linked), and their characteristic pattern of transmission has led to the identification of about 100 X-linked traits. Among the important gene loci on the X are those for the protan and deutan types of red-green color blindness, the two most common coagulation defects (hemophilia A and Christmas disease), the glucose-6-phosphate dehydrogenase variants and the Xg blood group system.

A gene on the X has no allele on the Y, so males are **hemizygous** for X-linked genes. There is one possible exception. The X and Y must pair at meiosis; otherwise they would not segregate regularly. Since in meiotic metaphase figures they show end-to-end pairing (Fig. 7–1), it is likely that they have some homologous loci at the pairing ends.

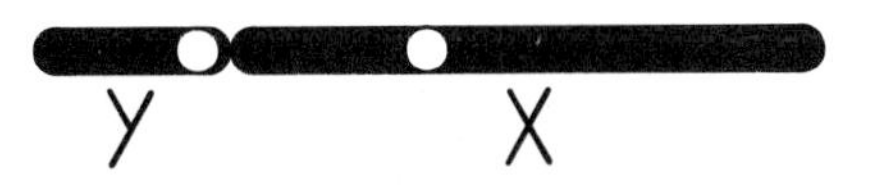

Figure 7-1 Diagrammatic representation of the X and Y chromosomes paired end to end by the tips of their short arms at metaphase of meiosis I. (See also Fig. 2-13.)

THE Y CHROMOSOME

The Y is a member of the small acrocentric G group, distinguishable from the other members of that group (21 and 22) by its slightly greater length, less widely flaring arms and absence of satellites on the short arm. The Y can often be distinguished in interphase as well as metaphase cells by its vivid fluorescence with quinacrine stains (page 15), though fluorescence in interphase nuclei is not restricted to the Y, and it is risky to attempt to diagnose sex by Y fluorescence in interphase cells. It was formerly thought that in man, as in the fruit fly *Drosophila*, the number of X's rather than the presence or absence of a Y determined the phenotypic sex, but it is now known that phenotypic sex in man depends chiefly on the presence of a Y, and that normally anyone who has a Y is a phenotypic male. However, modifications of the male phenotype are possible even when a Y is present.

SEX DETERMINATION

The gametes of the female all have an X chromosome; hence, the female is said to be **homogametic**. Males may transmit either X or Y; consequently, the male is said to be **heterogametic**. (In birds the situation is reversed and the female is the heterogametic sex.)

		Ova	
		X	X
Sperm	X	XX	XX
	Y	XY	XY

As the checkerboard shows, the sex of the child depends upon whether the father contributes an X or a Y. The mother, since she can contribute only an X, cannot determine the child's sex. The theoretical sex ratio is one male to one female, but at birth it is actually about 106 males to 100 females (a sex ratio of 1.06). The sex ratio of abortuses is about the same as the sex ratio of births (Carr, 1971).

Sex Chromosome Complements Other than XX and XY

A wide variety of aneuploids and other anomalies are known for the sex chromosomes. Some of the possible variants are:

Males	*Females*
XYY	X
XXY	XXX
XXXY	XXXX
XXYY	XXXXX
XXXYY	XY
XXXXY	and even XYY
XX	

To this list of variants must be added mosaics and, in females, karyotypes having a structural variant of the X such as an isochromosome, deletion or ring; or, in males, karyotypes with short (presumably deleted) or long (presumably duplicated or partly duplicated) Y chromosomes. Mosaics for sex chromosomes are relatively common, and extra sex chromosomes appear to be better tolerated than extra autosomes.

Sex Chromatin

Since early in the twentieth century it has been known that male cells differ from female cells in their complement of sex chromosomes, but it was not until 1949 that a difference in the *interphase* cells of males and females was detected. In that year Dr. Murray Barr of the University of Western Ontario, and his graduate student, E. G. Bertram, noted that a previously recognized but little understood mass of chromatin in the nuclei of some nerve cells was frequently present in females but not in males (Fig. 7–2). This mass is now known as the sex chromatin or Barr body. Cells (or individuals) are said to be chromatin-positive if sex chromatin is present, and chromatin-negative if it is not.

Barr later found sex chromatin in the cells of most of the tissues of females of many species of mammals including humans. The Barr body can be studied in three types of cells: nerve cells (Fig. 7–2), epithelial cells (Fig. 7–3) and polymorphonuclear leukocytes (Fig. 7–4). Of these three, the epithelium of the buccal mucosa is the most convenient tissue for demonstration of sex chromatin. Nerve cell studies are usually limited to autopsy material.

The technique of obtaining a buccal smear consists of scraping a

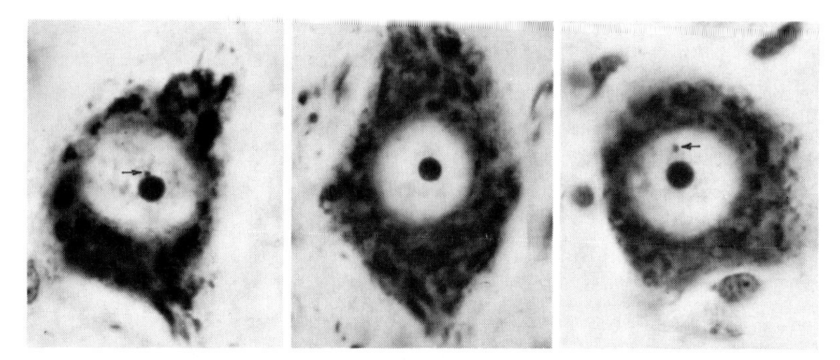

Figure 7–2 *Sex chromatin in the nerve cells of the cat. These are the illustrations used in the original description of sex chromatin. The sex chromatin (arrows) appears only in the female. The center cell, from a male, shows no sex chromatin. (From Barr and Bertram 1949.* Nature *163:676–677.)*

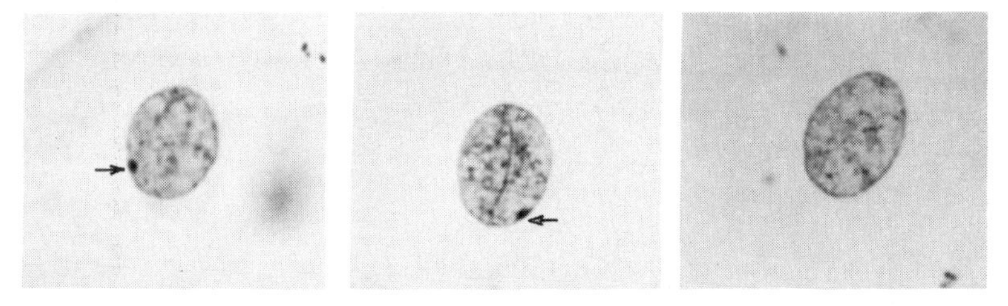

Figure 7–3 *Sex chromatin in the epithelial cells of the human buccal mucosa. Arrows point to the sex chromatin at the nuclear membrane. The male cell (right) shows no sex chromatin. These photographs appeared in the original paper describing the buccal mucosa technique for observation of the sex chromatin. (From Moore and Barr 1955.* Lancet *2:57–58.)*

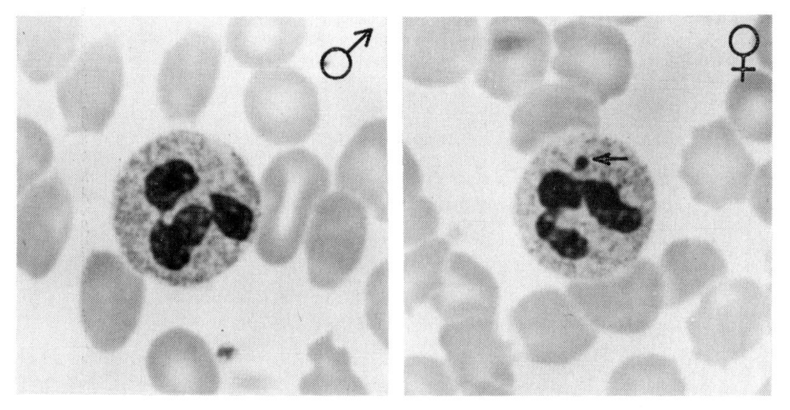

Figure 7–4 *Sex chromatin in human neutrophils. Note the "drumstick" (arrow) in the female cell on the right, and the absence of a drumstick in the male cell on the left. (From Barr 1957.* Progress in Gynecology *3:131–141. Grune & Stratton, Inc., New York.)*

few cells from the inside of the cheek, placing them on a slide, staining with any nuclear stain and examining the slide under the microscope.

Sex chromatin can be identified in amniotic fluid cells (fetal cells obtained by amniocentesis). The Y chromosome can also be identified in amniotic fluid cells by fluorescent staining techniques. However, when the sex of a fetus is to be determined prenatally, karyotyping to visualize the whole chromosome complement is much more reliable than testing for X and Y chromosomes by looking for sex chromatin and the fluorescent Y.

The association of the sex chromatin with the number of X chromosomes and its lack of association with the Y can be seen from the following table. The phenotypes of some of the individuals with anomalous chromosome complements are discussed later in this chapter.

Chromosome Complement	*Number of Barr Bodies*
45,X 46,XY 47,XYY	0
46,XX 47,XXY 48,XXYY	1
47,XXX 48,XXXY 49,XXXYY	2
48,XXXX 49,XXXXY	3
49,XXXXX	4

Note that sex chromatin is present if there are two or more X chromosomes, and that the number of sex chromatin bodies is one less than the number of X's. Not all cells show the characteristic number. Normal females exhibit recognizable Barr bodies in only about 50 per cent of the epithelial cells at any one time. Whether or not a Barr body can be seen in any one cell depends on the stage of the cell cycle and the orientation of the nucleus on the slide.

The Barr body is altered in size in individuals who have a structural aberration of one X chromosome, such as a deletion, ring or an isochromosome X. The abnormal X chromosome always forms the sex chromatin and the presence of unusually small or large Barr bodies in the buccal smear may indicate that a structural variant of the X is present.

At prophase of mitosis in normal females, one X chromosome is dark-staining (heteropyknotic) and forms a mass about the size of the sex chromatin. Autoradiographic studies with tritiated thymidine, a radioactive compound taken into the DNA of replicating chromosomes, show that one X chromosome completes its replication late in the DNA synthesis stage of the cell cycle and that this X is usually located peripherally, in the area of the nucleus where the sex chromatin is found. The late-replicating X, which is condensed and out of phase with the rest of the chromosome set, forms the sex chromatin.

THE LYON HYPOTHESIS OF GENE ACTION ON THE X CHROMOSOME

For many years the action of X-linked genes was a puzzle to geneticists. How could it be that females with two representatives of every gene on the X chromosome formed no more of the product of these genes than did hemizygous males with only a single X? And why were females homozygous for an X-linked mutant gene no more severely affected than were hemizygous males? Some mechanism of "dosage compensation" was indicated.

A hypothesis to explain dosage compensation in terms of a single active X chromosome was arrived at independently by several groups of workers in 1961; it is usually known as the "Lyon hypothesis," named after Mary Lyon, who was the first to state it explicitly and in detail. Lyon based her hypothesis (which after a decade is well accepted and should probably be referred to as a theory or principle) in part on genetic observations of X-linked coat color genes in the mouse and in part upon cytological data. She noted that in the female mouse heterozygous for X-linked coat color genes, the coat was neither like that of either homozygous phenotype nor intermediate between the two; instead, it was mottled, i.e., made up of patches of the two colors, the patches being random in arrangement and rarely crossing the midline, as shown in Figure 7–5. (Female tortoiseshell cats show the same effect.) On the other hand, males never had this patchy phenotype but instead had coats of a uniform color. Mice with one X and no Y chromosome also expressed X-linked coat color genes as a uniform, not mottled, coat color.

The chief cytological observation on which the Lyon hypothesis is based is that the number of sex chromatin bodies in interphase cells is always one less than the number of X chromosomes seen at metaphase. Additional cytological evidence is found in the observations that at prophase one X chromosome is heteropyknotic, that one X completes its replication later than the rest of the chromosome complement and that this late-replicating X is located in the area where the sex chromatin is found. Thus, it is clear that the sex chromatin is a heteropyknotic (condensed) X chromosome; only one X is active in cellular metabolism, and the second X (or in abnormal cells with extra X chromosomes, any extra X) appears as a sex chromatin body. In the male, the single X is uncoiled and active at all times, and consequently there is no sex chromatin.

The Lyon hypothesis states that:

1. In the somatic cells of female mammals, only one X chromosome is active. The second X chromosome is condensed and inactive, and appears in interphase cells as the sex chromatin.

2. Inactivation occurs early in embryonic life.

3. The inactive X can be either the paternal or maternal X (X^P or X^M) in different cells of the same individual; but after the "decision" as to

Figure 7–5 *Female mouse heterozygous for an X-linked coat color gene* tortoiseshell. *The mosaic phenotype of such females provided one of the first lines of evidence for the Lyon hypothesis (see text). (From Thompson, M. W. 1965.* Canad. J. Genet. Cytol. *7:202–213.)*

which X will be inactivated has been made in a particular cell, all the clonal descendànts of that cell will "abide by the decision"; i.e., they will have the same inactive X. In other words, inactivation is *random* but *fixed.*

In human embryos the sex chromatin has not been seen before the sixteenth day of gestation. More information is required about the timing of X inactivation in embryonic life. There is some evidence that it is not completed in all tissues at the same time. For example, Nadler's (1968) data suggest that the level of activity of G6PD is about twice as high in cells of female embryos as in cells of male embryos at very early stages, and declines to the male level by about the sixteenth week of development.

GENETIC CONSEQUENCES OF THE LYON HYPOTHESIS

The Lyon hypothesis has three principal genetic consequences: dosage compensation, variability of expression in heterozygous females and mosaicism.

Dosage Compensation

The amount of product of X-linked genes, such as G6PD or antihemophilic globulin (the substance deficient in classic hemophilia), is equivalent in the two sexes. For many years there was no satisfactory explanation as to how compensation for the dosage effect of the two X chromosomes of the female as compared with the male's single X was accomplished. X inactivation adequately explains this phenomenon. However, a number of problems remain. One of these is the abnormal phenotype shown by individuals with abnormal sex chromosome complements. If one and only one X is active regardless of the number present, it is difficult to understand why an XXY individual should show any phenotypic abnormality at all, and why individuals with more than one extra X chromosome should be progressively more abnormal.

Variability of Expression in Heterozygous Females

Since inactivation is random, females heterozygous for X-linked genes should have varying proportions of cells in which a particular allele is active. In other words, among such females there should be considerable phenotypic variability.

A demonstration that this really does happen in human females has been given by Nance (1964) for G6PD. In 30 women heterozygous for the A and B variants of G6PD, he estimated the relative proportions of the two enzyme variants and found a wide range and approximately normal distribution, as shown in Figure 7–6. For many, if not all, of the X-linked eye disorders, "manifesting heterozygotes" have been observed, and these patients may be regarded as extreme examples of "unfavorable Lyonization," in whom the deleterious allele is functional in a majority of the cells. Color blindness, classic hemophilia, Christmas disease and Duchenne muscular dystrophy are other conditions in which heterozygote variability has been noted (Lyon, 1962).

Mosaicism

If the Lyon hypothesis is true, it should be possible to demonstrate that women heterozygous for X-linked genes actually do have two populations of cells, one population with one X active, the other with the alternative X active. The first demonstration of mosaicism at the cellular level was provided by cloning of cultured fibroblasts from a woman heterozygous for two different G6PD alleles. Two different clonal populations were demonstrated, differing as to which G6PD allele was functioning (Davidson et al., 1963). This is direct evidence for mosaicism. More recently, mosaicism has been shown at the cellular level in women heterozygous for the X-linked type of mucopolysaccharidosis (Hunter's syndrome) and for HGPRT deficiency.

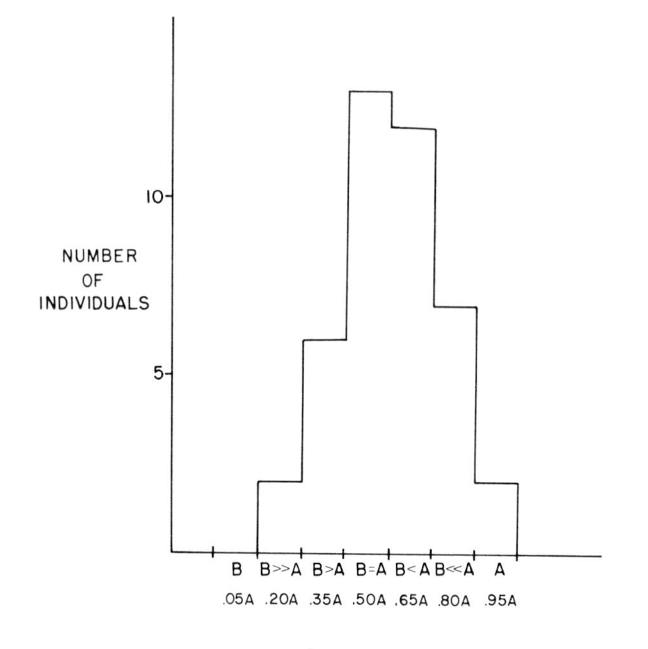

Figure 7–6 *G6PD phenotype distribution of 42 women heterozygous for two G6PD alleles,* A^+ *and* B. *(From Nance, W. E. 1968.* Genetic Studies of Human Serum and Erythrocyte Polymorphisms: Glucose-6-Phosphate Dehydrogenase, Haptoglobin, Hemoglobin, Transferrin, Lactic Dehydrogenase and Catalase. *Ph.D. Thesis, University of Wisconsin, p. 81.)*

Evidence Against the Lyon Hypothesis

The single active X hypothesis has been extensively criticized by Gruneberg. Though his criticisms are based chiefly on his studies of X-linked genes in the mouse, he has also examined 60 human X-linked conditions and has pointed out a number of difficulties in interpretation of their heterozygous expression on the basis of X inactivation (Gruneberg, 1967). As an alternative to the single active X hypothesis, he proposes the complemental-X hypothesis, which assumes that in the female mammal both X chromosomes are active, but jointly function to the same extent as the single X of the male. Gruneberg's criticisms have been useful in pointing out weaknesses in the Lyon hypothesis as originally stated, and in delineating the types of evidence that must be provided to substantiate it; but on the whole the data, particularly the demonstrations of two populations at the cellular level in heterozygous females, strongly support the Lyon hypothesis.

One locus on the X for which the evidence for inactivation is as yet equivocal is the Xg blood group locus. Usually there is no evidence of mosaicism in Xg^a/Xg women, but there is one exceptional family reported: an Xg(a+) woman heterozygous for an X-linked form of microcytosis, whose husband was Xg(a–) and normal; in their two daughters, half the red cells were microcytic (as would be expected on the basis of Lyonization). When the normal and microcytic red cells of the daughters were separated, the microcytic ones were found to be Xg(a+) and the normal ones Xg(a–) (MacDiarmid et al., 1967). It also appears that the Xg locus is inactivated if it is on a structurally abnormal X (Race and Sanger, 1969).

ANOMALIES OF THE SEX CHROMOSOMES

INCIDENCE

Anomalies of the sex chromosomes are not uncommon. In 15 newborn chromosome surveys combined, 31 of 11,039 live male births were found to have a sex chromosome abnormality (2.8 per thousand). The types of anomalies and their frequencies were as follows:

Type	*Number*	*Rate per 1000 ♂ births*
XYY	17	1.5
XXY	13	1.2
XX	1	0.1

Other types of aneuploids (XXYY, and so forth) were not represented in this large series; consequently they must be rare.

Sex chromatin surveys have shown a somewhat higher frequency of chromatin-positive males, usually about 2 per 1000 male births. The difference may be due to chance or to differences in ascertainment.

The incidence of sex chromosome anomalies in newborn females is less well known. About 0.45 infants per thousand female births are chromatin-negative; this group includes both 45,X (estimated at 0.40 per thousand) and 46,XY (0.05 per thousand). The incidence of XXX females is about 0.65 per thousand live female births.

MOSAICISM

A small proportion of chromatin-positive males are mosaics, usually 46,XY/47,XXY. Many mosaics are found among females with ovarian dysgenesis (see below). The frequencies of different chromosomal types of ovarian dysgenesis given by Hamerton (1971) are approximately as follows:

Ovarian Dysgenesis		***Per Cent of Subgroup***	***Per Cent of Total***
62.5% Chromatin-negative	45,X	90	56
	Mosaic	10	6
37.5% Chromatin-positive	Mosaic	80	30
	Structural aberration of X	16	6
	46,XX	4	2
			100

Among the structural aberrations of the X chromosome, the most common is an isochromosome for the long arm of the X.

TURNER'S SYNDROME (OVARIAN DYSGENESIS)

Turner's syndrome attracted special attention from geneticists when it was learned that typical patients, though phenotypic females, were chromatin-negative. Because of the discrepancy between their phenotypic sex and what was then called their "nuclear sex," Turner's syndrome was among the first conditions studied when chromosome analysis became possible. It was found that typical patients had 45 chromosomes with only a single X. It is now known that many patients with ovarian dysgenesis are chromatin-positive, but that such patients usually are mosaics with a 45,X cell line or have one normal and one structurally abnormal X (see previous section).

Patients with Turner's syndrome are usually short (under five feet), with webbing of the neck, low hairline at the nape of the neck and a characteristic facial appearance (Fig. 7–7). The chest is usually wide with broadly spaced nipples, the breasts are poorly developed, the external genitalia are juvenile and the internal sexual organs are female although the ovary is usually only a streak of connective tissue; however, it is arranged in the manner of ovarian stroma, and ovarian follicles may be present. Edema of the extremities, coarctation (narrowing) of the aorta and cubitus valgus (reduced carrying angle at the elbow) are frequent. Axillary and pubic hair are usually present. Primary amenorrhea is common, though not invariable. Many patients are not recognized until they seek medical advice at the age of puberty because of amenorrhea and failure of development of secondary sexual characteristics, but in the newborn period marked edema of the feet and webbing of the neck may give the clue that chromosome studies are indicated.

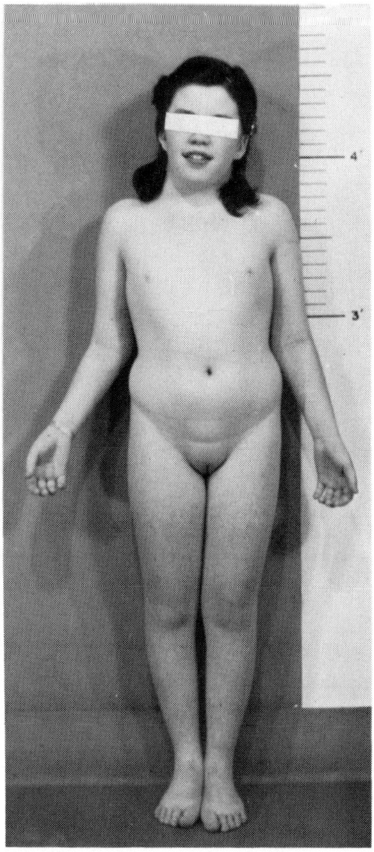

Figure 7–7 *Phenotype of 45,X Turner's syndrome. For additional details, see text. (From Barr 1960.* Amer. J. Hum. Genet. *12:118–127.)*

Because the 45,X constitution and other variants of Turner's syndrome are relatively benign defects, it is surprising that this chromosome complement is the most frequent one found in first-trimester abortions. About one-third of spontaneous first-trimester abortions are heteroploid; and of these, one-fourth are 45,X. It is estimated that not more than 2 per cent of all 45,X zygotes go to term (Carr, 1971).

Nance and Uchida (1964) report that there is a five- to tenfold increase in twinning in sibships of 45,X patients and that most of the twins are monozygotic (MZ). They speculate that 45,X individuals may originate in a postzygotic event which produces a chromosome loss. Eight sets of MZ twins, in which only one of each set was 45,X, have been described (summarized by Benirschke, 1972), and all of these must of course have arisen by postzygotic accidents.

The Xg blood groups in 45,X patients and their parents can sometimes reveal whether the single X is paternal or maternal. In about 75 per cent of patients the single X is from the mother (i.e., the error is in the paternal gamete), and in about 25 per cent it is from the father (Race and Sanger, 1969). There is no maternal-age effect as there is for Down's syndrome and other aneuploidies.

47,XXX (TRIPLE X) AND OTHER X POLYSOMIES

47,XXX females have two sex chromatin bodies in their somatic cells. These patients are usually quite normal in appearance, though perhaps they are more likely to be mentally subnormal or psychotic than are the controls. The incidence in the general population is about one in 500 female births. Some 47,XXX females have borne children, all of whom are normal. This is surprising, since one would expect that half their ova would have two X chromosomes.

Patients with four or five X chromosomes are also physically normal, but are severely retarded.

47,XXY (KLINEFELTER'S SYNDROME)

The incidence of chromatin-positive Klinefelter's syndrome is approximately one in 800 live male births. The only constant characteristics are small testes and hyalinization of the seminiferous tubules. Usually secondary sexual characters are poorly developed, gynecomastia may develop and many patients are tall and eunuchoid. Subnormal mentality is common and perhaps one in 50 institutionalized retardates is 47,XXY. Figure 7–8 illustrates the phenotype of a typical patient with Klinefelter's syndrome.

Like Turner's syndrome, Klinefelter's syndrome shows discrepancy between phenotypic sex and sex chromatin. The patients are males but are sex chromatin-positive.

Whereas the error in 45,X Turner's syndrome is usually in the paternal gamete, in Klinefelter's syndrome it is maternal in 64 per cent of cases and paternal in 36 per cent. The accident of gametogenesis can occur at either the first or second meiotic division, or at both (as in the rare and more severely abnormal variants of Klinefelter's syndrome with up to four X's).

47,XYY

Males with a second Y chromosome have aroused great interest since they were found to be frequent among males in a maximum security prison. About 3 per cent of such subjects are XYY, and among the group over six feet tall the incidence is much higher (over 20 per cent). In the

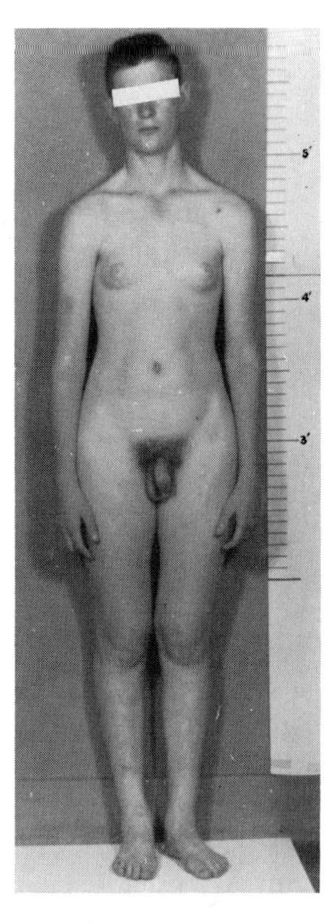

Figure 7–8 *Phenotype of 47,XXY Klinefelter's syndrome. Gynecomastia, present in this patient, occurs in only about 20 per cent of cases. (Courtesy of M. L. Barr.)*

general population XYY is rarely identified, though in newborn surveys the incidence is about 1.5 per 1000 male births. Thus, a very large proportion of all XYY males must be indistinguishable from normal males on the basis of their phenotype and behavior.

The relationship between XYY and aggressive, psychopathic or criminal behavior has aroused great public interest. It is estimated that XYY males are perhaps six times as likely to be imprisoned as are normal XY's. When XYY infants are identified in newborn surveys, or even prenatally, it is difficult to know how to handle the problem. Should an XYY fetus be aborted? Should the parents of an XYY infant be told that the child has an increased risk of behavioral difficulties? If so, how should the child's upbringing be modified? The answers to these questions await better understanding of the range of variation of XYY's, since the majority of those who have been studied so far have been ascertained because of their antisocial behavior.

XYY males originate by paternal nondisjunction at the second meiotic division, which produces YY sperm. The less common XXYY and XXXYY variants also probably originate in the father, by a series of nondisjunctional events.

ORIGIN OF SEX CHROMOSOME ABNORMALITIES

The causes of chromosomal aberrations have been discussed in a general way (page 144), but because the sex chromosomal abnormalities permit more detailed explanations, we will consider them separately here.

Anomalies of Number

The basis of all numerical anomalies is nondisjunction (page 135). In most instances, except when mosaicism is present, the nondisjunctional event probably occurred at a meiotic division (either first or second) in one of the parents. In the simplest case, we may assume nondisjunction of the X chromosomes at the first meiotic division in a female, followed by a normal second meiotic division:

Oocyte	Meiosis I	Result	Meiosis II	Ova
XX	Nondisjunction of X's	→ XX	Normal	→ XX
				→ XX
		→ No X	Normal	→ No X
				→ No X

Only one of the four cells indicated at the right becomes an ovum (the other three will be polar bodies). Probably any of the four types can form an ovum. Thus the ovum may have either of two constitutions, XX or X-less. Assuming fertilization by a normal sperm, the sex chromosome complement of the zygote could be any one of the following.

		Ova	
		XX	–X
Sperm	X	XXX	X only
	Y	XXY	Y only

XXX	Triple X syndrome
XXY	Klinefelter's syndrome
X	Turner's syndrome
Y, no X	Inviable

Clearly, this one relatively straightforward mechanism, nondisjunction at meiosis I in the female, can produce three common types of sex chromosome abnormalities. Nondisjunction of sister chromatids at meiosis II can produce these types also, and it is known that nondisjunction can take place at either of the two meiotic divisions. Nondisjunction can also take place in both meiotic divisions in succession, or after fertilization.

If nondisjunction happens at an early division of the zygote before the actual embryo has begun to develop, one abnormal cell might give rise to all the cells of the embryo, while extra-embryonic tissue might be normal. This mechanism could theoretically give rise to XXY, XYY or XXYY chromosome complements originating from a normal XY zygote, but it may be an unusual occurrence. If nondisjunction takes place shortly after rather than before the true embryo begins to differentiate, an XY zygote could develop into a person with, for example, XXY, Y only, and XY cell lines. The line with a Y but no X is presumably inviable, so the individual would be an XY/XXY mosaic. The same type of mosaicism could arise in an XXY zygote if one X was lost from one cell line early in development. This is only one example of the many types of sex chromosome mosaicism that have been recorded.

Source of the Abnormal Gamete

It is of some scientific interest to determine in which parent the nondisjunctional event took place, and this sometimes can be done by determining certain X-linked marker genes of the parents and abnormal child. For instance, if a patient with 45,X Turner's syndrome is color blind, her X must carry a color blindness allele. If her father is not color blind, his X cannot have a color blindness gene. Thus, the patient must have received her single X from her mother; her father gave her neither an X nor a Y, so the nondisjunctional event was an accident of spermatogenesis (Fig. 7–9).

The Xg blood group system is also used to discover in which parent nondisjunction occurred. Two examples are shown in Figure 7–10 for two patients with Klinefelter's syndrome. The patient on the left has received both an X and a Y from his father, since he has an Xg^{a} allele and his mother could only have provided him with *Xg*. This indicates that nondisjunction of X and Y must have occurred in the father at meiosis I. In the patient on the right both X's must be maternal, but nondisjunction could have occurred either at the first or the second meiotic division or even in an early postzygotic division.

INTERSEX

An intersex may be defined as a patient with ambiguous genitalia. The term is often wrongly used in medical writing to describe patients

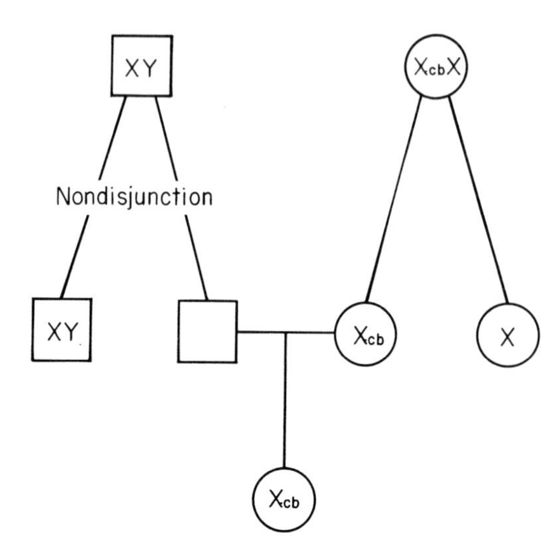

Figure 7–9 *Source of the abnormal gamete. For description, see text. The X chromosome carrying the color blindness gene (Xcb) must have come from the carrier mother since the father is not color blind.*

with Klinefelter's or Turner's syndrome, in which the genitalia are not ambiguous.

Aberrations of the sex chromosomes are not always involved in intersex conditions. Common examples of intersex conditions due to single mutant genes are also considered in this chapter.

An intersex phenotype does not show all the criteria normally associated with his or her chromosomal sex. "It" may be sterile, but not necessarily. It may be infantile in sexual development and may never

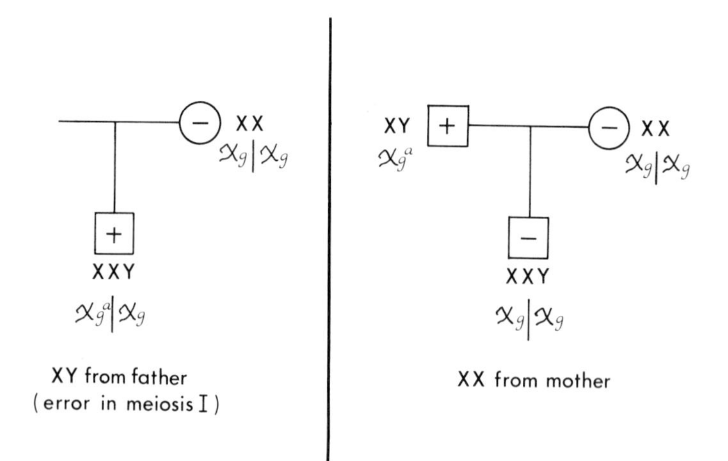

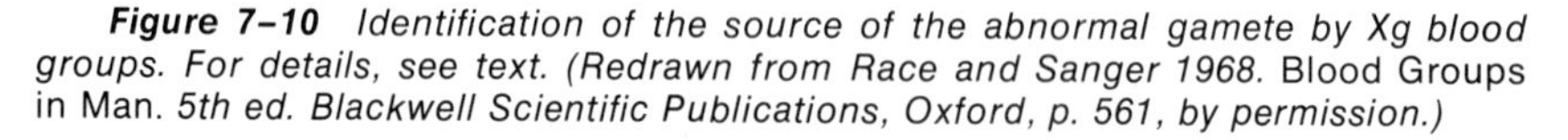

Figure 7–10 *Identification of the source of the abnormal gamete by Xg blood groups. For details, see text. (Redrawn from Race and Sanger 1968.* Blood Groups in Man. *5th ed. Blackwell Scientific Publications, Oxford, p. 561, by permission.)*

mature properly. It may have such ambiguous genitalia that sex assignment is impossible, or it may have characteristics of both sexes.

Homosexuality, transvestitism and other sexual psychological disturbances do not as a rule have their basis in abnormality of the sex chromosomes or in single-gene defects.

INTERSEX CONDITIONS WITH NORMAL CHROMOSOMES

True Hermaphroditism

True hermaphroditism is a very rare condition, only 84 cases having been reported (Hamerton, 1971). A true hermaphrodite possesses both testicular and ovarian tissue, either as two separate organs or as a single ovotestis. For a diagnosis of true hermaphroditism the gonads need not be functional, but only histologically identifiable. Internal and external sexual organs are very variable and not in any way diagnostic. Sex hormone studies are not helpful.

The great majority of hermaphrodites are 46,XX, but some are 46, XY and others are 46,XX/46,XY chimeras or mosaics.

Pseudohermaphroditism

Pseudohermaphrodites, unlike true hermaphrodites, have gonadal tissue of only one sex:

1. *Male* pseudohermaphrodites are 46,XY and have only testicular tissue.
2. *Female* pseudohermaphrodites are 46,XX and have only ovarian tissue.

The sexual organs of all pseudohermaphrodites are ambiguous, i.e., have anomalies which tend to make them resemble those of the opposite sex. These anomalies may range from hypospadias or enlarged clitoris to variants in which assignment of the individual to either sex is virtually impossible.

Male Pseudohermaphroditism

Testicular Feminization. In typical testicular feminization the external genitalia are female but the patient is 46,XY and has testes which may be in the abdominal cavity, inguinal canal or labia. Often these patients are well-developed females who seek medical advice because of amenorrhea, sterility or inguinal hernia (in which at operation a testis may be found). In an "incomplete" form of the disorder, the patients are males who develop breasts at puberty or who have ambiguous external genitalia. In some families all affected patients are phenotypic females, but in other families a spectrum of abnormality may be seen.

The basic defect in testicular feminization appears to be target organ unresponsiveness to testosterone, possibly due to inability to convert testosterone to its active form, dihydrotestosterone.

The genetic aspects of testicular feminization are described elsewhere (page 81). As in other X-linked genetic lethals, one-third of all cases are due to new mutation, i.e., have no family history of the disorder.

Female Pseudohermaphroditism

Congenital Adrenal Hyperplasia. This is the most common cause of female pseudohermaphroditism. The incidence is perhaps about one in 25,000 births. Several distinct genetic and clinical forms are known, all inherited as autosomal recessives and each characterized by a block in a specific step in cortisol biosynthesis, resulting in increased secretion of ACTH (adrenocorticotropic hormone) and hyperplasia of the adrenal glands. This in turn leads to masculinization of female fetuses. Affected baby girls frequently have major anomalies of the external genitalia, often to the point that sex assignment may be impossible. The clitoris may be enlarged and the labia majora rugose and even fused. In males the same genotype produces premature virilization, but there is no difficulty in identifying the sex. In one form of congenital adrenal hyperplasia, salt-losing crises may be very severe.

Other Forms of Female Pseudohermaphroditism. The external genitalia of a female fetus may be masculinized if the fetal circulation contains excessive amounts of either male or female sex hormones. These hormones reach the fetal circulation from the maternal circulation, and may originate either endogenously or exogenously. Androgenic hormones may be present in excessive amounts if the mother's adrenal cortex is overactive or if she has received hormone therapy, e.g., some progestins used to prevent spontaneous abortion. [When therapy itself produces a disorder, the disorder is said to be ìatrogenic (*iatros*, physician).]

The Treatment of Intersex Disorders

There are certain general principles of treatment that can serve as useful guidelines in the management of intersex disorders.

The chromosomal sex is *not* the final indicator of the sex to which the patient should be assigned. Much more important factors in the decision are the presenting phenotype and the sex of rearing. If by medical or surgical intervention the patient can be allowed to lead a relatively normal though infertile life as a member of one sex, this is usually the best option. Altering the sex of rearing is usually unwise, except in very early infancy.

The diagnosis should be made as early as possible, for several reasons:

1. To prevent or forestall problems developing in the future. For instance, if a "girl" is found to be chromatin-negative, complete karyotyping should be done. If a Y is found, the possibility of virilizing changes occurring at puberty must be anticipated; if there is no Y, the problems of growth and development associated with Turner's syndrome can be anticipated and to some extent ameliorated.

2. To avoid the severe psychological trauma of possible change of sex in later life.

3. To prevent the crises resulting from severe upsets in electrolyte balance that happen in one type of congenital adrenal hyperplasia.

Intersex disorders cause great concern and embarrassment to the family and eventually to the patient. They must be handled very carefully and tactfully. For example, it is usually disastrous to suggest to a patient with testicular feminization that she is "really a boy."

GENERAL REFERENCES

Federman, D. D. 1967. *Abnormal Sexual Development. A Genetic and Endocrine Approach to Differential Diagnosis.* W. B. Saunders Company, Philadelphia.

Hamerton, J. L. 1971. *Human Cytogenetics.* Vol. II. *Clinical Cytogenetics.* Academic Press, Inc., New York.

McKusick, V. A. 1964. *On the X Chromosome of Man.* Amer. Inst. Bio. Sciences, Washington.

Moore, K. L. 1966. *The Sex Chromatin.* W. B. Saunders Company, Philadelphia.

Ohno, S. 1967. *Sex Chromosomes and Sex-Linked Genes.* Springer-Verlag New York, Inc., New York.

PROBLEMS

1. A color blind woman with 45,X Turner's syndrome has a color blind father. From which parent did she receive a chromosomally abnormal gamete?

2. In a color blind male with Klinefelter's syndrome:
   (a) What is the genotype?
   (b) If his parents have normal color vision, what are their genotypes? In which parent did nondisjunction occur?

3. In a triple-X female:
   (a) What kinds of gametes would be produced?
   (b) If each type is equally likely to be fertilized and to produce a living child, and if fertilization is by a normal sperm, what are the theoretical karyotypes and phenotypes of her possible progeny? (Actually, XXX women appear to have normal offspring only.)

8

Immunogenetics

Immunogenetics is concerned with the genetic aspects of antigens, antibodies, and their reactions. An **antigen** is a substance (usually but not necessarily a protein) that can stimulate the formation of an **antibody** and **react specifically** with the antibody so produced. The antigen may be a genetically determined substance on the surface of a red blood cell, nucleated cell or bacterium; or it may be something not immediately related to any living cell. The only common characteristic of antigens is their ability to elicit the formation of antibody. In immunogenetics, the most important antigens are those which are genetically determined and located on body cells.

The reaction to an antigen is first elicited in the lymphoid tissue (such as lymph node, tonsil, or spleen) and is termed an **immune response**. Lymphoid cells, as described on page 180, may respond to the antigenic stimulus in one of two ways: by production of antibodies or by production of **killer cells**. In general, antibodies effect the rejection of bacteria, viruses and so forth, while killer cells effect the rejection of transplanted cells or tissues.

Antibodies are serum proteins of the gamma globulin class, referred to as **immunoglobulins** (Ig). The antibody-antigen reaction depends upon mutually interlocking sites, which vary in number depending on the particular type of antibody, and which are specific for each antigen-antibody pair.

When the antigen-antibody complex is formed on the surface of cells, a naturally occurring component of the blood, called **complement**, may cause lysis of the cells concerned.

One of the mysteries of genetics is how the body is able to respond to each of the thousands of antigenic substances by making a specific antibody against it. The problem is reviewed by Edelman (1970); according to the theory most widely accepted at present, antibody production depends on multiple somatic mutations or, alternatively, somatic recombinations of the genes that control antibody production (page 235). It is postulated that somatic mutations or recombinations occur constantly in that part of the DNA which controls antibody specificity. As a result, many thousands of individual new antibodies are constantly being formed. As it is formed, each antibody appears on the cell surface; and when an antigen is encountered which has combining sites matching those of the particular antibody, the cell which produces the antibody "locks in" on its production and undergoes repeated divisions. The progeny of this particular cell is geared to produce only this one antibody.

A similar mechanism appears to underlie the cellular type of immune response. When a foreign **histocompatibility antigen** matches an antibody on the surface of a lymphoid cell, the cell commences differentiation and proliferation to produce killer cells (page 180).

The ability of the body to reject cells with histocompatibility antigens different from its own allows it to reject the malignant cells which probably arise *de novo* from time to time in each of us. This defense mechanism of course conveys a biological advantage which ensures the evolutionary survival of the mechanism itself.

Two kinds of genetically determined antigens are of major importance in transplantation. These are the tissue antigens (histocompatibility antigens) and the ABH antigens of the ABO blood group system. Although A, B and H are also histocompatibility antigens in the general sense of having importance in transplantation, here we shall restrict the term histocompatibility antigen to the antigens determined by the alleles at the HL-A and similar loci (see later).

The histocompatibility antigens are glycoproteins on nucleated cells, including the white blood cells. The A, B and H antigens on the red blood cells are glycolipids. The *secreted* A, B and H substances in genetically "secretor" persons are glycoproteins.

Four areas of immunogenetics are of medical significance: (1) blood groups and problems related to blood group incompatibilities, (2) transplantation, (3) immune deficiency diseases and (4) autoimmune diseases. The importance of the blood group systems in genetics and in clinical medicine entitles them to a separate chapter (Chapter 9).

TRANSPLANTATION

In the earliest attempts at transplanting tissue from one individual to another, the graft usually survived a few days and then died. The only exception to the rule that grafts die within a few days involves grafts between monozygotic twins, which normally are accepted. This exception reflects the genetic basis of antigenic specificity. Today, because of our knowledge of the immune reaction and methods of circumventing or suppressing it, "spare-parts surgery" has become an important area in medicine. Surgeons can now replace damaged parts or organs by transplants from suitable donors with reasonable prospects of success, though much remains to be learned.

The mechanism of rejection is the immune reaction. Implanted tissue is antigenic to the host, which reacts by the production of lymphocytes antagonistic to the graft. These lymphocytes invade the tissues surrounding the graft and cause graft rejection. The exact mechanism is not fully understood. The prerequisite for further advances in surgical transplantation is not improved surgical technique, but further improvement in our understanding of the genetic and pharmacological possibilities of circumventing or suppressing the immune reaction.

LYMPHOID TISSUE AND THE IMMUNE REACTION

The regional lymph nodes respond to a graft of foreign tissue by cellular proliferation. The important cells in this response are lymphocytes, which, although uniform in appearance (white blood cells with a large nucleus of small amounts of cytoplasm), play two different roles (Roitt et al., 1969; Moller, 1971).

The two lines receive their names because of their similarity to two types of cells in birds. The two types of cells are called **T-cells** ("thymus-dependent" cells) and **B-cells** ("bursa-equivalent" cells, which are thymus-independent). In addition to the thymus, birds have the bursa of Fabricius, which is located at the distal end of the alimentary canal. Cells derived from the thymus are important in cell-mediated immune reactions such as the rejection of graft tissue, and cells from the bursa are important in the production of antibodies. In mammals, including man, where no bursa exists, a common stem cell found in fetal liver and adult bone marrow appears to give rise to both a T-cell line and a B-cell line. In both series there are cells which are antigen-sensitive.

Comparison of B- and T-cells

B-cells, which are sensitive to antigens such as those carried by bacteria, viruses and red blood cells, respond to antigenic stimulation by differentiation and proliferation. The result of this process is the production of mature **plasma cells** which are capable of synthesizing various types of antibodies. Plasma cells of any one clone are able to synthesize only one specific antibody. The different types of antibodies and their structures are discussed on page 234; for understanding the basics of immunogenetics, however, detailed information concerning the types of antibodies is not required.

T-cells are sensitive to antigens such as those found on implanted tissue and those which may appear on malignant cells in the body. They respond by differentiation and proliferation to form blast cells. Some blast cells produce the killer cells which travel to the graft site and bring about rejection of the implanted tissue. These killer cells, which appear to be lymphocytes, seem to release factors which are mitogenic to other lymphocytes and which also, in some way, stimulate cells of the B series to produce antibodies. Other blast cells produce "primed" antigen-sensitive cells which survive for long periods and are ready, if they meet the same antigen in the future, to react immediately; this type of cell is responsible for the second set reaction described on page 186.

The role of B- and T-cells, as presently understood, may be shown diagrammatically as follows:

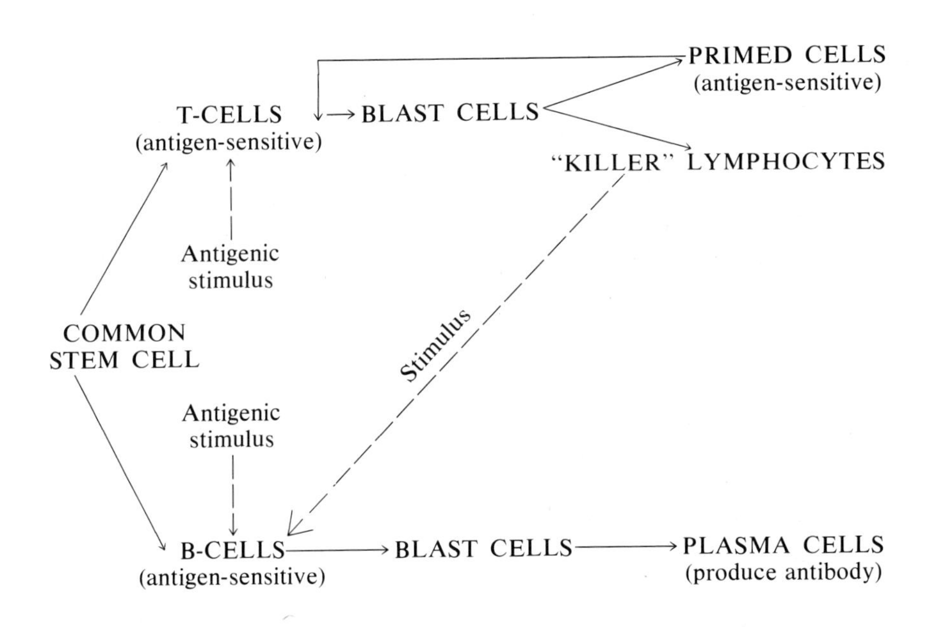

Some of the circulating lymphocytes of the blood are able to respond to antigenic stimulation. This can be demonstrated by the addition of antigens to cultures of lymphocytes, which respond by undergoing mitosis. In other words, antigens have a mitogenic action on lymphocytes. The strength of the antigenic stimulus determines the amount of mitotic activity produced in the culture. Lymphocytes of one individual are normally mitogenic to lymphocytes of an antigenically different person, as can be shown by mixing cultures of lymphocytes from two different people (Bain and Lowenstein, 1964). Mixed cultures of lymphocytes of monozygotic twins show little or no mitotic activity, but those of genetically different individuals show an amount of activity which reflects the genetic disparity between the two. This characteristic is the basis of one of the techniques for determining the degree of antigenic similarity of prospective graft donor-host combinations. (See page 192 and the table on the following page.)

THE REACTION OF LYMPHOCYTES IN CULTURE TO ANTIGENS

Antigenic Agent	Resulting Mitotic Activity
Bacterial toxin	Strong
Lymphocytes of person with very different genotype	Strong
Lymphocytes of person with more similar genotype	Moderate
Lymphocytes of identical twin	None
Lymphocytes from self	None

TRANSPLANTATION GENETICS AS SEEN IN INBRED STRAINS OF ANIMALS

The principles of transplantation genetics as they apply to man and other mammals have been uncovered largely through experimental work with highly inbred strains of animals, particularly mice. Because inbred mice, some of which have been inbred by brother-sister matings for well over a hundred generations, are genetically virtually identical (**isogenic**), animals of an inbred strain do not reject tissue from other animals of the same strain. Because their identical genes produce identical antigens, the host does not recognize the graft as foreign and therefore does not respond to it by antibody production.

Tissue from one mouse of an inbred strain can (with very rare exceptions) be successfully implanted into the F_1 of a cross between a mouse of the donor strain and another inbred strain (Figs. 8–1 and 8–2). Because an inbred mouse is homozygous at all (or virtually all) loci, its progeny receive one representative of each of its antigen-producing genes. Since these genes are codominant, all the antigen-producing genes of each parent find phenotypic expression in the F_1 hybrid. The hybrid, having all the antigens of each parent, does not regard tissue from either parent as foreign, and therefore does not produce killer cells against it. The principle may be illustrated as follows:

INBRED STRAIN A × INBRED STRAIN B ⟶ HYBRID AB

Donor	Host		
	A	B	AB
A	+*	−	+
B	−	+	+
AB	−	−	+

*+, acceptance; −, rejection.

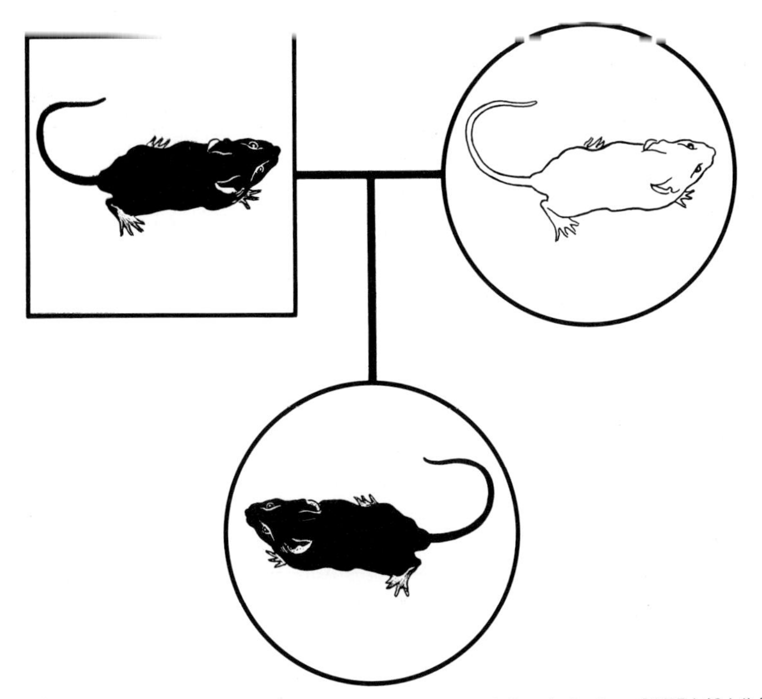

Figure 8–1 *Cross between mice of two different inbred strains, C57BL/6J (black) and A/HeJ (white), producing black F_1 progeny having all the histocompatibility antigens of both parents. See also Figure 8–2.*

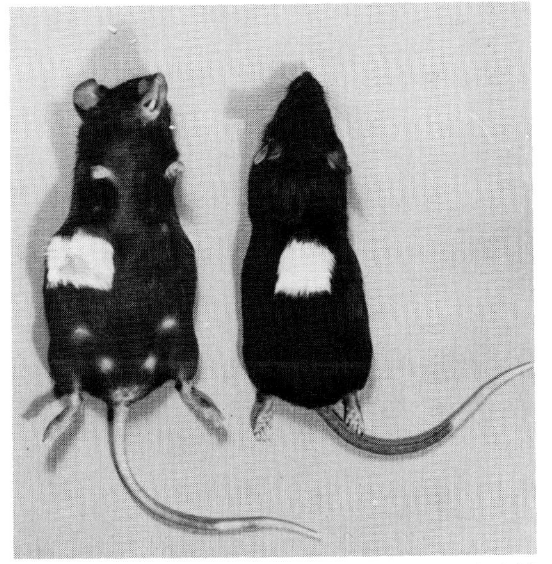

Figure 8–2 *Two F_1 progeny of the cross shown in Figure 8–1. The animal on the left has received a mammary gland graft (complete with skin and nipple) and the animal on the right has received a dorsal skin graft, both from donors of the white parental strain. Note that several months after transplantation both grafts are still flourishing.*

CLASSIFICATION OF TYPES OF GRAFTS

The nomenclature for different types of grafts is now fairly well standardized (Snell, 1964). An **isograft** is a graft between two genetically identical members of a species. Monozygotic twins, members of an inbred strain and the F_1 of crosses between two inbred strains fulfill these requirements. An **allograft** is a graft between two genetically different members of a species. These two types of grafts are also described as **isogeneic** and **allogeneic**, respectively.

Grafts between members of different species, e.g., ape to man, are referred to as **heterografts** or **xenografts**. As a rule they are rejected more quickly than are allografts.

The term **homograft** as first defined is often used incorrectly and does not convey precise information. Originally, a homograft referred to a graft of nucleated cells from one animal to another of the same species. Normally such a graft is rejected, and the term **homograft reaction** refers to this response. Some authors now restrict the term homograft to refer only to a graft between genetically different members of the same species. In at least two instances, homografts, as originally defined, are not rejected:

1. Identical twins accept tissue from each other.
2. Animals of highly inbred strains accept tissue from other members of the same strain.

Since homografts may or may not be accepted, it is preferable to use the more precise terminology: *isograft* if donor and host are genetically identical, and *allograft* if they are within the same species but genetically different. The classification of grafts is summarized in Table 8–1.

THE ALLOGRAFT (HOMOGRAFT) REACTION

Rejection of an allogeneic graft by a normal host involves a process called the allograft (homograft) reaction, which has also been termed the **primary response** to distinguish it from the **second set** (secondary) response. The typical sequence of events in this reaction is as follows. The example chosen is a skin graft in man or mouse, but the general principles apply to all forms of allogeneic grafts.

1. For the first few days the graft appears to "take," and a blood supply continuous with that of the host is established, allowing the skin to appear pink.
2. After about the sixth to ninth day (the exact time depending on the degree of incompatibility of the tissue), the graft shows inflammation and infiltration by cells of the lymphoid series, which also invade the graft bed.
3. The blood supply shuts down and the graft becomes cyanotic and swollen.

TABLE 8–1 CLASSIFICATION OF TYPES OF GRAFTS

Type of Graft	Synonym(s)	Relationship of Donor to Host	Result	Example
Autograft	Autogenous graft	Individual to another area of his own body	Acceptance	Skin graft from one area of body to another
Isograft	Isogeneic graft Syngeneic graft	Members of a pair of MZ twins Two members of one inbred strain	Acceptance Acceptance	MZ twin to co-twin Mice: strain A to strain A
Allograft	Allogeneic graft Homograft*	Host and donor of same species but of different genotypes	Rejection	Man to man
Heterograft	Xenograft	Host and donor of different species	Rejection	Ape to man

*If the term homograft is used without modification, it usually refers to an allograft.

4. The graft dies, sloughs and is lost.

Different tissues in the same animal show different strengths of antigenic action. Skin is perhaps the most difficult tissue to implant. At the other end of the scale, endocrine tissue is the most easily accepted epithelial tissue. As a rule, connective tissue structures produce the least immune response, and therefore rejection of connective tissue structures is on the average more easily avoided than that of epithelial structures.

SECOND SET RESPONSE

After rejection of an implant the host will reject a second graft from the same donor even more rapidly; in other words, the host has become **sensitized** to the donor tissue. In skin grafting, the second implant is usually rejected as a "white graft"; that is, before it is vascularized and pink. The more rapid response to the same antigenic stimulus on second exposure is the second set (secondary) response. The second set response results from the fact that the body now has primed antigen-sensitive cells of the T series (page 181). In the presence of the antigen to which they are sensitive, these cells quickly react by differentiating and dividing to produce blast cells, which in turn proliferate and differentiate to become the killer cells upon which graft rejection depends. Since the second set response seems to involve "memory" on the part of the host tissue, it is sometimes called the **anamnestic** response.

SECOND SET RESPONSE

Step	*Host-Donor Relationship*	*Type of Reaction*	*Result*
1	Same species, different genotype	Allograft reaction	Rejection
2	Same individuals as in Step 1	Second set response	Faster rejection

GRAFT-VERSUS-HOST REACTION

Not only does the host react against donor cells but donor cells, if they are immune-competent, may react against the antigens of the host. This is the **graft-versus-host** reaction.

An illustrative but discouraging facet of the graft-versus-host reaction is the failure of what might appear to be a logical treatment for radiation damage to the hematopoietic (blood forming) tissue of the body. After large doses of whole-body radiation, the irradiated individual loses the ability to produce immune-competent cells and so loses the immune response. The victim is then easy prey for bacterial infections which he would otherwise reject with no difficulty. Not only does he fail to react to infections, but he becomes tolerant to implants of foreign tissue. It would therefore seem logical to give him a transplant of hematopoietic tissue to replace his own damaged hematopoietic cells. Unfortunately these very cells, instead of being therapeutic, react to the antigens of the host and produce antibodies to the host cells, often with fatal consequences. If the reaction occurs in a young animal, it may not be fatal, but it inhibits growth and renders the host weak and sickly (Fig. 8–3), therefore producing **"runt disease."**

Recent reports indicate that in certain disorders, for instance, Swiss type agammaglobulinemia, in which the individual lacks immune-competent cells of his own (de Köning et al., 1969), or acute lymphatic leukemia, in which immune-competent cells had been killed by drugs or radiation (Graw et al., 1970), certain marrow grafts may take. They do not produce a fatal graft-versus-host reaction if the donor and the recipient are identical for all, or nearly all, HL-A antigens (page 189).

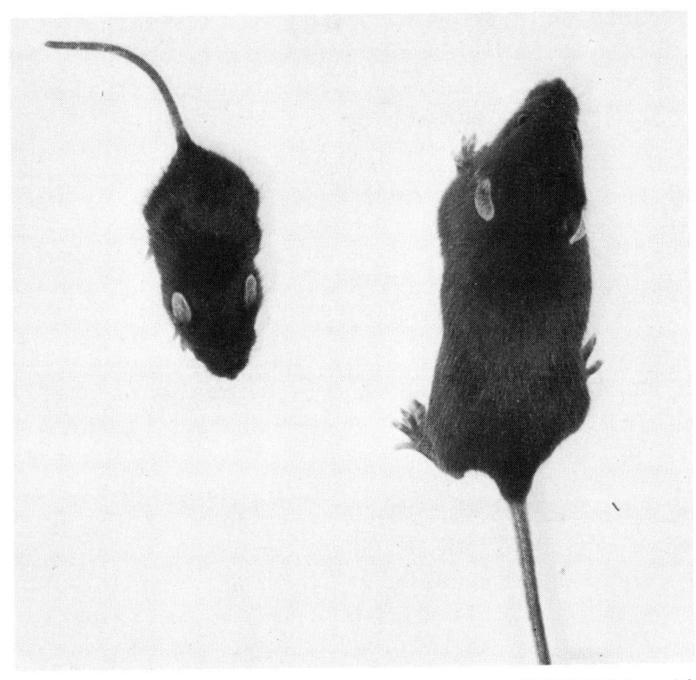

Figure 8–3 *Runt disease in F_1 hybrid of cross between C57BL/6J ♀ and A/HeJ ♂. Within 24 hours of birth, the animal on the left received an intraperitoneal injection of 10^7 spleen cells from a C57BL/6J adult. These immune-competent cells formed antibody against the A antigens of the host, producing runting. The animal on the right is a normal littermate control. (Photograph courtesy of R. Escoffery.)*

THE DEVELOPMENT OF IMMUNE COMPETENCE

Tissue antigens are formed at various times during fetal development, but the ability to react to antigens does not develop until shortly before birth. By this time antigens which might cause an immune response are all normally present, so that the immune mechanism does not produce antibodies against them. (Certain proteins which develop later in life, such as those on spermatozoa, do not normally alert the immune mechanism.) The condition of tolerance to one's own antigens is **immunological homeostasis**; it is dependent on the fact that in fetal life the individual is tolerant to any antigens he encounters *and this tolerance normally lasts throughout his lifetime* provided these antigens continue to be present. Indeed, a fetus (or in some species, a newborn) will accept cells from a genetically different donor and allow the cells to proliferate and persist, making him a **chimera** (page 229) with life-long tolerance to tissue from that donor. This phenomenon, shown in the following table and known as **acquired tolerance**, can be used to enable mice of one strain to accept tissue from mice of another strain, thereby allowing study of the same tissue in different natural environments. Similarly, human dizygotic twins may (though only rarely) exchange blood cells in utero, and thereafter such cells may persist, making each twin a chimera who will accept tissue from his co-twin. A twin chimera may also have a population of red cells with antigens indicating that their origin is from the exchanged stem cells of the co-twin.

Host		*Donor Tissue*		*Result*
Neonatal animal of Strain A	+	Mature cells of animal of Strain B	→	Acquired tolerance

Acquired tolerance and graft-versus-host disease can both be produced by injecting cells from animals of one strain into neonatal animals of a genetically different strain. If the transplanted cells are not immune-competent and the antigenic difference between the strains is not great, the grafted tissue survives and acquired tolerance results. On the other hand, if immune-competent cells are transplanted, graft-versus-host reaction will occur. If the antigenic differences between the two strains are great, such as a difference at the H-2 locus in two strains of mice, the only tissue likely to be accepted is connective tissue, which contains immune-competent cells and so produces graft-versus-host disease. Thus, acquired tolerance without runt disease may be impossible to produce in strains that differ at strong histocompatibility loci.

An exception to immunological homeostasis is the mechanism of **autoimmunity** (see Autoimmune Disease, page 197). In this condition individuals produce antibodies to their own antigens.

HISTOCOMPATIBILITY GENES

Although it is always risky to extrapolate information gained in one species to elucidate the situation in a different species, the **histocompatibility genes** that produce tissue antigens in the mouse appear to have general significance for mammals, including man. These genes are referred to as *H* (histocompatibility) genes, and the antigens they produce are called H antigens.

Histocompatibility Genes in the Mouse

There are, very approximately, 15 gene loci for histocompatibility in the mouse, each locus having 20 or more alleles. The loci are called H-1, H-2 and so forth; and alleles at a locus (say the H-2 locus) are H-2^a and *H-2*^{b}, with corresponding antigens H-2^a and H-2^b. The H-2 locus is a complex locus, with at least two closely linked regions determining perhaps 30 antigenic specificities (Klein and Shreffler, 1971). Information on the different histocompatibility series is incomplete, but it is clear that the antigens vary greatly in strength, with those determined by the H-2 locus being particularly potent in their ability to stimulate antibody production. It is also known that the H-2 antigens occur in different strengths in different cells of the body: skin or spleen cells show great antigenic strength and kidney cells show considerably less. H-2 antigens can also be demonstrated on the red blood cell. The antigen is a component of the cell membrane and its active principle appears to be a glycoprotein.

There is one main exception to the general rule that animals of an inbred strain usually accept tissue from other animals of that strain. In some but not all strains of mice, females will reject tissue from males. Since the antigen involved in this reaction is determined by a gene on the Y chromosome, it is called the H-Y antigen. H-Y is probably the same for all inbred strains of mice in which it is demonstrable, and is relatively weak compared to H-2 antigens.

Histocompatibility Genes in Man

In man, as in the mouse, one histocompatibility system is of particular importance in the rejection of implanted tissue, although its exact role is still not fully understood. This is the HL-A (human leukocyte A) system, which was formerly known as the Hu-1 system. It is determined by a single complex locus with at least two subloci. The HL-A locus is believed to be homologous to at least part of the H-2 locus (Snell et al., 1971). Knowledge about it is accumulating so rapidly that any attempt to describe it is likely to be out of date even by the time of publication.

Two HL-A subloci, termed LA and 4, are well established (Mann et al., 1969; Dausset, 1971). There are at least eight known alleles at the LA sublocus and 13 at the 4 sublocus. Because recombination between these subloci is very rare, the two are probably located very close together on the same chromosome. The two subloci form a **haplotype** which moves as a unit from parent to child (except in the rare cases where there is a crossover within the locus). Since there are at least 8 × 13 possible combinations of the alleles at the two subloci, the number of haplotypes theoretically possible is at least 104. In all probability other undiscovered alleles exist at each sublocus. There is evidence for at least one other sublocus (Kissmeyer-Nielsen et al., 1971), and possibly even more. As in the H-2 system in the mouse, the antigens are on the cell membrane and appear to consist of glycoprotein; the exact nature of the antigen has not been determined, although it may be that the antigenic site itself resides in the protein fraction. The molecular weight of the glycoprotein is probably between 30,000 and 50,000. By papain digestion it is possible to separate the two antigenic products of each haplotype, LA and 4.

Certain HL-A antigens are now recognized and classified in a manner accepted around the world. Others are under investigation but have not yet been accorded world-wide recognition. Those that are universally recognized are described as A-1, A-2 and so forth, the designation A applying to antigens whose production is controlled by genes at either the LA or the 4 sublocus.

Since each person has two haplotypes representing his four subloci and since he is usually heterozygous for each sublocus, he typically has four different HL-A antigens. (See Fig. 8–4.) In the unlikely event that he is homozygous at one sublocus, then only three different HL-A antigens are formed. To ensure the best match for organ transplantation, both donor and recipient should match with respect to all HL-A antigens as well as the antigens of the ABO blood group system.

TRANSPLANTATION IN MAN

Technically there are no insurmountable surgical problems to the transplantation of tissues in man. Kidney, heart, liver and skin have all been transplanted with respectable periods of survival (Bergan, 1971; Rake et al., 1970; Batchelor and Hackett, 1970). When transplants fail the cause is usually not poor surgical technique but rejection, which is genetically inevitable unless the challenge of the immune mechanism can be successfully met. In what ways has the immunological barrier been overcome in the successful cases?

The main problem is to avoid rejection of the transplant. In current practice there are three general approaches in attempting to ensure success:

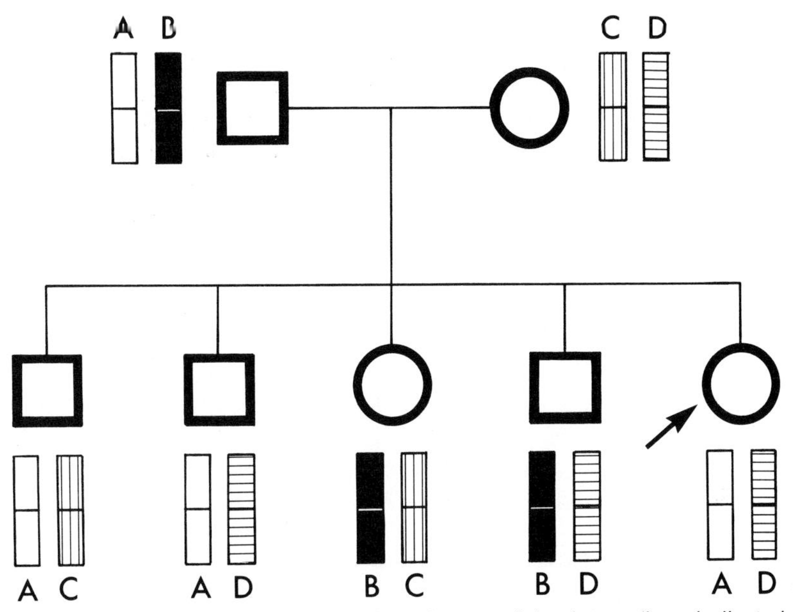

Figure 8–4 *The inheritance of HL-A haplotypes. A haplotype (here indicated as A, B, C or D) consists of two subloci, each with different antigenic specificities. Thus, there are normally four HL-A antigens per individual. Since the haplotypes are transmitted as units, a child has only one haplotype in common with either parent, but there is a one in four probability that two sibs match exactly. The theoretical distribution of haplotypes in a family of five sibs and their parents is shown. The child indicated by an arrow has one matching sib.*

1. Selection of a donor antigenically identical to the host or antigenically as compatible as possible.
2. Avoidance of the patient's immune mechanism.
3. Suppression of the patient's immune mechanism.

It is sometimes feasible to graft skin or bone from one part of the body to another; since this is an autograft, no immune reaction occurs. If a whole organ is needed there is only one possible genetically identical donor. This is the patient's monozygotic twin, if he is fortunate enough to be the one person in 200 who has a living monozygotic twin. Renal transplantations between identical twins have been successfully performed since 1956, but there is always a certain amount of risk to the donor as well as to the recipient and such operations are never undertaken except as a last resort. Unfortunately, in identical twins, successfully grafted donor kidneys may be subject to the disease process that necessitated the transplant; for example, glomerulonephritis may develop in a kidney transplant just as it did in the original kidney.

In certain types of grafts, the immune reaction is not important. Corneal transplants will survive, but there is debate about the reason for this survival. Since the cornea is not vascularized, it may be that killer cells do not come into contact with the implanted tissue. Another possible explanation is that because of the peculiar lymphatic drainage of the cornea, the rejection mechanism is not alerted. Bone grafts appear to be accepted, but actually the donor tissue merely acts as a framework upon which new host growth can occur; consequently the rejection phenomenon is not a factor. Blood vessels are a third major group of structures in which homografts serve a useful purpose, but here also the form and structure of the graft, rather than its living cells, appear to be important; and, for the replacement of blood vessels, plastic prostheses are usually more satisfactory. For both bone and blood vessel grafts, frozen tissue serves the need very adequately.

If, however, the tissue must live and function in the host and there are no special circumstances whereby the immune reaction may be avoided, the best hope is to choose a tissue which will elicit as mild a reaction as possible, and to find a way to suppress even this reaction to the extent that the living tissue will be allowed to persist.

Donor Selection

Donor selection now becomes of major importance, since the less antigenicity donor tissues have for the host, the less immune response will have to be suppressed. The methods used for selection of donors as outlined below may well be modified in the future, but certain principles are emerging which make donor selection less a matter of chance and more a matter of science.

Relatives as Donors. With rare exceptions, everyone is heterozygous for two different haplotypes, and unless the parents are consanguineous there is only a very small chance that they will have haplotypes in common. Thus, if a parent is used as a donor, the recipient will have one haplotype in common with the donor, but there is only a very slight chance that his other haplotype will also match the donor's. Since as a rule each HL-A haplotype is transmitted from parent to child as a unit, the probability that any two sibs will be identical with respect to the HL-A locus is one in four. This is why, until matching techniques are perfected, sibs are preferable to parents as tissue donors (Fig. 8–4).

Matching Tests. Various tests have been devised to determine if a prospective donor and recipient have matching histocompatibility antigens. Tests are done *in vitro* to remove the possibility of causing a second set reaction in the recipient. There are two general types:

1. **Mixed leukocyte culture test.** This test takes advantage of the property of lymphocytes to divide in the presence of foreign antigen (page 181). Lymphocytes of the prospective donor and recipient are mixed and cultured in short-term tissue culture, and the mitogenic stimulation

is measured, usually by measuring the amount of radioactivity after the incorporation of tritiated thymidine (a measure of DNA synthesis). Since the vital question is usually whether the donor has antigens which will stimulate a reaction by the recipient, the test is usually made "one-way," i.e., donor cells are inactivated by an inhibitor of DNA synthesis. Any cells which are then stimulated to divide are therefore those of the recipient. As an additional precaution, the "one-way" test is also performed with recipient cells inactivated to warn of possible graft-versus-host reaction.

2. **Lymphocytotoxicity test.** This test, used in tissue typing, depends on the ability of living lymphocytes to remain impervious to certain dyes (e.g., trypan blue) that stain dead lymphocytes. An antibody to an antigen borne by a group of lymphocytes will, in the presence of complement, kill those lymphocytes. The test makes use of a battery of different sera in which lymphocytes from the recipient and from each of a group of prospective donors are incubated, after the addition of complement. Following incubation, the dye is added to each tube and the number of dead lymphocytes counted. If donor and recipient lymphocytes react differently to any given serum, they differ in one or more antigenic specificities.

Tissue typing techniques can be used to determine at how many subloci a donor and recipient differ. If the recipient is a close relative of the donor and has all the donor antigens, and the donor has all or all but one of the host antigens, rejection should be minimal; if the donor has one antigen which is lacking in the host, the result is usually fairly good; but when the donor differs from the host in two or more antigens, the result will probably be poor. Cadaver tissue, even with good HL-A matching, may be rejected; as a result there is controversy as to the value of HL-A testing of cadaver donors. A sib is the best donor, especially a sib who matches the recipient for all testable HL-A and ABH antigens. Antigens of the ABO blood group system are very important in determining graft acceptance, because most tissues carry these antigens and a person who does not possess antigen A or B has the antibody against that antigen (page 206). Therefore, a potential donor who does not match the recipient on the ABO system is usually not employed, because the recipient's existing anti-A or anti-B antibodies may themselves produce rejection of the graft. (There are exceptions to this principle.)

Immunosuppression

After selection of the donor with the most suitable genotype, a way must be found to ensure further that the transplanted tissue is not rejected. Certain drugs known as **antimetabolites** inhibit the immune reaction, probably by interference with RNA production, but in so doing they make the patient susceptible to various infections which otherwise he would overcome with no difficulty. The dosage of antimetabolites must be kept as low as possible so that the immune mechanism may remain at least partially functional. Drugs such as azathioprine or methotrexate can be given in doses large enough to suppress the immune reaction completely; but if the graft is antigenically not greatly dissimilar to the host tissue, smaller amounts may suffice. If tissue from a donor of very different genotype is used, large dosages of drugs are necessary.

Radiation and steroid compounds may also be used to assist the suppression of immune response. A recent development has been the use of **antilymphocyte serum** (ALS), which contains antibodies to human lymphocytes. Since lymphocytes are essential to the rejection of transplants, the antilymphocytic globulin (ALG) in ALS should prolong the survival of implanted tissue. The results are as yet equivocal and the place of ALG in clinical immunosuppression is not clear (Mee and Evans, 1970; Woodruff, 1971).

Why Are Some Grafts Rejected?

Even when all the knowledge available to him and all the tools of his armamentarium are applied, the surgeon still cannot guarantee success in allografts in man. There can be little doubt about the importance of using sib pairs with matching HL-A antigens to enhance the possibility of successful skin grafting or renal transplantation. However, in unrelated recipient-donor pairs, results are less encouraging. This may be due to the imprecise nature of testing procedures and to lack of knowledge of other possible *LA* and *4* alleles or even alleles at other (sub) loci not yet fully determined.

One possible cause of graft rejection is that the recipient may have primed antigen-sensitive cells against the antigens of the donor tissue. If the recipient has previously had a blood transfusion, he may have received white cells with antigens identical to some donor antigens; if so, he will then react by a second set response to antigens whose effect might otherwise have been overcome by low doses of immunosuppressive drugs. The same problem may occur in women who have borne children; the cells of the fetus may have crossed the placental barrier and sensitized the mother to the HL-A antigens of the fetus.

The practical value of genetic knowledge concerning transplantation of human tissue is perhaps best illustrated in kidney transplantation. Of 3646 renal transplants registered with the Human Kidney Transplant Registry (Murray et al., 1971), 40 per cent were performed in 1968 and 1969. (More recent implants have not yet survived long enough for useful statistical investigation.) The two-year survival rate of kidneys from consanguineous donors is 75 per cent, but for unrelated donors only 41 per cent. The 1971 report indicates that sibs and parents are equally satisfactory as kidney donors despite theoretical considerations (page 190).

This unexpected finding is probably explained by increased use of HL-A and ABO typing and by the use of sib donors who did not match perfectly but who were used because there were no well-matched sibs available.

IMMUNE DEFICIENCY DISEASES

Until the discovery and development of antibiotics, infections were a major cause of death in infancy and childhood, and genetic defects formed only a small part of childhood morbidity and mortality. However, morbidity from infectious disease is now a much less important problem, and when it occurs it is often due to the child's inability to cope with infection. This inability is often genetically determined and may be caused by an **"immune deficiency disease,"** which reflects defects in the origin and function of the immune-competent cells.

As described on page 180, at some stage of development there is a stem cell ancestral to both thymus-dependent and bursa-equivalent cells. If such a stem cell or any of its progeny is abnormal or missing, specific forms of immune deficiency will result. In most immune deficiency diseases the gamma globulin levels are reduced, but the term agammaglobulinemia often applied to these conditions is inaccurate, since there is invariably some gamma globulin present.

1. X-LINKED RECESSIVE AGAMMAGLOBULINEMIA (BRUTON'S DISEASE)

In this condition the different immunoglobulins are greatly reduced in amount. There are virtually no plasma cells and the tonsils are very small. However, the thymus appears to be normal and the individual is able to reject allografts. The disease therefore depends upon a deficiency of B-cells and is almost identical to that seen in chickens from which the bursa of Fabricius has been excised. Total plasma gamma globulin levels are very low and the power to form antibodies is very limited. As a result, affected individuals have many acute infections, especially of the pneumococcal type.

2. DIGEORGE SYNDROME

In this condition, which apparently has no genetic basis, T-cells are missing. There are normal plasma cells, and all gamma globulins are at normal levels. DiGeorge (1968) reports one patient who did not reject skin grafts.

3. AUTOSOMAL RECESSIVE (SWISS TYPE) AGAMMAGLOBULINEMIA

In this disorder patients cannot produce a normal amount of circulating antibody, immunoglobulins are absent or decreased, and there is no cell-mediated response. There is a very low lymphocyte level in the blood and few or no plasma cells are seen. The thymus is vestigial and there are none of the germinal centers in the lymph nodes that normally indicate the presence of B-cells. The affected individual is susceptible to many types of infection and will not reject skin grafts. Affected children usually die within the first six months of life; almost invariably, death occurs before the end of the first year. The logical explanation is that there is a deficiency in the stem cell. Total absence of the red cell enzyme adenosine deaminase (ADA) has recently been noted in two children with this disorder (Giblett et al., 1972). In each case the presumptively heterozygous parents had reduced amounts of ADA.

4. X-LINKED THYMIC APLASIA

This condition has been described as "X-linked Swiss" immune deficiency, but it represents a different genetic defect. It is less severe than the Swiss type, the patients usually surviving up to two years. Again the defect appears to be in the stem cell.

THE FETUS AS AN ALLOGRAFT

A human mother will not accept tissue from her child, nor will an inbred female mouse accept tissue from its offspring if the father is of a different inbred strain; but in both these cases rejection may take longer than usual. Though the trophoblast (fetal tissue) certainly acts as a graft on the mother, it is not rejected. At present the mechanisms that prevent rejection of the fetus are not fully understood.

It is known that red and white cells from the fetus invade the maternal circulation, so the mother has ample opportunity to make antibodies and/or killer cells directed against fetal tissue. There is no doubt that some antibodies (IgG) cross the placental barrier from mother to fetus, but these do not appear to harm the fetus except in certain cases of blood group incompatibility, and some of them confer on the newborn a passive immunity to infections against which the mother has circulating antibody.

Anderson (1971) has recently reviewed the matter of fetal-maternal immune relationships. He concludes that the depression of the immune response of the mother to fetal tissue resembles other forms of immune tolerance, and that trophoblast cells do not exhibit the expected strength of tissue antigens. Fetal antigens appear to develop very slowly as the products of conception grow, and their slow development may allow a gradual reduction of function of the maternal immune apparatus for these particular antigens.

AUTOIMMUNE DISEASE

Though the body normally exists in a state of immunological homeostasis (page 188), occasionally it produces antibodies (autoantibodies) to some specific antigen of its own. When this happens a state of **autoimmunity** is said to exist. Theoretically, a change in immunological homeostasis may result from a change in either side of the antigen-antibody balance. Both types of change may be involved in human autoimmune disease:

Altered Antigen + Normal Antibody ⟶ Autoimmune Disease
or
Normal Antigen + Altered Antibody ⟶ Autoimmune Disease

An *antigen* may change as a result of an alteration in the tissue that produces it (the **target tissue,** so called because it is the target of a specific antibody). Such an altered tissue could be produced by somatic mutation or by viral transformation. In either case, an altered protein would be produced.

A cell not present at birth may have foreign antigenic properties when it appears, but normally such a cell does not alert the immune mechanism. For example, injection of its own spermatozoa produces an immune reaction in the guinea pig, but normally spermatozoa do not reach the lymphoid cells. Injury or disease may release from a target tissue an antigen that has existed for many years, secluded from the rest of the body. In the condition known as **sympathetic ophthalmia,** severe injury to one eye may result, weeks later, in blindness to the other eye, because sequestered antigens in the injured eye were released by the injury and incited the production of antibodies, which then attacked the same tissue of the uninjured eye.

Homeostatic balance might be affected on the *antibody* side by somatic mutation of a gene controlling the production of an antibody, leading to the formation of a slightly altered antibody which might then react with some existing antigens. This is believed to be the mechanism of acquired hemolytic anemia. Somatic mutation is thought to result in a **forbidden clone** of cells, capable of producing an **autoantibody** which would react with normal target tissue. To many investigators this possibility explains most autoimmune phenomena. It is possible that by transformation, viral infection might produce a similar change in an antibody-producing cell which could then form a forbidden clone.

If there is exposure to an antigen which is very like some normal antigen of the host, the new antibody formed in response to this abnormal antigen may be capable of combining with the normal antigen as well. For example, the antigen produced by streptococci may be so similar to some of the histocompatibility antigens of human heart muscle that immunoglobulin formed in response to a streptococcal infection may also adhere to human heart cells. Similar phenomena may account for rheumatoid arthritis and some other autoimmune disorders, such as certain kinds of nephritis, which appear to be related to a streptococcal infection.

At one time it was thought that inflammation of the thyroid gland could be produced immunologically because thyroglobulin was a sequestered antigen that somehow was released to alert the immune mechanism. This interpretation is no longer accepted. It is now believed that slightly altered thyroglobulin, in structure similar to thyroglobulin from a different species, can cause the production of antibodies which react with the normal as well as with the altered thyroglobulin. It is assumed that if, by mutation, a spontaneous modification of thyroglobulin occurred, antibodies would be produced which would also combine with normal thyroid tissue and cause autoimmune thyroiditis.

In order to establish that a disease is autoimmune in nature, it is necessary to demonstrate in the patient autoantibodies which react specifically with his own tissues. Such autoantibodies have been shown, for example, in patients with lymphocytic thyroiditis (Hashimoto's thyroiditis), pernicious anemia, idiopathic Addison's disease, ulcerative colitis, rheumatic fever and systemic lupus erythematosus. If the antigenic substance is from a single organ the disease is said to be **organ-specific** (e.g., Hashimoto's thyroiditis). If the antigenic substance is found in many tissues the disease is said to be **systemic** (e.g., systemic lupus erythematosus).

In rheumatoid arthritis, autoantibodies can be demonstrated that unite with gamma globulin of normal human plasma; in other words, these autoantibodies are "anti-antibodies." The autoantibody, **rheumatoid factor,** can be demonstrated in affected individuals, particularly in the plasma cells around involved joints and in the germinal centers of lymph nodes.

The tendency to form forbidden clones may have a genetic basis. Immunological homeostasis is based upon immune tolerance, which is not genetic in character but depends on tolerance for the antigens that are exhibited by each individual during his embryonic existence and that persist throughout his lifetime. However, rheumatoid factor may be found in relatives of patients with rheumatoid arthritis or in relatives of patients with systemic lupus erythematosus. The familial incidence of rheumatoid factor suggests that autoimmune disease has a genetic basis.

A remarkable feature of autoimmune disease is the tendency of patients suffering from one such disease to show a high incidence of another; for example, pernicious anemia is common in patients with Hashimoto's thyroiditis. Moreover, patients with one autoimmune disease may show a very high incidence of autoantibodies to antigens of organs not affected by the original disease, without actually having a second autoimmune disease. For example, patients with diabetes, particularly of the juvenile type, have an unusually high incidence of autoantibodies to organs other than the pancreas (Irvine et al., 1970). The significance of these observations is not yet clear, but they too suggest a genetic tendency toward the production of autoantibodies.

CHROMOSOMAL ABERRATIONS AND AUTOIMMUNE DISEASE

Fialkow (1969) has reviewed evidence that the presence of autoimmune disease of the thyroid gland, even at a subclinical level, may predispose to the development of major chromosomal aberrations. The first indications of this association came when it was found that some patients with Turner's syndrome (page 168) showed thyroid autoantibodies. Elevated thyroid autoantibody levels may also be found in patients with Down's syndrome and sometimes in their mothers, but not in their fathers. For instance, Dallaire and Flynn (1967) report that thyroid antibodies are twice as prevalent in mothers of patients with Down's syndrome as in a group of matched control mothers. It is suspected that the autoantibody may be a factor which increases the probability of nondisjunction during gametogenesis. This association should be considered in the genetic counseling of women with autoimmune disease.

THE THYMUS IN AUTOIMMUNE DISEASE

Since the thymus plays a major role in the immune mechanism of the body, it is not surprising that in autoimmune disease the thymus is abnormal. Histological abnormalities of the thymus are seen in myasthenia gravis, thyrotoxicosis, Addison's disease, systemic lupus erythematosus, rheumatoid arthritis and autoimmune hemolytic anemia. In myasthenia gravis and systemic lupus erythematosus there is enlargement of the gland. Perhaps the thymus itself is a victim of autoantibodies and because of impaired function does not fulfill its (postulated) function of eliminating forbidden clones.

SOME AUTOIMMUNE DISEASES

It should be noted that autoimmune diseases have a predilection for connective tissue; many of them, which show quite similar histopatho-

TABLE 8–2 AUTOIMMUNE DISEASES

Systemic Diseases

Chronic discoid lupus erythematosus	Rheumatoid arthritis
Dermatomyositis	Sjøgren's syndrome
Polyarteritis nodosa	Systemic lupus erythematosus

Organ-Specific Diseases

Disease	*Organ Involved*
Acquired hemolytic anemia	Blood
Addison's disease (idiopathic)	Endocrine system—adrenal gland
Certain forms of encephalitis	Central nervous system
Hashimoto's thyroiditis	Endocrine system—thyroid gland
Infantile eczema	Skin
Multiple sclerosis	Central nervous system
Myasthenia gravis	Muscle
Pernicious anemia	Blood
Some forms of chronic gastritis	Digestive system
Some types of glomerulonephritis	Genitourinary system
Some types of orchitis	Genitourinary system
Sympathetic ophthalmia	Eye
Ulcerative colitis	Digestive system

logical changes, constitute the so-called **collagen diseases.** Table 8–2 lists some diseases generally considered to have an autoimmune component. The great variety of tissue and organs affected and the diversity of the resultant symptoms indicate the importance of autoimmunity to modern medicine.

GENERAL REFERENCES

Fialkow, P. J. 1969. Genetic aspects of autoimmunity. *Progr. Med. Genet.* 6:117–167.

Good, R. A., Peterson, R. D. A., Perey, D. Y., Finstad, J., and Cooper, M. D. 1968. The immunological deficiency diseases of man. In: *Immunological Deficiency Diseases in Man.* (Bergsma, D., ed.) Birth Defects Original Article Series 4:17–34. The National Foundation, New York.

Holborow, E. J. 1967. An ABC of modern immunology. *Lancet* 1:833–835, 890–893, 942–944, 995–997, 1049–1052, 1098–1101, 1148–1150, 1208–1210.

Humphrey, J. H., and White, R. G. 1970. *Immunology for Students of Medicine,* 3rd ed. Blackwell Scientific Publications Ltd., Oxford.

Rapaport, F. A., ed. 1971. Proceedings of the Third International Congress of the Transplantation Society. *Transplantation Proceedings* 3:1.

PROBLEMS

1. An F_1 hybrid mouse produced by crossing a female of the C3H strain with a male of the C57BL/6 strain is called a C3B6F_1.

(a) From which of the following will a C3B6F$_1$ normally tolerate skin grafts?
  (1) C3H
  (2) C57BL/6
  (3) C3B6F$_1$
  (4) Wild mouse
  (5) Mouse of another inbred strain

(b) Which of the following will normally accept a skin graft from a C3B6F$_1$?
  (1) C3H
  (2) C57BL/6
  (3) C3B6F$_1$
  (4) Wild mouse
  (5) Mouse of another inbred strain

2. Arrange the following relatives of the recipient in order of their suitability as donors of a kidney transplant:

   Sib, father, MZ twin, uncle, cousin, mother, DZ twin.

9

Blood Groups and Other Genetic Markers

Some 40 genetic polymorphisms in the antigens of red blood cells, the red cell enzymes and the serum proteins are now known. Because of their ready classification into different phenotypes, their simple mode of inheritance and their different frequencies in different populations, these characteristics are known as **genetic markers.** In addition to their own intrinsic interest and medical significance, they have useful applications in family studies, linkage analysis and population genetics. Though all the genetic polymorphisms of blood might appropriately be called "blood groups," the term is restricted in usage to the red cell antigens.

The usefulness of a genetic trait as a genetic marker depends upon the following characteristics:

1. A simple and unequivocal pattern of inheritance. Most of the blood groups, red cell enzyme variants and serum protein variants are determined by systems of codominant genes, i.e., both alleles of heterozygotes are phenotypically expressed, permitting the genotype to be inferred directly from the phenotype. (Exceptions include the *O* allele of the ABO blood group system, which has no detectable product.)

2. Accurate classification of the different phenotypes by reliable techniques.

3. A relatively high frequency of each of the common alleles at the locus. If the rarer of a pair of alleles is encountered in only a small percentage of a population, it is not likely to be very useful for family studies and linkage analysis. Equal frequency of the two alleles is the ideal, since this situation provides the best opportunity for segregation to be observed at the locus.

4. Absence of effect of environmental factors, age, interaction with other genes or other variables on the expression of the trait.

The study of the blood group systems has permitted many important contributions to human genetics, and to the elucidation of genetic principles. Multiple allelism at a locus was first demonstrated in man by the ABO blood group genes. In linkage studies, the first four autosomal linkages to be found all involved blood groups. The X-linked blood group locus, Xg, is the standard point of reference on the X chromosome, and has been shown to be within measurable distance of at least four other loci: those for ichthyosis, ocular albinism, angiokeratoma and retinoschisis (page 286). The Xg locus has also been useful in studies of X-chromosome aneuploids. Population genetics has made extensive use of the blood groups to demonstrate Hardy-Weinberg equilibrium and to study the effects of genetic drift and gene flow in altering blood group frequencies in populations. Interaction of nonallelic genes is demonstrated by the ABO-Lewis-secretor relationship. The controversy over whether the C, D and E antigens of the Rh system are determined by different, closely linked alleles, or whether they are different expressions of a single allele, illustrates the difficulties of deciding between these possibilities in a human system.

BLOOD GROUPS

The human blood groups occupy a special place in medical genetics both because of their many contributions to the development of genetic principles and because of their clinical importance in blood transfusion and obstetrics.

The first successful transfusion of human blood was performed in 1818, but transfusion for therapeutic purposes did not become reasonably safe until the discovery of the ABO blood group system by Landsteiner in 1900. Even when the donor's blood group is the same, with respect to the ABO system, as that of the patient, careful matching must be carried out to detect antibodies reacting with other red cell antigens.

In 1971, the last year for which figures are available at this writing, transfusions were given to more than 235,000 persons in Canada. In other words, in a single year, one in 100 Canadians received one or more blood transfusions and thus had an opportunity to form antibodies against antigens on the donor's red cells.

The clinical significance of the blood groups in maternal-fetal incompatibility results from the fact that sensitization of a pregnant woman to antigens of her fetus (usually, though not always, antigens of the Rh system) can produce hemolytic disease in her newborn child. (It should also be remembered that the survival of transfused cells may be curtailed by the "immune" antibodies formed by multiparous women.) Until recently, hemolytic disease of the newborn ranked among the first 10 causes of perinatal death, and led to physical or mental handicap in surviving children who were inadequately treated. The development of a means of prevention of sensitization of the mother has been one of the most significant advances in obstetrics and pediatrics in recent years. It is described in a subsequent section (page 222).

BLOOD GROUP NOTATION

The notation for blood group genes has been developed in a piecemeal fashion, and is riddled with inconsistencies. Three chief kinds of notation are used:

1. Alternative alleles may be designated by letter sequences (*A* and *B*, *M* and *N*).

2. Alternative alleles may be designated by large and small letters (*S* and *s*, *K* and *k*, *C* and *c*). Note that use of a lower-case letter does not here imply that the trait is recessive.

3. Alternative alleles may be designated by a symbol with a superscript (Lu^a and Lu^b, Fy^a and Fy^b).

A blood group system is now usually named for a person in whom the antibody was first recognized (e.g., Duffy and Kidd), and some part of this name is used to designate the gene locus concerned (Fy for Duffy, Jk for baby J. Kidd). The Lutheran system, on the other hand, was named for the donor of the provoking antigen. For the more recently discovered blood groups, there is a standard notation which can be described with reference to the Kidd groups. The symbol Jk is used for the locus, and the known genes are Jk^a and Jk^b. A third allele has been postulated and is symbolized by merely *Jk* without a superscript, to show that the antibody defining the corresponding antigen has not yet been discovered. The two known antigens are called $\mathrm{Jk^a}$ and $\mathrm{Jk^b}$, and they are detected by antibodies anti-$\mathrm{Jk^a}$ and anti-$\mathrm{Jk^b}$. The reactions of red blood cells are described as Jk(a+b+), Jk(a+b–), Jk(a–b+) or Jk(a–b–), depending upon the results of testing the cells with anti-$\mathrm{Jk^a}$ and anti-$\mathrm{Jk^b}$.

The following description of the best-known blood group systems makes extensive use of Race and Sanger's *Blood Groups in Man* (5th edition, 1968), to which readers are referred for additional information and references. Though most of the known blood group systems receive at least a brief mention in the following pages, not all the known antigens have been included; complications of some systems have been omitted, especially when they do not elucidate new principles.

Fourteen blood group systems are listed in Table 9–1 in chronological order of their recognition. The first example of the antibody defining each system was found as follows: naturally occurring in healthy subjects (ABO, Lewis); in immunized animals (MN, P, Rh); in the mothers of infants with hemolytic disease of the newborn (Rh, Kell, Kidd, Diego); or during cross-matching tests (Lutheran, Duffy, I, Cartwright, Xg, Dombrock).

In addition to the 14 blood group systems shown in Table 9–1, there

TABLE 9–1 SOME OF THE KNOWN BLOOD GROUP SYSTEMS

System	Discovered by	Main Antigens*
ABO	Landsteiner, 1900	A_1, A_2, B, H
MNSs	Landsteiner and Levine, 1927	M, N, S, s
P	Landsteiner and Levine, 1927	P_1, P_2, P^k
Rh	Landsteiner and Wiener, 1940	C, C^w, c, D, D^u, E, e and many others
Lutheran	Callender, Race and Paykoc, 1945	Lu^a, Lu^b
Kell	Coombs, Mourant and Race, 1946	K, k, Kp^a, Kp^b, Js^a, Js^b
Lewis	Mourant, 1946	Le^a, Le^b
Duffy	Cutbush, Mollison and Parkin, 1950	Fy^a, Fy^b
Kidd	Allen, Diamond and Niedziela, 1951	Jk^a, Jk^b
Diego	Layrisse, Arends and Dominguez, 1955	Di^a, Di^b
I	Wiener, Unger, Cohen and Feldman, 1956	I, i
Cartwright	Eaton, Morton, Pickles and White, 1956	Yt^a, Yt^b
Xg	Mann, Cahan, Gelb, Fisher, Hamper, Tippett, Sanger and Race, 1962	Xg^a
Dombrock	Swanson, Polesky, Tippett and Sanger, 1965	Do^a, Do^b

*Not a complete list.

are several others whose rank as independent systems is still uncertain. An example of the discovery of a "new" system which eventually was shown to be part of an old system is provided by "Gonzales." The antigen Go^a was originally defined in 1962 by an antibody which caused mild hemolytic disease of the newborn, and was later found to be restricted to Negroes, among whom its incidence was about 2 per cent. It was eventually recognized that Go^a belongs to the Rh system, in which it is one of several types of unusual D antigen.

Public and Private Antigens. A number of other antigens are known which are either very common (public) or very rare (private). Public and private antigens probably differ from ordinary blood group systems only in their incidence. It is hard to know whether an ultrarare or ultracommon antigen belongs to some known blood group system or is actually part of a new system, but in either case its frequency detracts from its utility as a genetic marker, because there are so few families in which segregation is demonstrable at the locus concerned.

THE ABO BLOOD GROUP SYSTEM

Landsteiner's discovery of the ABO blood groups is the first notable event in the history of the blood groups. He and his colleagues found that the blood of any individual belongs to one of four different types. These types are distinguished by an agglutination reaction; if the serum of the recipient is mixed with cells from an incompatible donor, the cells are agglutinated. The clumped red cells, or "agglutinates," can usually be observed with the naked eye.

There are four major ABO phenotypes, known as O, A, B and AB. The reaction of each phenotype with anti-A (−A) and anti-B (−B) is as follows:

Red Cell Phenotype	***Reaction With*** *−A*	*−B*
O	−*	−
A	+	−
B	−	+
AB	+	+

*−denotes no agglutination.
+denotes agglutination.

Group A individuals possess antigen A on their red cells, group B individuals possess antigen B, group AB individuals possess both A and B and group O individuals possess neither. The inheritance of the ABO blood groups has been outlined in Chapter 4 as an example of multiple allelism.

A remarkable feature of the ABO groups is the reciprocal relation between the antigens present on the red cells and the antibodies in the serum of the same individual, summarized in Table 9–2. (Serum and plasma are interchangeable terms in this connection. Serum is more satisfactory than plasma for laboratory tests.) Anti-A is found in the serum of persons whose cells do not contain A, and anti-B is found in persons whose cells do not contain B. The reason for this reciprocal relation is uncertain, although it is believed that formation of anti-A and anti-B is probably a response to A and B substances occurring

TABLE 9–2 THE ABO BLOOD GROUPS

Blood Group *(Phenotype)*	***Genotype***	***Antigens on Red Cells***	***Antibodies in Serum***
O	*O/O*	neither	anti-A, anti-B
A	*A/A* *A/O*	A	anti-B
B	*B/B* *B/O*	B	anti-A
AB	*A/B*	A, B	neither

naturally in the environment. However they originate, the regular presence of anti-A and anti-B explains the failure of many of the early attempts to transfuse blood.

An antibody present in the plasma of the recipient will cause agglutination of donor red cells if the donor cells carry the corresponding antigen (Table 9–3).

TABLE 9–3 AGGLUTINATION REACTIONS WITHIN THE ABO BLOOD GROUP SYSTEM

Recipient		***Blood Group of Donor's Red Cells***			
Blood Group	*Antibodies (Serum or Plasma)*	*O*	*A*	*B*	*AB*
O	anti-A, anti-B	–*	+	+	+
A	anti-B	–	–	+	+
B	anti-A	–	+	–	+
AB	neither	–	–	–	–

*– indicates no agglutination; + indicates agglutination of donor cells by antibodies in recipient's serum (or plasma).

Hence, there are "compatible" and "incompatible" combinations, a "compatible" combination being one in which the red cells being tested (or the red cells of a donor) do not carry an antigen corresponding to the antibodies in the test (or recipient) serum. (Antibodies in the donor's plasma are not usually taken into account in transfusion, presumably because they are greatly diluted in the recipient's circulation.) Although there are theoretically "universal donors" (group O) and "universal recipients" (group AB), a patient is always given blood of his own ABO group, except in real emergencies.

Subtypes of the ABO groups can be recognized. The most important is the separation of group A into A_1 and A_2, with a corresponding separation of AB into A_1B and A_2B. About 85 per cent of group A bloods are A_1. Other variants of both A and B are known, but they are rare.

H Substance

H substance (H antigen) is the substrate from which the A and B substances are made by the action of the *A* and *B* genes. The *O* gene is thought to be an "amorph," i.e., a completely inactive gene; thus group O cells contain unaltered H substance. Anti-H is found in the serum of some subjects of group A_1, and in all subjects of the "Bombay" phenotype (see below).

The Bombay Phenotype

The Bombay phenotype, which is very rare, is produced by interaction between the ABO genes and a rare mutant at a different locus. It was first discovered in 1952 in Bombay when two men needed transfusion, one after stab wounds and the other after a railway accident. Their cells were not agglutinated by anti-A, anti-B or anti-H; their serum contained all three antibodies.

Figure 9–1 is a pedigree showing the Bombay phenotype. In this family a woman whose red cells were grouped as O produced an AB

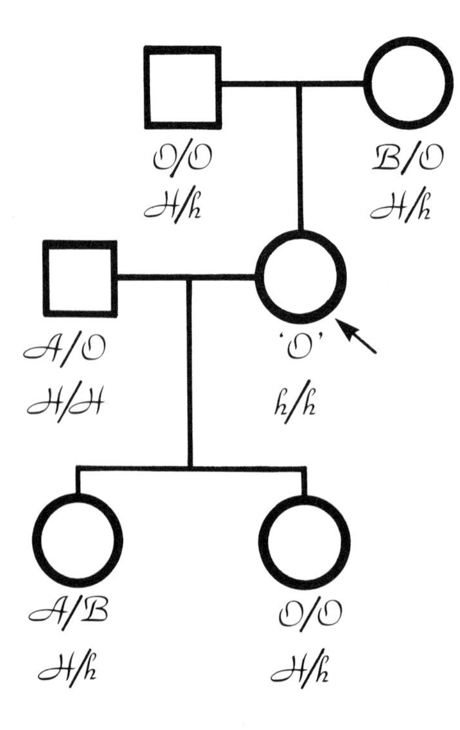

Figure 9–1 *A pedigree of the Bombay phenotype (genotype* h/h*) described in the text. (After Levine et al. 1964.* Blood *10:1100–1108, by permission.)*

child. Family studies led to the conclusion that the woman's "true blood group" was B, but she could not form the B antigen. (It is noteworthy that her parents were consanguineous.) Most persons possess a gene *H* (*H*/*H* or *H*/*h*) required for the development of the H substance from a precursor. Persons of the Bombay phenotype are homozygous for an inactive allele (amorph) *h*. When H is not formed, the enzymes determined by *A* and *B* genes have no substrate on which to act, so that *h*/*h* persons cannot make the A or B antigen even if they have the *A* or *B* gene.

Secretion of Blood Group Substances

The A, B and H antigens of the ABO blood group system occur in other tissues as well as on the red cell, and in most people are also secreted in a water-soluble form in a number of body fluids. The ability to secrete these substances is determined by a secretor gene, *Se*; its allele, *se*, has no known function. Secretors of H, A and B are either *Se*/*Se* or *Se*/*se*; nonsecretors (about 22 per cent of the population) are *se*/*se*. The ABH antigens on the red cells and in the saliva of secretors and nonsecretors are shown in Table 9–4.

The ABO-secretor relationship is an example of nonallelic interaction (page 87). For example, in a family where both parents are *A*/*O*, *Se*/*se*, the ratio of secretors of A substance to nonsecretors of A substance in the progeny is 9:7. This ratio is a modification of the 9:3:3:1 Mendelian dihybrid ratio, and it results from interaction of the two nonallelic genes in such a way that one of the genes depends for its expression on the product of the other gene. *Epistasis* is the general term used to describe this type of interaction.

Genetic Pathways of Synthesis of ABO Blood Group Substances

The biosynthesis of the ABO blood group substances has been unraveled chiefly by the work of Watkins and Morgan. Only one of their

TABLE 9–4 ABH ANTIGENS IN SECRETORS AND NONSECRETORS

Blood Group	Antigens of Secretors: Red cells	Antigens of Secretors: Saliva	Antigens of Nonsecretors: Red cells	Antigens of Nonsecretors: Saliva
O	H	H	H	–
A	A	A and H	A	–
B	B	B and H	B	–
AB	A and B	A, B and H	A and B	–

many papers on the subject is cited in our references (Morgan and Watkins, 1969). They have studied the biosynthesis of secreted blood group substances, but it is assumed that the biosynthesis of red cell antigens is similar. The precursor molecule, itself presumably the product of many biosynthetic steps, forms the "backbone" onto which each specific transferase determined by the *H, A* and *B* genes (but not *O*) adds a specific sugar. The steps may be summarized as follows:

SUGAR TRANSFERASES

Gene	***Transferase***	***Acceptor***	***Sugar Added****	***Blood Group Substance***
H	*H*-transferase	Precursor	L-fucose	H
A	*A*-transferase	H substance	Gal*N*Ac	A
B	*B*-transferase	H substance	D-Gal	B
O	–	–	–	(Unchanged H)

*Abbreviations: Gal*N*Ac, *N*-acetylgalactosamine
D-Gal, D-galactose

The relationship between the blood group genes, their products and the blood group substances is shown in Figure 9–2. The observation that

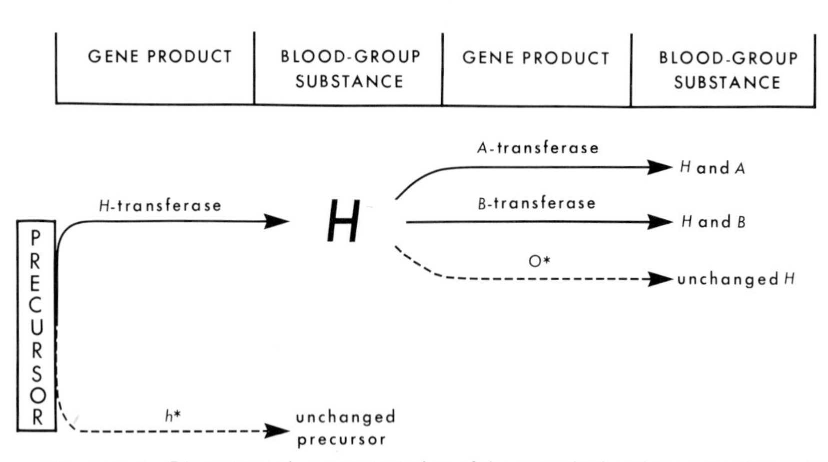

Figure 9–2 *Diagrammatic representation of the genetical pathways leading to the formation of the H, A and B blood group substances.**

*Genes *h* and *O* are amorphs, i.e., they have no detectable effect on the precursor. (Based on the work of Morgan and Watkins, and of Ceppellini.)

red cells of the subtype A_2 have fewer A receptors than those of subtype A_1 fits well with the recent finding that the A^1 and A^2 transferases are quantitatively and qualitatively different, although both enzymes add the same sugar (Gal.*N*Ac) to the same acceptor (H) (Schachter et al., 1973).

In the interests of simplicity, the role of the Lewis genes has been omitted from this discussion. The Lewis (*Le*) gene is involved in the synthesis of Le^a and Le^b substances in secretions and in the plasma. Unlike other blood group antigens, the Lewis phenotype of the red cells depends on the adsorption of Lewis substances from the surrounding plasma.

Frequency of the ABO Groups and Genes

Because the relative proportions of the ABO groups vary widely in different populations, frequency figures are valid only for the specific population from which they are derived. In white North Americans, groups O and A usually have frequencies in the 40 to 50 per cent range, group B about 10 per cent and group AB about 3 to 4 per cent. The following figures will serve as an illustration of a population in Hardy-Weinberg equilibrium (see Chapter 11).

Blood Group	Frequency
O	0.46
A	0.42
B	0.09
AB	0.03
	1.00

For the example given above, the calculated frequencies of the alleles *O*, *A* and *B* are 0.68, 0.26 and 0.06, respectively. In other populations, the frequencies might be quite different; for example, group B is much more common in Eastern Europe and in Asia than in Western Europe or areas of Western European immigration.

Linkage Relations of the ABO Groups

The first linkage described for the ABO locus is that of the "nail-patella syndrome" (NP), a rare autosomal dominant disorder characterized by dystrophy of the nails, absence or reduction of the patella and other bone dysplasias and, occasionally, nephropathy. The recombination fraction between the two loci is about 15 per cent.

Adenylate kinase (AK) is a polymorphic red cell enzyme (which also occurs in muscle, where it is known as myokinase). There are two common alleles, AK^1 and AK^2, the rarer AK^2 having a gene frequency of about 0.05 in Britain. This locus is very closely linked to the NP locus. The three loci, ABO, NP and AK, were the first triplet linkage group to be described in man. The probable order of the genes and their distance apart is believed to be as follows:

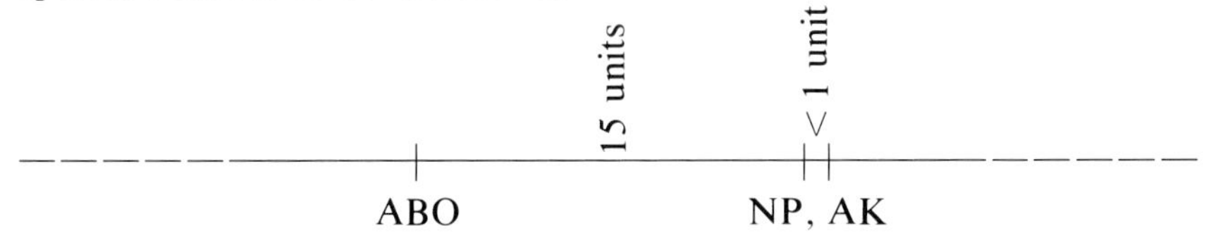

The ABO Groups and Disease

Theoretically, if a gene is to be maintained in a population at a frequency higher than can be explained by recurrent mutation, it must confer some advantage. For the ABO blood groups, however, three alleles are maintained at quite high frequencies even though there is no major advantage conferred by any genotype. When a population contains two or more phenotypes in such proportions that the rarest of them cannot be maintained merely by recurrent mutation, it is said to exhibit genetic polymorphism (Ford, 1940). Though research has so far failed to demonstrate that any particular blood group is advantageous (Reed, 1969), the ABO blood groups may not be genetically neutral. On the contrary, they may actually have a bearing on selective perinatal mortality (see discussion of hemolytic disease of the newborn, page 216). Their relation to disease susceptibility is well known, but may not be of great genetic significance because the diseases concerned usually affect people in middle and later life, and only rarely during the peak reproductive period.

Disease Susceptibility. The idea that certain blood groups may be associated with susceptibility or resistance to certain categories of disease was ridiculed for many years, but by now numerous studies all over the world, with highly consistent results, make it impossible to doubt that certain associations are real (even though false ones are still reported from time to time). The first strong evidence was produced in 1953 by Aird et al., who described an excess of group A among patients with gastric cancer, an observation that has since been repeatedly confirmed. Another even closer association involves duodenal ulcer, group O and nonsecretion of ABH substances. Group O persons are about 1.4 times as likely as group A persons to develop duodenal ulcer. Nonsecretors are at quite a large additional risk; Clarke reports that 37 per cent of 973 duodenal ulcer patients are nonsecretors, compared with 24 per cent of a control series. Considering both the ABO and the secretor status, group O nonsecretors are about twice as likely to have duodenal ulcer as are group A secretors.

Several other associations have been suggested, particularly pernicious anemia and diabetes mellitus with group A. The diseases which show an association with a particular blood group all appear to be diseases of the upper part of the gastrointestinal tract. Conditions that have been investigated for association with the ABO blood groups but have not shown any significant deviation include cancer of the esophagus and pancreas, ulcerative colitis and leukemia. A recent suggestion of an association between carcinoma of the breast and absence of the S antigen (see next section) has failed to be confirmed.

THE MNSs BLOOD GROUP SYSTEM

After the discovery of the ABO blood group system in 1900, no other blood group antigens were found until 1927, when Landsteiner and Levine used a different approach. They injected human blood into rabbits, which developed antibodies against it. The "immune serum" formed by the rabbits could distinguish between different human red cell samples. The antigens so recognized were called M and N.

As originally described, the MN groups were models of genetic simplicity. They appeared to depend upon a pair of codominant alleles, *M* and *N*, roughly equal in frequency, which produced three genotypes, *M*/*M*, *M*/*N* and *N*/*N*, and three corresponding phenotypes, M, MN and N.

Blood Group (Phenotype)	Genotype	Reaction With Anti-M	Reaction With Anti-N	Approximate Frequency (European)
M	*M*/*M*	+	−	0.28
MN	*M*/*N*	+	+	0.50
N	*N*/*N*	−	+	0.22

The Ss subdivisions of the MN groups, which make the genetic pattern less simple, were discovered some 20 years later when the first example of anti-S was recognized in Sydney. It now appears that combinations of MN and Ss are inherited as units, i.e., *MS, Ms, NS* and *Ns*. The genetic interpretation of these combinations is not yet clear; the situation is somewhat similar to that for the Rh groups (see below). Rarely in Blacks and frequently in Congo Pygmies, red cells have been found which lack both S and s. Further complications of the MNSs system are discussed in Race and Sanger.

The MN groups have little importance in blood transfusion or in maternal-fetal incompatibility. Their major significance in medical genetics is that their relative frequencies and codominant pattern of inheritance make them useful in the solution of identification problems.

Many population surveys show a greater number of MN heterozygotes than would be expected on the basis of the Hardy-Weinberg equilibrium. The reason for this excess is not known; there is no obvious "heterozygote advantage" that would explain it.

THE Rh BLOOD GROUP SYSTEM

The Rh blood groups (discovered by Landsteiner and Wiener in 1940) rank with the ABO groups in clinical importance because of their relation to hemolytic disease of the newborn (HDN) and their importance in transfusion. An Rh-negative woman may form anti-Rh when pregnant with an Rh-positive fetus, and Rh-negative persons may form anti-Rh in response to transfusion with Rh-positive blood. Unlike anti-A and anti-B, anti-Rh antibodies do not occur naturally.

The Rh antigen was so named because Rhesus monkeys were used in the experiments through which it was discovered. Anti-Rh will react with the red blood cells of about 84 per cent of white North Americans. The 84 per cent who have the Rh antigen are Rh positive, and the 16 per cent who lack it are Rh negative.

Genetically, the Rh groups are complex, but as an introduction the Rh system can be described simply as follows: A pair of alleles *D* and *d* produce three genotypes, *D/D, D/d* and *d/d*. *D/D* and *D/d* are phenotypically Rh positive, and *d/d* is Rh negative. Anti-D is a synonym for anti-Rh. Anti-d has never been detected. We now know that the human anti-D is not identical to the anti-rhesus antibody formed by guinea pigs against Rhesus monkey red cells; that particular antibody has been named anti-LW after Landsteiner and Wiener.

To discuss the genetics of the Rh system with any degree of sophistication, it is necessary to become familiar with not just one, but two complicated sets of symbols.

According to the Fisher-Race hypothesis, the Rh blood groups are determined by a series of three closely linked genes, *C, D* and *E*, with allelic forms *c, d* and *e*. There are eight different possible chromosomal combinations of these six genes. Their frequencies are very different in different parts of the world; for example, a high frequency of Rh-negative individuals (about 30 per cent) is found in the Basques and a Berber tribe, and close to 100 per cent of Orientals and North American Indians are Rh positive. The frequencies of eight Rh gene complexes in an English population are:

CDe	0.41	*cdE*	0.01
cde	0.39	*Cde*	0.01
cDE	0.14	*CDE*	low
cDe	0.03	*CdE*	low

Sometimes these genes are listed in the order DCE, since there is some suggestion that this is the order of the three genes on the chromosome.

Each person has two Rh gene complexes, so the total number of possible combinations of the eight gene complexes listed above is 36. Rh-negative persons are homozygous *d*/*d* (irrespective of the presence of *C* or *E*) but, because of the comparative rarity of *Cde*, *cdE* and *CdE* as compared with *cde*, the great majority of Rh-negative persons are *cde*/*cde*. Additional alleles (C^{w}, D^{u} and so forth) are known, so the total number of Rh genotypes is well over 100.

Wiener has not accepted the three-allele theory of Fisher and Race, but prefers to think of a single locus with a large series of alleles. His notation for these alleles is given below for each of the gene complexes commonly used:

Fisher and Race Gene Notation	Wiener Gene Notation
CDe	R^1
cDe	R^0
cDE	R^2
CDE	R^z
Cde	r'
cde	r
cdE	r''
CdE	r^y

The most frequent Rh genotypes in Caucasians are:

CDe/cde	R^1/r	0.33
CDe/CDe	R^1/R^1	0.17
cde/cde	r/r	0.15

The genetic interpretation of the Rh locus is still clouded, but by analogy with microorganisms, it seems probable that Rh is a complex locus within which there are many different mutational sites.

Rh_{null}

The Rh_{null} phenotype is a very rare one, first found in an Australian aborigine, in which there are no Rh antigens. Rh_{null} people also lack LW, the "true" Rhesus antigen mentioned above. They are thought to be homozygous for a rare gene X^0r, whose allele $X'r$ is responsible for synthesis of a precursor of both Rh and LW antigens.

Rh_{null} is comparable to Bombay and "null" phenotypes in other blood group systems. Probably the explanation for all these is failure to form a product which is a substrate for formation of the blood group antigens of that system. The absence of Rh antigens as in Rh_{null} is sometimes associated with a hemolytic anemia known as Rh_{null} disease.

The –D–/–D– Genotype

A very rare Rh blood type is known in which the C and the E series of antigens appear to be completely missing; only D is present. This blood type has been given the symbol –D–/–D– and was originally interpreted as a deletion of the part of the chromosome which carries the C and E loci. (If the order of the genes on the chromosome were *DCE*, it would be quite understandable that a deletion might involve *C* and *E* but not *D*.) A pedigree of a large Canadian Métis family in which this rare genotype occurs is given in Figure 9–3. Individuals with this rare Rh combination have a high frequency of parental consanguinity, and a surprisingly high proportion of homozygous sibs (significantly more than the 25 per cent expected). The cause of –D–/–D– is more likely to be a separate suppressor gene like X^0r than a chromosomal deletion.

Hemolytic Disease of the Newborn

Hemolytic disease of the newborn (HDN) may be defined as a condition in which the life span of the infant's cells is shortened because of

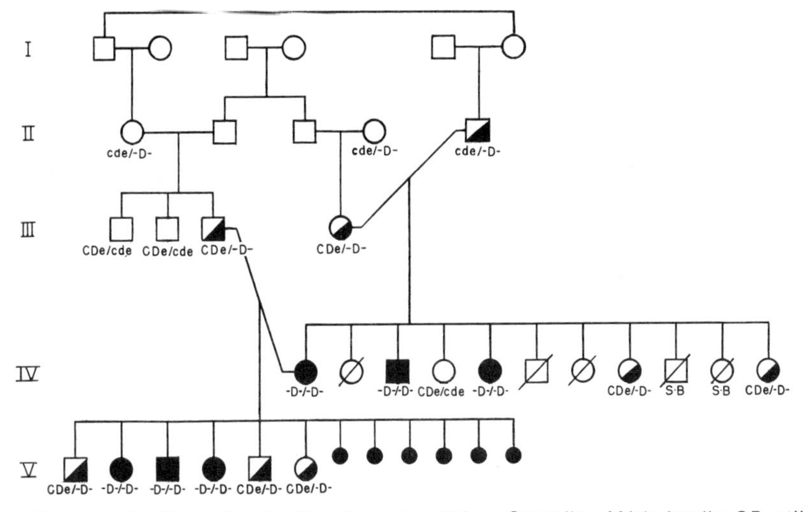

Figure 9–3 *The —D—/—D— "genotype" in a Canadian Métis family. SB, stillborn. (The pedigree of Buchanan and McIntyre 1955.* Brit. J. Haemat. *1:304–307, with corrections provided by Buchanan. Personal communication, 1965.)*

the action of specific antibodies derived from the mother by placental transfer. The disease thus begins *in utero* and continues for only as long as some maternal antibody remains in the infant; in practice, an increased rate of red cell destruction can seldom be detected for more than about six weeks after birth, although maternal antibody may be detected in the infant for as long as three months. Thus, HDN is an *acquired* hemolytic anemia, and must be clearly distinguished from *inherited* anemia such as that caused by G6PD deficiency or hereditary spherocytosis.

Two main types of HDN may be recognized: one due to anti-D (anti-Rh) and the other due to anti-A. Most of the cases of HDN recognized clinically are due to anti-D; the cases due to anti-A are often difficult to diagnose but tend to be so mild as to require no treatment. HDN is only rarely caused by other blood group antibodies such as anti-K and anti-Fy^a. Anti-Le^a and anti-Le^b are often present in the serum of pregnant women, but these antibodies are almost invariably IgM globulins and do not pass the placenta. In a Caucasian population, HDN occurs approximately once in 100 births.

Mechanism of Hemolytic Disease of the Newborn

Normally the maternal and fetal blood streams are completely separated by the placental barrier, but when breaks occur in this thin membrane small amounts of fetal blood reach the maternal blood stream. In an Rh-negative mother, Rh-positive fetal cells may stimulate the formation of anti-Rh, which is then transferred to the fetal circulation (Fig. 9–4).

In the fetus, red cells which are heavily "coated" with anti-Rh are rapidly removed from the circulation. The fetus becomes anemic and responds by releasing large numbers of erythroblasts (nucleated immature red cells) into its blood. The presence of erythroblasts in the fetal blood accounts for the name sometimes used to describe this disease: *erythroblastosis fetalis*. Hydrops (edema due to heart failure secondary to severe anemia) may cause intrauterine death.

After birth, the rapid destruction of cells produces a large amount of bilirubin, which causes jaundice to appear during the first 24 hours of life. If hyperbilirubinemia is not promptly prevented by "replacement transfusion" (i.e., replacing the infant's cells with Rh-negative cells), deposition of unconjugated bilirubin in the brain (kernicterus) may produce cerebral damage, which few infants survive. Those who do survive kernicterus may have high-frequency deafness, mental retardation or the athetoid type of cerebral palsy.

A detailed account of the diagnosis and treatment of HDN may be found in Mollison's *Blood Transfusion in Clinical Medicine* (5th edition, 1972). The most important aid in the rapid diagnosis of HDN was provided by Coombs et al. (1946) who showed that the maternal antibody (globulin) on the infant's cells could be detected by a **direct antiglobulin** test immediately after birth.

Factors Influencing the Development of Hemolytic Disease of the Newborn

Hemolytic disease of the newborn develops only when the infant has inherited from the father an antigen which the mother lacks. How-

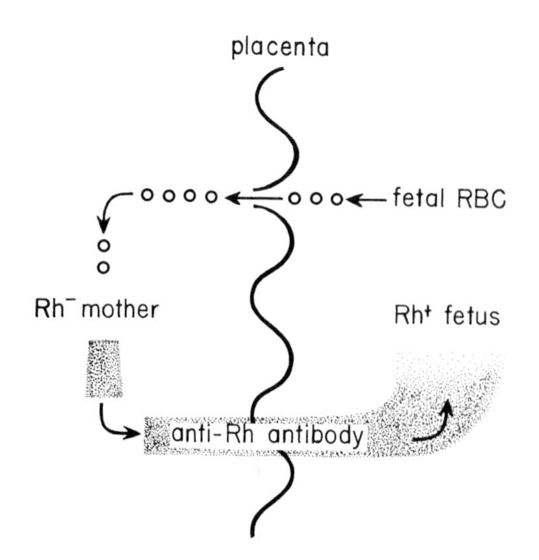

Figure 9–4 *Diagrammatic representation of the immunization of an Rh-negative mother by the cells of her Rh-positive fetus, followed by passage of the maternal antibody to the fetal blood stream, where it causes hemolysis of the fetal cells.*

ever, the number of genetic opportunities for HDN is far in excess of the number of infants who actually develop HDN, since the probability that a given infant will develop HDN is also determined by the following:

1. The mother's previous exposure to the antigen by transfusion or earlier incompatible pregnancy.
2. The effectiveness of the placenta as a barrier to fetal red cells.
3. The potency of the antigen.
4. The maternal capacity to form antibody.
5. The mother's ABO group.
6. The immunoglobulin class of the maternal antibodies (since IgG globulin is transferred to the fetus and IgM globulin is not).
7. The titer of the maternal antibodies.
8. The effect of maternal antibody on the fetal red cells.

Genetic Aspects. The probability that a fetus will develop HDN because of anti-Rh depends in part upon the frequency of Rh-negative persons in the population. Rh hemolytic disease, for example, should occur much more frequently in Berbers of the Atlas Mountains, among whom about 30 per cent are Rh negative, than in Orientals, who are almost entirely Rh positive.

PROBABILITY OF AN RH-POSITIVE FETUS. In a population where the mating is random, the probability that any fetus carried by an Rh-negative woman is Rh positive must be the same as the frequency of the *D* gene. Thus, one can readily calculate the genetic risk of HDN due to anti-Rh, as follows:

Frequency of Rh-negative genotype (d/d) = 0.16 (approximately)

Frequency of $d = \sqrt{0.16}$ = 0.40 (p)

Frequency of $D = 1.00 - 0.40$ = 0.60 (q)

Frequency of Rh-positive genotype (D/D or D/d) = 0.84 ($q^2 + 2pq$)

Risk that *any pregnant woman* in the population is Rh negative and carrying an Rh-positive fetus = frequency of $d/d \times$ frequency of $D = 0.16 \times 0.60 = 0.096$ (approximately 10 per cent.)

Risk that *any pregnant* Rh-*negative woman* is carrying an Rh-positive fetus = frequency of $D = 0.60$ (60 per cent).

One can be more specific about the risk if one can determine the father's genotype as well as the mother's. Of course, if both parents are Rh negative, the infant will always be Rh negative, so HDN due to anti-Rh cannot develop. If the father is Rh positive, the probability that he is heterozygous is:

$$\frac{2pq}{q^2 + 2pq} = 0.57\ (57\%).$$

If he is heterozygous, one-half of his children (on the average) will receive his *d* gene and will be Rh negative. Thus if the Rh-positive father's genotype is not precisely known, the probability that any one child will be Rh positive is:

$$\frac{0.36 + \frac{1}{2}(0.48)}{0.84} = 0.71\ (71\%)$$

The mathematics used above depends upon the Hardy-Weinberg law (page 277).

PROBABILITY OF A NON-O FETUS. The probability that a group O pregnant woman carries a non-group O child can be calculated in the same way, using the frequency figures for the ABO blood groups and genes given on page 211.

The probability that a pregnant group O mother is carrying a non-group O child is the same as the sum of the frequencies of the non-*O* alleles (i.e., the *A* and *B* alleles) and is thus 0.32 (32 per cent).

The probability that any pregnancy is group O mother/non-group O child is the frequency of group $O \times$ the frequency of non-*O* alleles = $0.46 \times 0.32 = 0.15$ (15 per cent).

In summary, the chance that an *O/O* woman carries a non-O fetus = 32 per cent and the chance that a *d/d* woman carries a D-positive fetus = 60 per cent. Nevertheless, HDN due to anti-Rh is a far more frequent clinical problem than is HDN due to anti-A (because the naturally occurring anti-A in group O mothers is not transferred to the fetus; see later).

Other Factors Affecting the Risk of HDN

1. *Relative Potency of the Antigen.* In Caucasians the frequency of the *c* and *D* genes of the Rh complex is almost identical, yet anti-c is rarely formed by *C/C* women during pregnancy. The antigens Fy^b, Do^b and Xg^a are even less efficient provokers of antibodies.

2. *The Maternal Capacity to Form Antibody.* There is great individual variation in the ability to form antibody, and possibly this variation is itself in part genetically determined. For example, an Rh-negative woman may (very rarely) have a child affected with HDN owing to anti-Rh in a first pregnancy, even though she has not been previously transfused. At the other end of the spectrum, an Rh-negative woman who has not developed anti-Rh in a second or third "incompatible" pregnancy is relatively unlikely to develop anti-Rh in subsequent "incompatible" pregnancies.

3. *The Immunoglobulin Class of the Maternal Antibodies.* Since the plasma of group O women always contains anti-A and anti-B, one might expect ABO hemolytic disease to be very common. However, "naturally occurring" anti-A is predominantly Ig M globulin (which is not transferred to the fetus). By contrast, anti-Rh is always predominantly

of the Ig G class and, if present in the mother, can always be found in her fetus.

4. *Protective Effect of ABO Incompatibility Against Rh Hemolytic Disease.* Rh-negative women who are "incompatibly" mated with respect to both ABO and Rh (i.e., who are group O, Rh negative and have non-group O, Rh-positive husbands) are much less likely to have infants affected with HDN than are women with group O husbands. One explanation offered for this important observation is that fetal red cells which carry antigen A (or B) are destroyed by maternal anti-A (or anti-B) before they can provoke the formation of anti-Rh. This protection is afforded only to nonimmunized women, i.e., women who have not already formed anti-Rh.

PREDICTION OF SEVERITY IN HDN DUE TO ANTI-RH. Fortunately, one-third of those infants born alive with HDN due to anti-Rh are only mildly affected (as judged by the hemoglobin concentration in their blood at birth). Subsequent Rh-positive sibs of such infants also tend to be only mildly affected. However, after the birth of a severely affected infant (anemic and requiring two or more replacement transfusions), the next Rh-positive sib has a high risk of intrauterine death.

The outlook for later pregnancies in families in which an infant has been affected with Rh hemolytic disease has been calculated by Walker et al. (Table 9–5). For example, in England in 1957, when a woman had had one stillbirth due to anti-Rh, there was a 70 per cent chance that her next Rh-positive child would also be stillborn. In order to decrease the number of stillbirths due to anti-Rh, a technique of treating the anemic fetus in utero was devised by Liley in 1962: early in the third trimester, Rh-negative red cells are injected into the peritoneal cavity of the fetus.

TABLE 9–5 RISK OF STILLBIRTH OF Rh-POSITIVE FETUS OF IMMUNIZED Rh-NEGATIVE WOMAN*

History of Rh-Immunized Woman	*Risk of Stillbirth in Present Pregnancy (%)*
No previously affected infant	7
Previous mildly affected infant	2
Previous moderately or severely affected infant	20
Previous very severely affected infant	55
One previous stillbirth	70
More than one previous stillbirth	80

*Data from Walker et al., 1957.

PREVENTION OF RH IMMUNIZATION. Clinical trials carried out in several countries over the past 10 years (i.e., since 1963) have established that primary immunization to the Rh antigen on fetal cells can be prevented by the injection of a small dose (300 micrograms) of potent anti-Rh given to the Rh-negative mother of a newborn Rh-positive infant. If the maternal circulation contains more than 10 ml. of fetal cells, the dose of Rh-immune globulin (RhIG) must be increased.

Abortion has been incriminated as a means of Rh-sensitization, and it has been suggested that all nonimmunized Rh-negative women who abort should receive RhIG as a prophylactic. There is general agreement that the suppressive effect of RhIG is dependent on the binding of antibody molecules to the antigen determinants on the red cells *in vivo,* but the mechanism of this suppression has yet to be elucidated.

OTHER BLOOD GROUP SYSTEMS

The P Blood Group System

The P blood group system was discovered by Landsteiner and Levine in 1927, in the same series of experiments that led to the discovery of the MN system. There are four recognizable phenotypes: P_1, P_2, p and P^k.

Genotype	Phenotype	Approximate Frequency (Caucasian)
P^1/P^1	P_1	0.29
P^1/P^2		0.50
P^1/p		Very rare
P^2/P^2	P_2	0.21
P^2/p		Very rare
p/p	p	Very rare; formerly known as Tj (a–)

The rare phenotype P^k, found in only 12 families, nine of Finnish extraction, may prove to be comparable to the Bombay phenotype of the ABO system. P^k people appear to have normal P_1 and P_2 genes, but lack an independent, very frequent, dominant gene.

The Lutheran Blood Group System

The Lutheran blood groups were discovered by Callender et al. in 1946. The phenotypes Lu (a+b–), Lu (a+b+) and Lu (a–b+) depend upon a pair of codominant alleles, Lu^a and Lu^b. Lu^a is much less common (0.04) than Lu^b (0.96). About 8 per cent of Caucasians are Lu (a+b+). Lu (a–b–) persons have been described, some of whom are homozygous for a "silent" *Lu* gene, and some of whom carry a dominant suppressor gene which affects the expression of all genes at the Lu locus.

Genotype	**Phenotype**	**Approximate Frequency** (*Caucasian*)
Lu^a/Lu^a	Lu (a+b–)	0.001
Lu^a/Lu^b	Lu (a+b+)	0.08
Lu^b/Lu^b	Lu (a–b+)	0.92

The Lutheran blood groups are historically interesting because they provided the first example of autosomal linkage and autosomal crossing over in man, the Lutheran-secretor linkage. Figure 9–5 illustrates the Lutheran and secretor genotypes in a family described by Sanger and Race. The father is homozygous for Lu^b and *se*, and the mother is heterozygous at both loci; this is the classical "back cross," in which the children's phenotypes reveal the genotype of the heterozygous parent. Only the maternal contribution to the children's genotypes is shown, since the father can transmit only Lu^b and *se*. In this family, the gene Lu^a and *se* are obviously on one maternal chromosome, and Lu^b and *Se* are on the other. This particular pedigree does not show crossing over between the two loci, but other family studies show a crossover frequency of about 15 per cent.

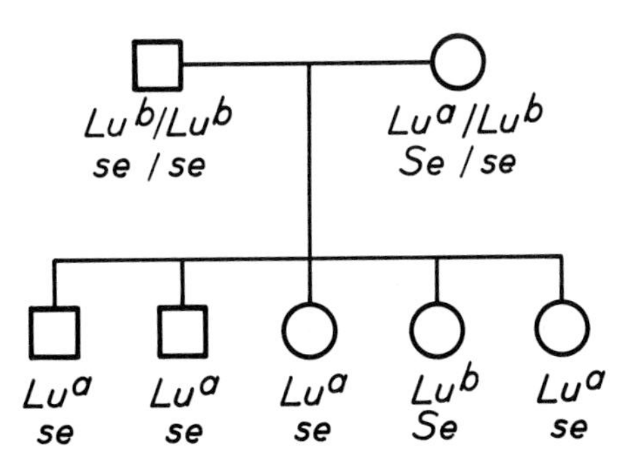

Figure 9–5 *A pedigree demonstrating Lutheran-secretor linkage, described in the text. (From Sanger and Race 1958.* Heredity *12:513–520.)*

The locus for myotonic dystrophy is also linked to the Lutheran and secretor loci. This linkage relationship can be exploited in suitable families for prenatal diagnosis of myotonic dystrophy, because it is possible to determine the secretor phenotype of the fetus from amniotic fluid. If the fluid contains H substance, the fetus is a secretor.

The Kell Blood Group System

The Kell blood group system was originally described in terms of two alternative phenotypes, Kell positive (K+) and Kell negative (K–), governed by a pair of alleles *K* and *k* (Coombs et al., 1946). Kell-positive people (*K*/*K* or *K*/*k*) make up about 9 per cent of the Caucasian population, but only about 1 to 2 per cent of the Negro population.

Like other blood group systems, the Kell system has become more complicated as new antibodies have been found to detect antigens other than K and k. Most of the complications are beyond the scope of this discussion, but one bears mention: it has been found that the two "Sutter" antigens, Js^a and Js^b, described by Giblett in 1958, are part of the Kell system. The antigen Js^a is noteworthy because, like the phenotype Fy(a–b–) and the Rh complex cDe, it is found commonly in Negroes but rarely in Caucasians.

The Kell system is sometimes implicated in hemolytic disease of the newborn. A K-negative woman whose child is K-positive may form anti-K, which causes severe hemolytic disease of the newborn. If there is also incompatibility on the ABO system (i.e., if the mother is O,K– and the fetus A,K+), the mother is less likely to form anti-K (see discussion of the protective effect of ABO incompatibility, page 221). In some boys with chronic granulomatous disease (an X-linked recessive), the red cells have "weakened" antigens as judged by their reactions with antibodies of the Kell system (Giblett et al., 1971).

The Lewis Blood Group System

The Lewis blood group system was first described by Mourant in 1946. Two antibodies, anti-Le^a and anti-Le^b, are known; they define two antigens, Le^a and Le^b. Almost everyone's red cells are either Le(a+) or Le(b+), so the common phenotypes are Le(a+b–) or Le(a–b+). Le(a–b–) is relatively rare in Caucasians but quite common in West Africans. The phenotype Le(a+b+) is not observed in adults.

Unlike other blood group systems, the Lewis groups did not at first lend themselves to a simple genetic explanation because of their complicated interactions with the ABO and secretor systems.

Secretors (persons of genotype *Se*/*Se* or *Se*/*se*) are those individuals who secrete ABH substances in their saliva. Persons whose red cells are Le(a+b–) *never* secrete ABH substances in the saliva, and persons whose red cells are Le(a–b+) *always* do so. Persons who have Le(a–b–) red cells are usually secretors of ABH substances.

It appears that the Lewis substances are not primarily red cell antigens but antigens of the body fluids, such as plasma and saliva, and that they are only secondarily adsorbed onto the red cells from the plasma. The presence of Le^a and Le^b antigens in the serum and saliva is governed by a pair of allelic genes, *Le* and *le*. The interactions of the Lewis and secretor genes produce the following *red cell* phenotypes:

Lewis Genotype	Secretor Genotype	Red Cell Phenotype	Approximate Frequency (Caucasian)
Le/Le or *Le/le*	*Se/Se* or *Se/se*	Le(a–b+)	0.69
Le/Le or *Le/le*	*se/se*	Le(a+b–)	0.26
le/le	*Se/Se, Se/se* or *se/se*	Le(a–b–)	0.05

The Duffy Blood Group System

The Duffy blood group system was first described by Cutbush et al. in 1950. There are two known allelic genes at the Duffy locus, Fy^a and Fy^b, and a third allele, *Fy*, has been postulated on the basis of family studies.

Genotype	Phenotype	Approximate Frequency (Caucasian)
Fy^a/Fy^a Fy^a/Fy	Fy(a+b–)	0.20
Fy^a/Fy^b	Fy(a+b+)	0.46
Fy^b/Fy^b Fy^b/Fy	Fy(a–b+)	0.33
Fy/Fy	Fy(a–b–)	<0.01

The Fy(a–b–) phenotype, or in other words the genotype *Fy/Fy*, is very common in West Africans (over 80 per cent *Fy/Fy*) and in American Negroes (nearly 40 per cent *Fy/Fy*), but very rare in Caucasians (about one in 1000 *Fy/Fy*). *Fy*, like the Kell gene Js^a, makes a sharp distinction between Negroes and Caucasians. The *gene* frequency of *Fy* is about 0.03 in Caucasians, and this means that a relatively large number of people are heterozygous Fy^a/Fy or Fy^b/Fy. One-eighth of all Fy(a+b−) and one-sixth of all Fy(a−b+) people carry a silent *Fy* gene. This can lead to errors in family testing. For example:

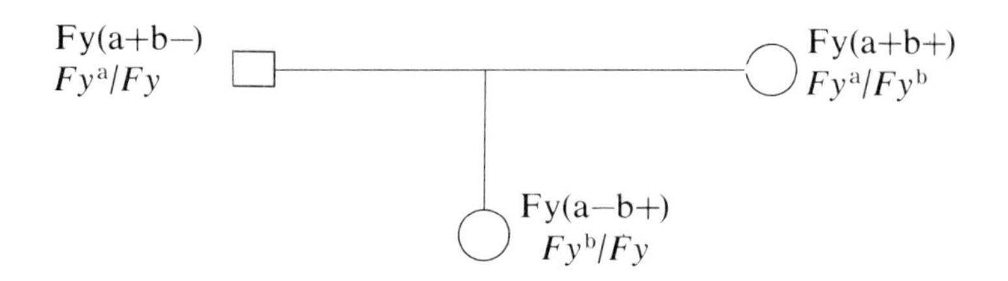

In this pedigree, failure to recognize that the father is Fy^a/Fy rather than Fy^a/Fy^a, and that the child is Fy^b/Fy rather than Fy^b/Fy^b, could lead to the false interpretation that the man is not the biological father of the child.

The Duffy locus has the distinction of being the first locus to be assigned to a particular autosome, chromosome no. 1. This assignment resulted from linkage studies of a family in which one of the pair of no. 1 chromosomes had an unusually long secondary constriction, which looked like a partly uncoiled segment (Fig. 9–6). Once Duffy had been assigned, markers were rapidly added to chromosome 1, and by 1972 it had 10 markers, more than twice as many as any other chromosome.

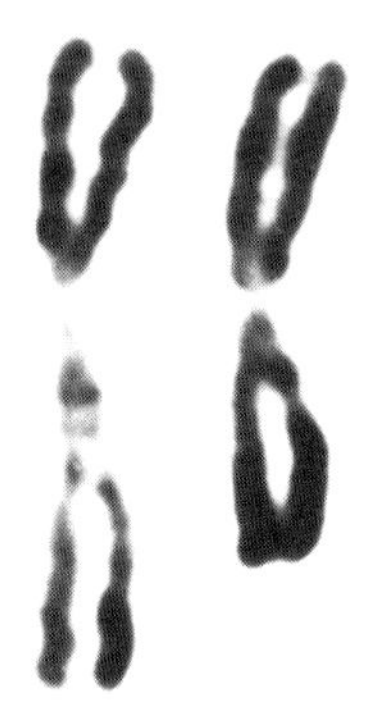

Figure 9–6 *A pair of no. 1 chromosomes from a metaphase cell of an individual heterozygous for a variant in which there is a long secondary constriction. Analysis of the Duffy blood groups in a family in which this variant occurred showed that the Duffy locus is closely linked to the uncoiled segment. (From Donahue et al. 1968.* Proc. Nat. Acad. Sci. *61:949–955, 1968.)*

The Kidd Blood Group System

The Kidd blood group system was described in 1951 by Allen et al. There are three known allelic genes, Jk^a and Jk^b, almost equal in frequency, and a rare, so far "silent" allele, *Jk*. The symbol $\mathrm{Jk^a}$ is used for the antigen detected by anti-$\mathrm{Jk^a}$, and $\mathrm{Jk^b}$ is used for the antigen detected by anti-Jk^b. A small number of blood samples do not react with either anti-Jk^a or anti-Jk^b, and they are described as Jk(a–b–).

There are great differences in the racial distribution of the Jk^a antigen. Over 90 per cent of American Blacks, 75 per cent of Caucasians and 50 per cent of the Chinese are Jk(a+). Anti-Jk^a has been known to cause hemolytic disease of the newborn and transfusion reactions.

Genotype	Phenotype	Approximate Frequency (Caucasian)
Jk^a/Jk^a Jk^a/Jk	Jk(a+b–)	0.25
Jk^a/Jk^b	Jk(a+b+)	0.50
Jk^b/Jk^b Jk^b/Jk	Jk(a–b+)	0.25
Jk/Jk	Jk(a–b–)	Very rare

The Diego Blood Group System

The Diego groups were discovered in 1955 by Layrisse et al. The symbol Di^a is used for the antigen, and the phenotypes are written as Di(a+) and Di(a–). The major interest of the Diego system is that the Di^a antigen appears to be a specific marker for Mongolians. Its frequency varies from about 36 per cent in some South American Indian tribes, to 10 per cent in Japanese, to practically zero in most Caucasians. Surprisingly, Canadian Eskimos, whose ethnic origin is clearly Mongolian, are entirely Di(a–); this observation is of great anthropological interest. The antigen Di^b has recently been detected.

Genotype	Phenotype	Approximate Frequency	
		Mongolian	Caucasian
Di^a/Di^a Di^a/Di	Di (a+)	0.10	0
Di/Di	Di (a–)	0.90	1.00

The Xg Blood Group System

The Xg blood group system, discovered by Mann et al. in 1962, is of special genetic importance because its locus is on the X chromosome.

The antibody which first defined it, anti-Xg^a, was found in a man who had been transfused many times because of a familial vascular disorder. Only a small number of other examples of anti-Xg^a have been found so far. The genotypes and phenotypes of the Xg system are shown below (and also on page 71, where they are discussed as an example of X-linked inheritance):

Genotype		Phenotype	Approximate Frequency (Caucasian)
Males	Xg^a/Y	Xg(a+)	0.67
	Xg/Y	Xg(a−)	0.33
Females	Xg^a/Xg^a Xg^a/Xg	Xg(a+)	0.89
	Xg/Xg	Xg(a−)	0.11

Because of its importance as an X-linked marker, the Xg system is often used as an example throughout this book of X-linked dominant inheritance (page 72) and in connection with X-chromosome aneuploidy (page 173).

The I Blood Group System

The I system, which was first described by Wiener et al. in 1956, appears to be quite different from other blood group systems in the development of the antigens. Practically all adults are I positive; in fact, the antigen I was at one time thought to be a "public" antigen possessed by all members of the human species. In the great majority of individuals, i is strong at birth and decreases in strength thereafter, whereas I is poorly developed at birth and increases as i decreases. A few rare adults (five of 27,000 in New York) are I negative, and this trait is inherited.

In many blood diseases, notably thalassemia major, the red cells are strongly agglutinated by anti-i, even though they continue to react strongly with anti-I. It has been suggested that the strong reaction with anti-i might be related to the erythropoietic stress placed on early erythroid precursors in hematological disorders (Giblett and Crookston, 1964). The I and i antigens are not useful genetic markers because, unlike other blood group antigens, they are influenced by age and environment.

The Yt (Cartwright) Blood Group System

The Yt system was discovered by Eaton et al. in 1956, when a previously unknown antibody, now known as anti-Yt^a, caused cross-matching difficulty. A second antibody, anti-Yt^{b}, was later identified. The genotypes, phenotypes and frequencies are as follows.

Genotype	Phenotype	Approximate Frequency (Caucasian)
Yt^{a}/Yt^{a}	Yt(a+b−)	0.91
Yt^{a}/Yt^{b}	Yt(a+b+)	0.07
Yt^{b}/Yt^{b}	Yt(a−b+)	0.02

Because 98 per cent of the population are Yt(a+), Yt is not very useful in family and linkage studies. About 98 per cent of the first 1000 persons tested had the antigen Yt^{a} and only 2 per cent lacked it.

The Dombrock Blood Group System

The Dombrock blood group system was first described by Swanson et al. in 1965. Anti-Do^{a} was detected in the serum of a transfused patient, Mrs. Dombrock. Approximately 64 per cent of Northern Europeans are Do(a+). The antithetical antibody, anti-Do^{b}, has now been discovered. Swanson et al. state, "The reactions of anti-Do^{a} alone place the Dombrock system sixth in the order of potential usefulness of blood group systems as markers of autosomes of white people."

BLOOD GROUP CHIMERAS

Monozygotic twins frequently have a common prenatal circulation, but dizygotic twins, even when they have fused placentas, do not usually develop anastomoses between their circulations. It is therefore very rare indeed in man (though quite usual in cattle) for any mixture of blood to occur between dizygotic twin fetuses. Only about 12 cases have been recorded in which one or both members of a dizygotic twin pair have a mixture of two kinds of blood. In the first such case to be described, a woman was observed to have a mixture of A and O cells. It was later found that she was a twin; her twin brother had died 25 years earlier. A person of this kind, with blood cells of two different genotypes derived from two different zygotes, is called a **blood group chimera.** "Whole-body" chimeras, probably usually formed by fusion of DZ twin zygotes or early embryos, are also known.)

In blood group chimeras, the exchange of hematopoietic cells takes place in fetal life, prior to the development of immune competence. The foreign cells do not stimulate the host to form antibodies against them, so they are accepted in the host's bone marrow, where they can persist and multiply throughout life. Since only hematopoietic cells and not germ cells have been exchanged, the host can, of course, transmit to his offspring only his own blood group genes (i.e., his "true" blood group).

GENETIC MARKERS IN SERUM

The technique of starch gel electrophoresis, introduced by Oliver Smithies in 1955, stimulated an interest in genetic markers of serum that shows no sign of waning. It has led to the discovery in blood of many polymorphisms of great interest to geneticists and biochemists.

The fluid portion of the whole blood is the **plasma.** When blood coagulates, the cells and other formed elements, together with fibrinogen, form the clot. The remaining fluid is the serum. Thus, in the present context, the terms plasma proteins and serum proteins are interchangeable.

A standard biochemical technique for the study of serum proteins is electrophoresis, a process by which the migration of ions in an electric field may be observed. In paper electrophoresis, a strip of filter paper is moistened with buffer and a drop of the solution containing a mixture of proteins to be tested is placed at one point. When an electric field is applied along the paper, the proteins migrate at characteristic rates which are determined by their net electrical charge, size and shape. After several hours, the positions of the various components can be determined by special staining methods.

In 1955 Smithies developed a technique in which the filter paper was replaced by a starch gel, which gave sharper zones and better resolution than did other methods. With this technique he and his colleagues soon discovered polymorphism in two different serum proteins, the haptoglobins and the transferrins.

Starch gel electrophoresis of a normal serum sample reveals a pattern of bands in which the different serum proteins assume characteristic positions as shown in Figure 9–7. When the usual appearance of the serum protein fractions has been established, variants can be detected. They are usually first noticed by differences in their speed of migration, which result from substitution in their polypeptides of one or more amino acids that differ in charge from the "usual type." Occasionally a serum is found in which one factor is completely absent.

Giblett's invaluable and authoritative textbook *Genetic Markers in Human Blood* is the source of much of the information in this section (as well as parts of the previous section).

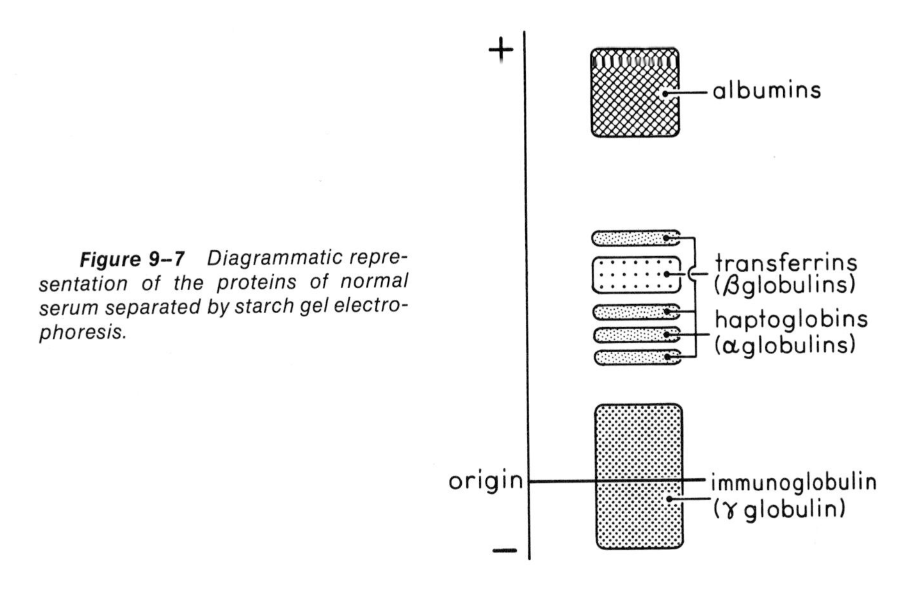

Figure 9–7 *Diagrammatic representation of the proteins of normal serum separated by starch gel electrophoresis.*

HAPTOGLOBINS

Haptoglobins are α globulins with the property of binding hemoglobin. There are three main types, Hp 1-1, Hp 2-1 and Hp 2-2, determined by two allelic genes, Hp^1 and Hp^2.

Genotype	***Phenotype***
Hp^1/Hp^1	Hp 1-1
Hp^1/Hp^2	Hp 2-1
Hp^2/Hp^2	Hp 2-2

It is known that there are two closely similar alleles for Hp^1: Hp^{1F} (fast) and Hp^{1S} (slow). Since Hp^{1F}, Hp^{1S} and Hp^2 are allelic, several phenotypes are possible.

The three common types of haptoglobin form distinctive patterns when subjected to starch gel electrophoresis. Hp 1-1 forms a single zone, and Hp 2-2 forms a series of zones such as one might expect to find with a mixture of polymers of a basic unit similar to Hp 1-1. Hp 2-1 has a polymeric series different from Hp 2-2, and unlike a mixture of Hp 1-1 and Hp 2-2 (Fig. 9–8).

The polypeptides corresponding to Hp^{1F} and Hp^{1S} differ in only one

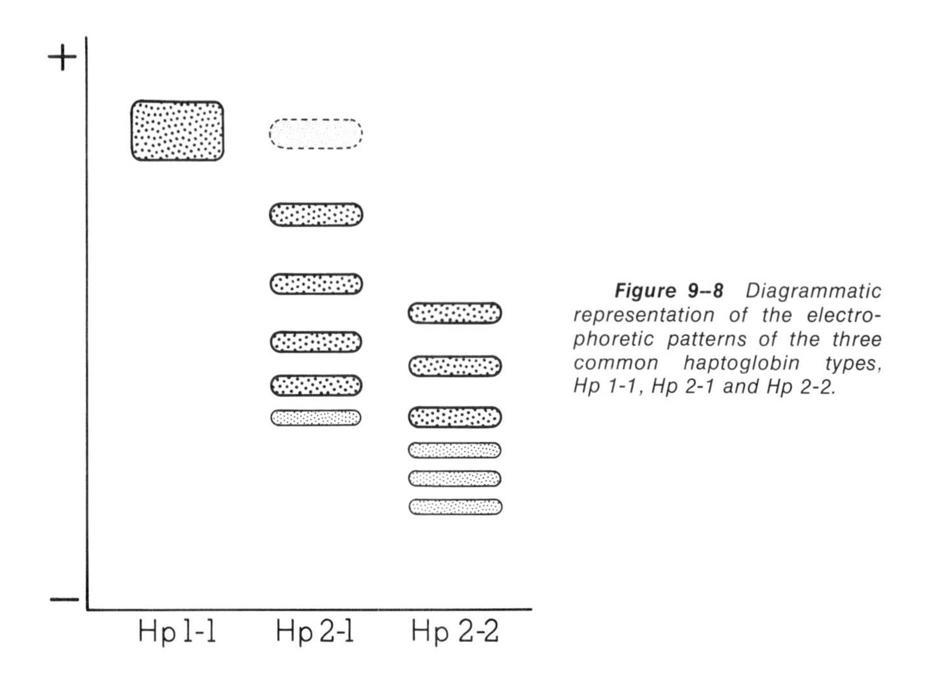

Figure 9–8 Diagrammatic representation of the electrophoretic patterns of the three common haptoglobin types, Hp 1-1, Hp 2-1 and Hp 2-2.

amino acid. Thus, a point mutation is responsible for the difference. Hp^2 is responsible for a different polypeptide chain which resembles a hypothetical fusion of parts of two Hp^1 chains. Smithies, Connell and Dixon (1962) have suggested that this similarity is the result of nonhomologous crossing over between a pair of chromosomes, so that part of one Hp^{1F} gene has been added to part of an Hp^{1S} gene. The result is duplication of some of the DNA and formation of a "new" polypeptide. The principle of nonhomologous crossing over with resulting duplication of portions of genes could be of major significance in evolution, since it could permit formation of new genes without actual chemical change, and could allow parts of two different genes to combine so that the resulting single polypeptide may have properties formerly divided between two polypeptides (Fig. 9–9). This unusual crossover event seems to have happened only once, probably early in human evolution. The frequency of the Hp^1 gene is about 0.35 to 0.40 in Europe, somewhat higher (about 0.55) in African and American Blacks and somewhat lower in Orientals (e.g., about 0.24 in Japan). Biochemically, it is curious that the new, longer Hp^2 polypeptide is stable and active. From the standpoint of evolution it is surprising that the new allele has spread until in many populations it has a frequency higher than that of the old Hp^1 allele. It is useless to speculate on what the selective advantage might be when the physiological role of haptoglobin is not very well understood. Total or near-total absence of haptoglobin, quite common in some Negro populations, does not seem to be deleterious.

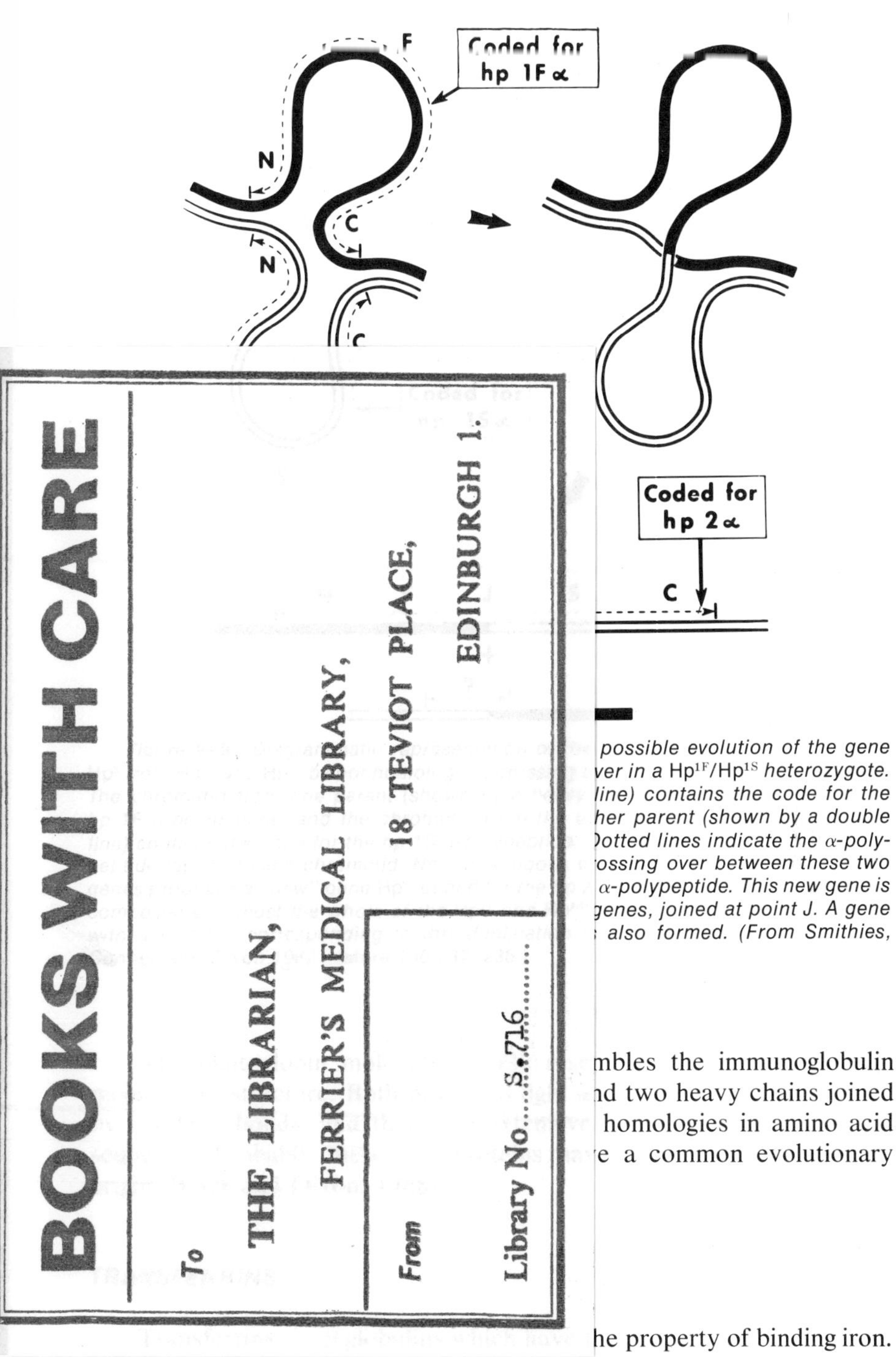

possible evolution of the gene ver in a Hp^{1F}/Hp^{1S} heterozygote. line) contains the code for the her parent (shown by a double Dotted lines indicate the α-poly- ossing over between these two α-polypeptide. This new gene is genes, joined at point J. A gene also formed. (From Smithies,

mbles the immunoglobulin nd two heavy chains joined homologies in amino acid e a common evolutionary

he property of binding iron. Smithies originally reported variant forms of transferrins in 1957, and by now some 20 variants have been identified electrophoretically. One form (C) (genotype Tf^{C}/Tf^{C}) is very common, and all the others are quite rare. It has been shown that the various transferrins do not differ in their ability to bind and transport iron. The different transferrins are probably produced by a series of codominant alleles. Transferrins are detectable by starch gel electrophoresis. Individuals homozygous for *C* have just one electrophoretic band, whereas all heterozygotes (e.g., B_1C) have two bands.

Transferrins show considerable racial variation, but the relative rarity of the abnormal types limits their usefulness in population studies. The genetical interest of the Tf locus was enhanced by the discovery that it is linked to the E_1 serum cholinesterase locus (Robson, 1966).

IMMUNOGLOBULINS

The variants of the Gm and Inv systems are located on polypeptide chains of the immunoglobulins (the serum proteins that are antibodies). The term **allotype** is often used for the different inherited antigens of the Gm and Inv series, in the sense of different products of allelic genes. This term is used more in immunology than in genetics. Before discussing the heritable variants of the immunoglobulins, a brief description of the immunoglobulins themselves is in order.

General Considerations of Immunoglobulins

Many synonyms have been used for gamma globulins. To avoid confusion, the following nomenclature will be used here for the five classes of immunoglobulins (Ig) currently recognized:

Ig Class	*Heavy Chain*	*Light Chain*
IgG	γ	κ or λ
IgM	μ	"
IgA	α	"
IgD	δ	"
IgE	ϵ	"

A gamma globulin molecule is composed of two identical light and two identical heavy chains held together by disulfide bonds (Fig. 9–10). The light (L) chains occur in two types, kappa (κ) and lambda (λ), differing in amino acid sequence. They are alike in all classes of immunoglobulins, but each class has its own characteristic heavy (H) chain. The molecular weight of IgG globulin and of IgA globulin is about 160,000 and the molecular weight of IgM globulin is about 900,000. The difference is that IgM is a pentamer, composed of five basic units. The

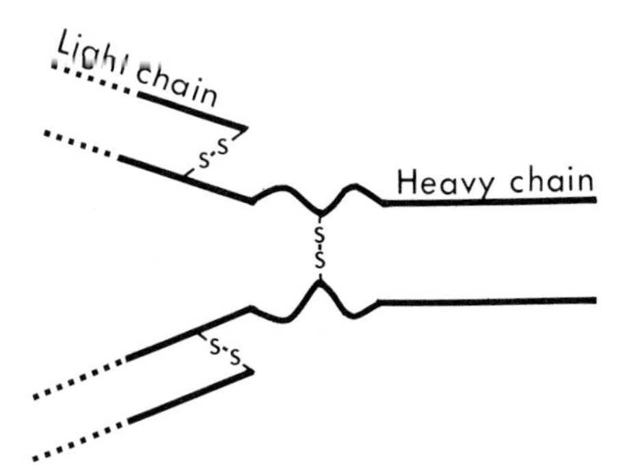

Figure 9–10 *Diagrammatic representation of an IgG molecule. There are two heavy chains and two light chains, each with a variable portion (broken line) and an invariant portion (solid line). The Gm locus is on the invariant portion of the heavy chain, and Inv is on the invariant portion of the κ type of light chain. (Adapted from Cohen, S., and Milstein, C. 1967.* Nature *214:449.)*

Inv antigen is on the kappa type of light chain, so Inv antigens are found in all classes of immunoglobulins. The Gm antigen is on the heavy chain of only Ig G.

According to the current concept, each chain of the gamma globulin molecule has a **fixed portion,** on which the Gm and Inv antigens are found, and a **variable portion,** which forms the antibody combining site. Thus, the Gm and Inv antigens are not involved in the antibody combining site of the molecule.

In addition to the two types of light chain, κ and λ, there are four subclasses of γ chains (designated γ_1, γ_2, γ_3, γ_4), two kinds of μ chains and two of α chains. With δ and ϵ, there are at least 12 structural genes for immunoglobulins. Each chain has an amino acid structure that begins with a variable portion and ends with an invariant segment. The individuality of the variable portion would require numerous genes, if each specific antibody is separately encoded in DNA.

There has been much speculation about the origin of immunoglobulin variability. One possibility is that variability depends upon somatic recombinations, but there are many theories as to the exact mechanism. Adjacent genes that had arisen by repeated tandem duplication might mispair and undergo crossing over, thereby allowing many different amino acid sequences in the variable portion of the chain (Edelman and Gally, 1967); or paired genes ("master" and "scrambler"), the one an inverted partial duplication of the other, might undergo crossing over (Smithies, 1967). Whatever the mechanism, it results in the synthesis of a wide variety of antibodies, each programmed to combine with a specific type of antigen molecule and each capable of being recalled promptly in large quantity if the same antigen is encountered a second time.

The Gm and Inv Allotypes

Several alleles at each of two independent loci, Gm and Inv, determine the Gm and Inv allotypes of immunoglobulins.

The initial discovery of Gm variants was made by Grubb in 1956. Detection of the Gm antigens is complicated. The serum from many patients with rheumatoid arthritis (RA) and a few normal sera will agglutinate globulin-coated red cells (e.g., Rh-sensitized cells). The agglutination reaction can be inhibited by serum from a normal person who has in his gamma globulin the same Gm antigen as that on the coating of the sensitized red cell. The explanation is that the "agglutinator" serum contains anti-Gm (e.g., anti-Gm^1), which reacts specifically with the antigen (e.g., Gm^1) on the sensitized red cell. Serum from a Gm (1+) person will inhibit the agglutination reaction, but other sera will not. Note that the detection of an unknown Gm antigen requires both a known agglutinator serum, and red cells coated with gamma globulin containing a known Gm antigen.

Antibody in Agglutinator	**Antigen on Coated Red Cells**	**Phenotype Deduced by Effect of Test Serum on Agglutination**	
		Inhibition	*No effect*
Anti-Gm^1	Gm^1	Gm(1)	Not Gm(1)
Anti-Gm^2	Gm^2	Gm(2)	Not Gm(2)

By this test, more than 20 different Gm phenotypes have been recognized so far. They are believed to be determined by a series of alleles (Gm^1, Gm^2 and so forth) at a single locus.

The Gm groups are genetically and immunologically complex and by no means fully understood. For example, the $Gm^{1,5}$ allele found in some Negro and Mongolian populations determines both Gm^1 and Gm^5 antigens, but in Caucasians Gm^1 and Gm^5 behave as alleles. Progress in working out Gm genetics is impeded because of the scarcity of reliable "agglutinators."

One feature of this "antigen on an antibody" is unique among autosomal loci: in an individual cell in a heterozygote, only one of the two Gm types is expressed.

The genetic variants at the Inv locus were discovered by Ropartz (1960). As with Gm, the phenotype was determined by tests involving the inhibition of the agglutination of cells coated with the appropriate antigen. There are three major codominant alleles:

Genes	**Frequency**
Inv^1	0.25
$Inv^{1,2}$	0.30
Inv^3	Common (0.70–0.90)

OTHER GENETIC MARKERS IN PLASMA

Beta Lipoprotein

β lipoprotein is the low-density class of serum lipoprotein, separable from the α, pre-β, and chylomicra classes of lipoprotein by ultracentrifugation and electrophoresis. β lipoprotein transports cholesterol in plasma. It has considerable genetical interest; for example, hyperbetalipoproteinemia is a rather common autosomal disorder in which homozygotes develop xanthomata (fatty tumors) and coronary insufficiency in childhood, and may die of cardiac infarction before reaching adult age. There are two systems of genetic variants in β lipoproteins, known as Ag and Lp. The Ag system is genetically complex, but two of its alleles, Ag^x and Ag^y, are useful in population studies because both alleles are common, so that segregation is often observed; however, their frequencies are strikingly different in European and Asian populations (e.g., Ag^x has a frequency of 0.23 in Sweden, and 0.73 in Japan).

The Protease Inhibitor (Pi) System

Alpha-1 antitrypsin is a serum protein of the α_1 fraction which can inhibit the activity of trypsin and other proteolytic enzymes. It is one of several protease inhibitors of serum. Several *Pi* alleles have been identified. The identification is complicated, requiring starch gel electrophoresis in one direction followed by "crossed electrophoresis" in agarose containing anti-Pi antibodies (Fig. 9–11). Here we will mention only the common allele Pi^M and the rare allele Pi^Z.

Pi^M/Pi^M homozygotes have α_1 antitrypsin of the common type; Pi^Z/Pi^Z homozygotes have α_1 antitrypsin deficiency. The particular clinical interest of the system is that α_1 antitrypsin deficiency is found in many patients with lung and liver disease. The extent and basis of this association are now under intensive study.

POLYMORPHISM

The examples discussed in this chapter indicate that many human proteins exist in more than one variant form. When at a given locus two or more alleles occur together in the same population, each with a frequency greater than 0.01, a polymorphism is said to exist. (A gene frequency of 0.01 produces a heterozygote frequency of 0.02, so for any polymorphism at least one person in 50 is a heterozygote).

Only about half of the 40 or so known polymorphisms of blood are mentioned in this chapter, or elsewhere in this book. In particular, red

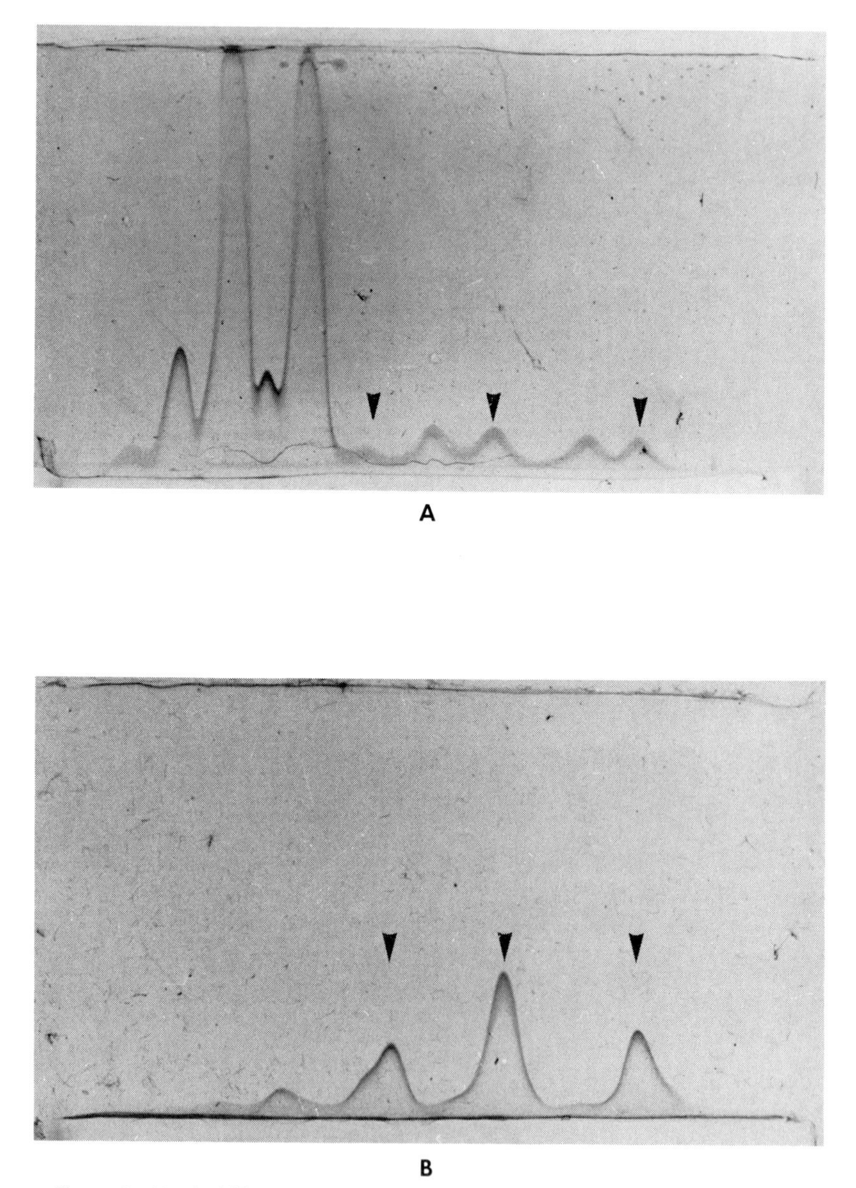

Figure 9–11 A, *MZ pattern on agarose with specific antiserum, at right angles to original electrophoresis in acid starch gel. Similar to MM but has extra Z bands, shown by arrows.*

B, *ZZ pattern on agarose; Z bands showing in MZ pattern indicated by arrows. (Photographs courtesy of D. Wilson Cox.)*

cell enzyme polymorphisms have been omitted here (except for adenylate kinase, noteworthy for its linkage to the ABO blood group locus).

The extent of polymorphism in human populations has been roughly estimated by Harris (1970). More than one-fourth (8 of 30) of the enzyme systems he examined by starch gel electrophoresis were found to be polymorphic. His estimates of average heterozygosity based upon observed incidence of heterozygotes for seven polymorphic enzymes is over 5 per cent. Assuming that only one-third of all such variants can be detected electrophoretically, the average heterozygosity per locus is about 16 per cent. (It is of interest that the average heterozygosity per locus is similar in Europeans and Blacks, though the incidence of heterozygotes at any one locus may differ widely between the two groups.) If at any one locus 16 per cent of the population are heterozygous, the total number of different combinations of protein variants in the populations must be enormously large. Perhaps everyone has a unique set of proteins.

The extent of protein polymorphism has led scientists to speculate about its significance in terms of evolution. It is not surprising that rare variants are found, because these could arise by mutation, would usually be deleterious and would be eliminated by selection. But it is surprising to find, for many proteins, two common variants instead of a single one.

If there are two common alleles at a locus in a population, one may be in the process of replacing the other, or the two may be maintained in equilibrium by heterozygote advantage. In either case, natural selection must be operating.

Kimura (1968) has challenged the classic concept that natural selection must be the essential process in maintaining polymorphisms. He argues that random genetic drift could account for the occurrence of two common alleles at many loci. In other words, there is no need to postulate that one or the other allele is selectively preferred or that the heterozygote is favored. "Neutral" mutations could become fixed by chance.

The controversy as to whether polymorphisms are maintained only by selection has not yet been resolved. Harris's demonstration that the majority of enzyme polymorphisms involve variants with different enzymatic activity scores a point for the selectionists, since a difference in activity might be a reasonable basis for selection.

GENERAL REFERENCES

Giblett, E. R. 1969. *Genetic Markers in Human Blood.* Blackwell Scientific Publications, Oxford.

Harris, H. 1970. *The Principles of Human Biochemical Genetics.* American Elsevier Publishing Co., Inc., New York.

Harris, H. 1971. Protein polymorphism in man. *Canad. J. Genet. Cytol.* 13:381–396.

Morgan, W. T. J., and Watkins, W. M. 1969. Genetic and biochemical aspects of human blood group A-, B-, H-, Le^{a}- and Le^{b}- specificity. *Brit. Med. Bull.* 25:30–34.

Race, R. R., and Sanger, R. 1968. *Blood Groups in Man.* 5th ed. Blackwell Scientific Publications, Oxford.

PROBLEMS

1. An Rh-negative woman has an Rh-negative child. What is her husband's Rh genotype?

2. An AB mother has an AB child. What are the possible genotypes of the father?

3. (a) A woman of blood group B, N has a child of blood group O, MN. She states that a certain man of blood group A, M is the father of the child. Can he be excluded as the father on the basis of these blood groups?
   (b) If the man and woman were both Rh positive and the child Rh negative, could he be excluded?
   (c) If the man were now found to be a secretor whereas the mother and child are nonsecretors, could he be excluded?
   (d) If his blood groups were A, N, Rh negative, what would your conclusion be?

4. Assign each of the following children to the right set of parents:

Children	Parents
O	AB x O
A	A x O
B	A x AB
AB	O x O

5. A woman who in infancy was affected with hemolytic disease of the newborn marries a man who did not have hemolytic disease but had several older sibs who were affected.
   (a) Give the probable genotypes of the couple and their parents.
   (b) If they have a large family, what is the risk that they will have a child with hemolytic disease?

6. (a) What blood groups can occur in the offspring of an AB x O mating?
   (b) Do the genes A and B segregate or assort independently?
   (c) What does this prove about genes A and B?

7. (a) Two parents are A/B, R/r and O/O, r/r respectively. What genotypes are possible in their offspring?
   (b) Do genes A and R segregate or assort?
   (c) What does this prove about genes A and R?

8. Two parents have Hp 2-1 haptoglobin. What haptoglobin types are expected in their children, and in what proportions?

10

Genes in Differentiation and Development

Now that the genetic code has been deciphered and protein synthesis is reasonably well understood, the genetics of development and differentiation is the next major genetic topic. All the cells of a multicellular organism have the same set of genes as the zygote from which they have descended, yet any one cell may differ greatly in phenotype and function from the zygote and from other differentiated cells. While changes are going on in the developing embryo, the business of the whole organism must proceed as usual; there is no closing down for alterations.

Differentiation does not depend upon differences in the genes actually present in the differentiating cells, for the genes can be altered only by mutation or chromosomal rearrangement, both of which are rare random events. Rather, differentiation depends upon variation in the activity of the genes. At any point in time, only a portion of the genes present in the chromosome set are active; at different times, any one gene may be active or inactive. Developmental genetics is concerned largely with the mechanisms which activate certain genes at appropriate times, and inactivate them when their products are not required.

The regulation of the expression of the genes can take place at several different levels: RNA synthesis or degradation, transfer of RNA from nucleus to cytoplasm, translation of RNA into protein or after the protein has been synthesized. In this brief discussion we can only touch upon a few examples of differential gene expression. Related topics of particular interest in medical genetics, including congenital malformations, pleiotropy and neoplasia as examples of abnormal development, will also be discussed.

RNA IN DEVELOPMENT

Most of our knowledge about the early development of the zygote comes from studies of the sea urchin egg, which, unlike the human egg, is readily available in quantity. Though the sea urchin is an echinoderm and not very closely related to man, the general principles of early development are in all probability the same in both organisms.

There is no protein synthesis in the sea urchin egg at the time of fertilization, but as soon as cell division begins in the zygote, protein synthesis also begins. Protein synthesis, of course, is dependent upon RNA.

In the first few cell divisions after fertilization, the RNA used for protein synthesis is probably of maternal origin, formed during oogenesis. In sea urchin eggs, only a small amount of RNA is synthesized immediately after fertilization, and this RNA is not essential to development (Gross et al., 1964). After the early blastula stage, RNA appears to be transcribed from the embryo's own DNA.

CHROMOSOMAL EVIDENCE OF GENE ACTIVITY

Evidence of gene activity and inactivity can be seen microscopically in the lampbrush chromosomes of amphibian oocytes and the giant salivary gland chromosomes of the larvae of *Drosophila* and of some other flies.

Lampbrush chromosomes are so named because they contain segments in which the chromatin appears as expanded pale loops (the lampbrush appearance). It is in these loops that RNA synthesis occurs. The condensed parts of the chromosome do not form RNA. Actinomycin D (which blocks the synthesis of RNA) or histones rich in arginine (which play some nonspecific part in genetic regulation) will suppress RNA synthesis in the loops and cause them to regress to the condensed form (Izawa et al., 1963). Thus, it is apparent that the genes are active in the lampbrush areas (in which the chromosome appears to be unwound) and inactive in the condensed areas. (See also Fig. 3–7.)

The giant salivary gland chromosomes of insects consist of many parallel chromosome strands, produced by a series of replications without separation. Homologous loci remain in apposition, so the whole of a chromosome pair shows a banded appearance (Fig. 10–1). In these chromosomes, active loci in which RNA is being synthesized can be seen as lightly staining puffs, and inactive loci as densely staining bands which contain arginine-rich histone and do not synthesize RNA. Any cell has only about 10 per cent of its genes switched on at one time (Beerman, 1963). Different patterns of puffing can be seen in different tissues or in the same tissue at different periods of differentiation.

In *Chironomus* (a small, two-winged insect commonly known as a midge or fish fly) there is a band that controls one type of salivary secretion. When certain salivary gland cells are actively forming this product, the band is puffed, but when the product is not being formed the band appears condensed. There is a species of *Chironomus* which never pro-

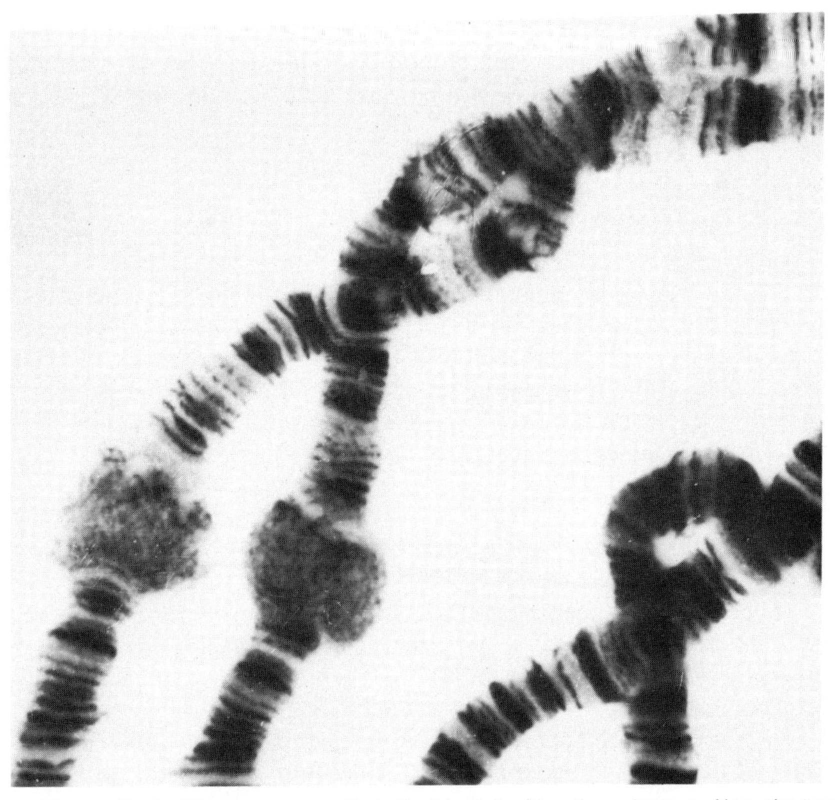

Figure 10–1 *Chromosome puffs in the black fly,* Simulium vittatum. *Note the two puffs on the strands at left, and the banded appearance of the rest of the strands. For details, see text. (Photomicrograph courtesy of J. J. Pasternak.)*

duces this secretion and never forms puffs at the site. A hybrid between secretor and nonsecretor species shows at appropriate times a puff on only one member of the pair of homologous chromosomes concerned.

SEQUENTIAL GENE ACTION

Ecdysone is a hormone of insects which brings about metamorphosis (molting). It has been shown to produce chromosomal puffs, suggesting that it acts by switching on certain genes (Clever, 1964). These genes in turn stimulate other loci which, in their turn, form puffs, but ecdysone itself does not affect these "secondary" genes. By switching on one gene, ecdysone sets in motion a chain of gene-directed events which effect metamorphosis. The puffing caused by ecdysone can be prevented by an inhibitor of nucleic acid metabolism (actinomycin), but not by an inhibitor of protein synthesis (chloramphenicol). Sequential activity of genes has not been demonstrated in human cells, but it could explain such reactions as the striking proliferation and differentiation of mammary gland epithelium when stimulated by estrogenic hormone.

PROTEIN SYNTHESIS DURING DEVELOPMENT

Since changes in cellular proteins reflect changes in the activity of specific genes, gene function during development can be described in terms of the protein composition of cells at successive stages. Lactate dehydrogenase and hemoglobin are examples of proteins that have been studied in connection with the genetic control of protein synthesis at different developmental stages.

LACTATE DEHYDROGENASE ISOZYMES

Lactate dehydrogenase (LDH) is an enzyme of vertebrates (and many invertebrates) which is important in cellular respiration. The LDH molecule is made up of four subunits (polypeptides) of two different kinds, designated A and B. Each kind is encoded by a separate structural gene. The two kinds of polypeptide chains combine randomly to form tetramers, so five different combinations are possible: four A chains (A_4B_0), three A chains and one B chain (A_3B_1), two of each (A_2B_2), one A and three B's (A_1B_3) or four B's (A_0B_4). The five different tetramers (isozymes) are distinguished by starch gel electrophoresis.

Isozyme	*Subunit Composition*	*Speed of Migration*
LDH-1	A_0B_4	Most rapidly migrating band
LDH-2	A_1B_3	↑
LDH-3	A_2B_2	│
LDH-4	A_3B_1	│
LDH-5	A_4B_0	Most slowly migrating band

In LDH synthesis the aggregation of subunits into tetramers occurs at random, in contrast to hemoglobin synthesis in which tetramers are always formed of two α and two non-α chains. The relative proportions of the different isozymes in any tissue, therefore, are determined by the relative proportions of A and B subunits present. The proportions of A and B subunits in turn depend upon the relative rate of action of the structural genes concerned.

If the two kinds of subunits are present in equal quantities, the five isozymes have the proportions 1:4:6:4:1, i.e., the binomial distribution obtained by the expansion of $(\frac{1}{2} + \frac{1}{2})^4$ (Fig. 10–2). If the subunits are produced in unequal amounts, the proportions of the five isozymes will vary accordingly.

The LDH isozyme pattern shows progressive changes during fetal and postnatal development. The *A* gene is more active than the *B* gene before birth, so embryonic tissues contain relatively more LDH-5 than do postnatal tissues. As development proceeds, the *B* gene becomes more active, more B subunits are formed and the whole isozyme pattern shifts toward the more rapidly moving components. The isozyme pattern is specific for any one mature tissue, but the ontogenetic changes are all in the same direction (Markert, 1964).

THE DEVELOPMENT OF HEMOGLOBIN

Hemoglobin is synthesized in erythroblasts, cells which are precursors of erythrocytes. Erythroblasts synthesize two main types of

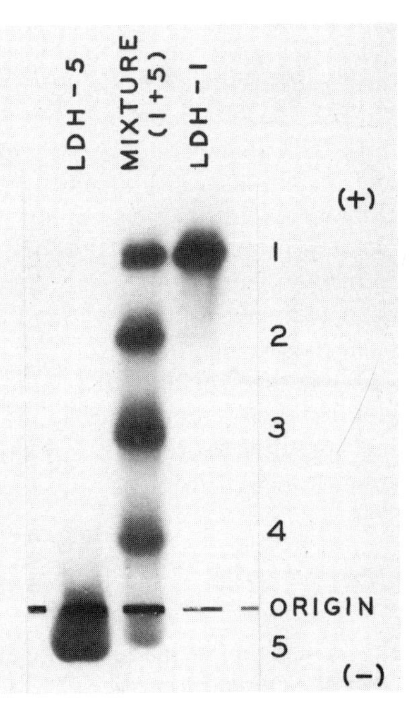

Figure 10–2 *Lactate dehydrogenase isozymes as distinguished by starch gel electrophoresis. A mixture of LDH-1 (A_0B_4) and LDH-5 (A_4B_0) will produce all five isozymes. For further discussion, see text. (From Markert 1963.* Science *140:1329–1330.)*

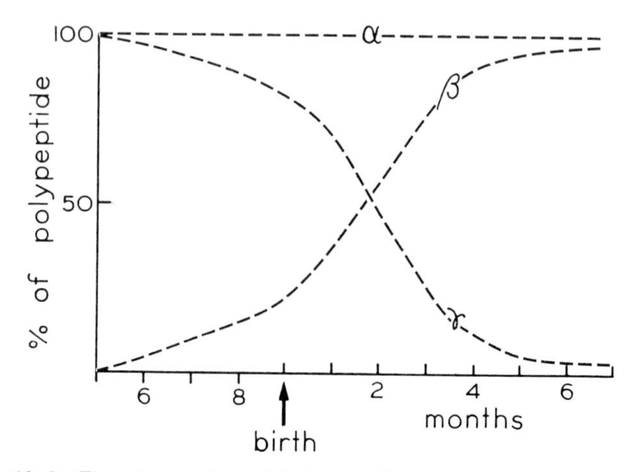

Figure 10–3 *The change from fetal to adult hemoglobin during late fetal and early postnatal life. During the last months of gestation, the production of γ chains begins to decline, and the production of β chains rises reciprocally, until by about six months after birth the production of the two types of polypeptide has reached the adult level. During the switch from γ chain to β chain production, the synthesis of α chains remains constant. (After Ingram 1963.* The Hemoglobins in Genetics and Evolution. *Columbia University Press, New York, p. 117.)*

hemoglobin: fetal hemoglobin, composed of two α and two γ chains ($\alpha_2\gamma_2$); and adult hemoglobin, composed of two α and two β chains ($\alpha_2\beta_2$). Each of the three types of polypeptide chains is encoded by a separate structural gene. The synthesis of α chains appears to continue at a steady rate from the time that hemoglobin production begins in the embryo, but the rate of synthesis of γ and β chains changes reciprocally during differentiation (Fig. 10–3). Gamma chains are formed by the fetus, but the rate of synthesis begins to decline during later fetal life, and after birth falls almost to zero. Meanwhile, the synthesis of β chains begins, and it continues at a steadily increasing rate until almost all the hemoglobin present is of the adult type. The minor variety of adult hemoglobin, Hb A_2, which has δ chains instead of β chains, is (like Hb A) formed only in the late fetal period and after birth. In very early development there is still another kind of hemoglobin chain, the ϵ chain, which forms embryonic hemoglobins (ϵ_4 and $\alpha_2\epsilon_2$).

The mechanism that regulates the activity of these different hemoglobin genes is not understood, though it is obvious from Figure 10–3 that a reciprocal relationship exists between the activity of the γ and β genes.

CONGENITAL MALFORMATIONS

A congenital malformation is a malformation present at birth. The term does not connote heredity. Conversely, heredity cannot immediately be excluded as a causative factor, it may or may not play a role in the appearance of the malformation. Since congenital malformations are present at birth, they are sometimes referred to as birth defects, although not all birth defects are malformations in the anatomical sense; on the contrary, many biochemical anomalies which are manifest at or near the time of birth are regarded as birth defects, even though they are associated with no actual malformation. For purposes of the present discussion, only gross morphological malformations present at birth will be considered.

Congenital malformations are highly variable both in kind and in causation, but all arise from maldevelopment during fetal life. In view of the great complexity of prenatal development and the interlocking roles played by genes and environmental factors, it is hardly surprising that developmental processes sometimes go awry. The delineation of specific malformations and the genetic and environmental factors that combine to produce them are important aspects of preventive medicine, since if the causes of a malformation can be discovered, prevention may become possible. Relatively few congenital malformations have clear-cut patterns of inheritance, yet the parents of a malformed child are usually deeply concerned about the risk that the malformation will recur in a later-born child; thus, congenital malformations as a group are a serious problem in genetic counseling.

Genetic couseling in congenital malformation is in an unsatisfactory state at present. To improve it, we need first to know more clearly the relative roles of genes and environment in the etiology of different types of malformations. As the following discussion will demonstrate, the recurrence risk of a specific malformation can be affected by many variables, such as geographic location, sex, maternal age, parity and the season of the year. For example, the risk of anencephaly is many times higher for girls born in Belfast in winter than for boys born in Lyons in summer (see later). The empiric risk figures available in the literature are usually (as most authors take care to emphasize) crude averages based on small and often heterogeneous samples. More accurate figures are urgently needed but are not likely to be available until the application of computer methods places much larger series of data at our disposal.

INCIDENCE

The total load of congenital malformations is difficult to estimate, but in the United States and Canada it is believed that one birth in 16 is that of a child with a defect so severe that it would, if unattended, result in lifelong handicap or death (O'Connor, 1963). About half of these are obvious at birth, and most of the remainder become apparent during the first year of life. This figure includes birth defects in the broad sense, not only congenital malformations. It is a curious fact that although the incidence of specific malformations varies considerably in different parts of the world, the total load is much the same everywhere.

In an Edinburgh study (Nelson and Forfar, 1969), 5.4 per cent of 8684 newborns had congenital malformations, of which 2.1 per cent were "major." On this basis at least one child in 50 has a medically significant congenital melformation.

CLASSIFICATION OF CONGENITAL MALFORMATIONS

There is no really satisfactory or simple way to classify congenital malformations. They are too variable in type and in causation and too incompletely understood to lend themselves to convenient docketing.

For purposes of genetic investigation, a classification by causes is a convenient way to approach the subject. A few malformations are clearly caused by single genes. Others are associated with chromosomal abnormalities. Still others are produced by known environmental factors. However, when all malformations of known etiology have been classified, a large and heterogeneous category remains. This category contains some of the most common and severe malformations, e.g., central nervous system malformations, most cases of cleft lip and palate, club foot, congenital dislocation of the hip and congenital malformations of the heart. Some of the genetic and nongenetic factors known to be involved in the production of these anomalies are discussed later in this chapter.

TERATOGENIC AGENTS

A teratogen is any agent which produces a malformation or raises the incidence of a malformation in a population. The best known kinds of teratogens in man are viruses, radiation and drugs. Some examples of each of these are given in Table 10–1.

A wide variety of agents has been shown to be teratogenic in experimental animals. To have a teratogenic effect, an agent must affect some *specific metabolic process* in the developing embryo. The *timing of application* of the teratogen is extremely important; most teratogens act during a rather limited period at the time the embryo is undergoing rapid organogenesis, and the extent of the damage produced by the teratogen varies with the timing of its application. The importance of timing is particularly well shown by the effect of rubella infections. Rhodes (1961) has found that the incidence of major defects in infants following maternal rubella in pregnancy falls from 50 per cent if the infection is in the first four weeks of pregnancy, to 17 per cent in the third month; thereafter it is close to zero. Studies with inbred strains of mice have shown that the *genotype* of the embryo may be critical in determining whether or not it

TABLE 10-1 SOME TERATOGENIC AGENTS AND THEIR EFFECTS IN HUMAN DEVELOPMENT

Agent	*Examples*	*Traits Produced*
Viruses	Rubella	Congenital heart defects, cataracts, deafness
	Cytomegalovirus	Mental retardation
Radiation		Microcephaly
Drugs	Thalidomide	Phocomelia
	Aminopterin	Variable malformations
	Progestin	Masculinization of female fetuses

will be susceptible to a given teratogen. In the A/J strain of mice, cleft palate occurs spontaneously in about 15 per cent of the young, whereas in the C57BL/6J strain, cleft palate is rare. Treatment with cortisone raises the incidence of cleft palate to 100 per cent in the A/J strain but only to 17 per cent in the C57LB/6J strain. Thus, the susceptibility of the embryo to the teratogenic effect of cortisone is determined at least partly by its own genotype.

GENETIC ASPECTS OF SOME COMMON CONGENITAL MALFORMATIONS

Cardiac Malformations

Congenital malformations of the heart are seen in many different conditions, including single-gene disorders (e.g., Marfan's syndrome, Ellis–van Creveld syndrome), chromosome disorders and syndromes of nongenetic etiology (e.g., rubella embryopathy). Many cardiac malformations fit the criteria for multifactorial inheritance (page 270).

If more than one member of a family has a congenital malformation of the heart, the type of malformation is often the same in all affected members, but it is not unusual for different anatomical defects to be present in members of the same family.

Central Nervous System Malformations

The neural tube defects, anencephaly and spina bifida, occur more frequently in association with one another than would be expected by chance. This suggests that they have a common origin. Anencephaly has been much studied because it is a very obvious congenital defect which is usually recorded when it occurs. Unlike some other congenital malformations, it shows striking variation in geographic distribution; e.g., it is more than 50 times as common in Belfast (0.671 per cent) as in Lyons (0.012 per cent). The sex ratio is one male to two females. The incidence is also said to vary with social class (lower classes are more susceptible), season of the year (higher frequency in winter months) and parity (first-born children more susceptible than later-born children).

Both anencephaly and spina bifida are more common in the later-born sibs of affected infants than in the general population. If one child in a family is born with anencephaly or spina bifida, possibly 5 per cent of the later-born children will have anencephaly and/or spina bifida. If two or more such children are born, the risk may be about 12 per cent. Most authors agree that in a few families the risk may be much higher, perhaps as high as 25 per cent, but mothers with such an unusually high risk of producing anencephalics cannot yet be recognized.

Very recently, a new dimension has been added to the neural tube defect problem by the suggestion of an association between the incidence of potato blight (a fungus disease) and the incidence of neural tube malformations a year or so later (Renwick, 1972). Experimental production of neural tube malformations in marmosets (New World monkeys) fed on blighted potatoes before and during pregnancy (Poswillo et al., 1972) supports this possibility.

Cleft Lip and Cleft Palate

Cleft lip with or without cleft palate is etiologically distinct from cleft palate alone (Fogh-Anderson, 1942). Both categories are heterogeneous. On the basis of etiology, four groups can be distinguished:

Cases due to single, mutant genes (30 known).
Cases due to chromosomal aberrations (especially D trisomy, E trisomy, XXXXY).
Cases due to teratogenic agents (rubella, thalidomide and so forth).
Cases of multifactorial origin (the great majority).

Cleft Lip With or Without Cleft Palate (CL or CLP). This disorder originates as a failure of fusion of the nasal processes of the frontal prominence with the maxillary process, at about the seventh week of development. The incidence is higher in males (60 to 80 per cent of cases). There is wide racial variation: one per 1000 in Caucasians, 1.7 per 1000 in Japan and 0.4 per 1000 in American Negroes. The recurrence risk for sibs when neither parent is affected is about 4 per cent after one affected child, and 9 per cent after two affected children. The risk for sibs varies with the severity of the defect in the proband: 5.7 per cent for bilateral CLP, 4.2 per cent for unilateral CLP and 2.5 per cent for unilateral CL. The pattern of distribution closely fits what is expected on the basis of multifactorial inheritance (page 270), unless a specific mutant gene, chromosomal aberration or teratogen is implicated.

Cleft Palate (CP). Failure of fusion of the secondary palate has a population incidence of 1/2500, and is more common in females than in males. There is little racial variation in incidence. The recurrence risk in sibs is about 2 per cent, apparently regardless of family history, though more data are needed. Microforms of CP include bifid uvula and congenital palatopharyngeal incompetence (a condition in which the individual speaks as though he has a cleft palate although there is no cleft).

Congenital Dislocation of the Hip

Congenital dislocation of the hip (CDH) has an overall incidence of about one in 1500, but it is far more common in females than in males. In one series the male:female ratio was approximately 1:7 (Cox, 1964). It shows geographical variations in frequency, probably related in part to genetic factors and in part to environmental factors. For example, its high frequency in some Canadian Indian groups and extremely low frequency in Eskimos may be related to the manner in which infants are dressed and carried, as well as to genetic differences in susceptibility.

Like other malformations of complex etiology, CDH is a heterogeneous category which includes CDH secondary to other defects, some of which themselves have a genetic basis. Late maternal age, primiparity and birth in winter are associated with an increased incidence.

Diagnosis and treatment of CDH involve numerous X-ray examinations of the pelvis. Cox has studied 91 women who underwent treatment for CDH, and found that the incidence of abnormalities in their offspring is increased as compared with the incidence in a control series. She points out that this observation raises the possibility that some of the increase is due to increased genetic susceptibility to malformation among the children of women with CDH, rather than to genetic damage resulting from the large gonadal doses of irradiation the mothers had received in childhood.

Fraser quotes a figure of 5 per cent as the recurrence risk of CDH in families in which both parents are normal and there is one affected child.

Club Foot

Like many other congenital malformations, the general category of club foot includes many cases which differ in etiology. Club foot may occur in association with spina bifida, cerebral palsy, arthrogryposis, congenital dislocation of the hip and other disorders, or it may occur as an isolated abnormality. There is no assurance that either type is etiologically uniform, and indeed it is more likely that both are heterogeneous.

There are three main types of club foot: talipes equinovarus (TEQ), talipes calcaneovalgus (TCV) and metatarsus varus (MTV). Each of the three types occurs about once in 1000 births. Wynne-Davies (1964) studied the genetic aspects of the three types.

For TEQ, the male:female ratio was 2:1. There was no association with consanguinity, maternal age or parity. The degree of concordance in monozygotic twins (in this series and in others) was about 35 per cent, whereas in dizygotic twins it was about 3 per cent. The recurrence risk in sibs is about 3 per cent; no data are given as to whether there is a difference in risk if a parent is also affected. The risk varies with the sex of the sib as follows:

Male sib of female patient	1 in 16
Female sib of female patient	1 in 40
Male sib of male patient	1 in 42
Female sib of male patient	0 in 187

Such a distribution suggests the possibility of multifactorial inheritance with a different threshold in males and in females. It is noteworthy that the risk is the same in a dizygotic twin of the propositus as in a later-born sib; this observation certainly suggests that twin pregnancy *per se* does not increase the risk of club foot.

For TCV and MTV, which are milder deformities than TEQ, the situation is quite different. Both are more common in girls, the male:female sex ratio being about 0.7:1. In TCV, but not in MTV, there is a pronounced tendency to primiparity and early maternal age. The recurrence risk in sibs for each malformation is about one in 20.

There is some evidence that a generalized defective formation of connective tissue may be the underlying genetic defect. Minor defects of connective tissue, such as joint laxity, hernia and congenital dislocation of the hip, were observed in nearly one-fifth of the patients. In another 4 to 5 per cent, there were abnormalities at the periphery of the limbs.

In summary, the recurrence risk of TEQ is about 3 per cent on the average, but it may be doubled in the male sibs of female patients. The recurrence risk of TCV and MTV is about 5 per cent.

PLEIOTROPY AND DEVELOPMENTAL DEFECTS

Pleiotropy is the term used to describe multiple effects of a single gene. An example is the gene *W* ("dominant spotting") in the mouse. Homozygous *W/W* mice or mice with other "double-dominant" genotypes of the W locus such as W/W^v, are black-eyed white mice with lifelong macrocytic anemia associated with a severe deficiency of hematopoietic stem cells and near total sterility corresponding to a deficiency of primordial germ cells. The triple deficiency of pigment cells, hematopoietic stem cells and germ cells probably has a common origin in some embryonic cell type ancestral to all three.

A different mutation in the mouse, *Steel* (symbol *Sl*), when homozygous or in a compound genotype (Sl/Sl^d) produces a phenotype indistinguishable from that of *W/W*. However, in this genotype the basic defect is not cellular but is related in some as yet unelucidated way to the environment in which the cells differentiate. W/W^v and Sl/Sl^d mice can be surgically combined in parabiosis because their genetic backgrounds are identical. When this is done, their circulations fuse, and under these conditions the anemia of both mice is "cured." If they are then separated, W/W^v remains "cured," because it has received normal hematopoietic stem cells from its Sl/Sl^d partner; but Sl/Sl^d slides back into its former anemic state, because it no longer has the environmental factor it had received from W/W^v. As well as being excellent examples of pleiotropy, these two mutations exhibit genetic heterogeneity: an identical phenotype produced by either of two different mutant genes, differing greatly in their basic mechanisms and therefore responding quite differently to treatment.

GENETICS AND CANCER

Cancer is basically a disorder of differentiation of some cells of the body. Instead of differentiating and assuming their normal roles, some of the cells (or more probably, one original cell) appear to escape from whatever factors produce differentiation, while at the same time they are no longer subject to the normal controls upon cell multiplication. As a result the undifferentiated or partly differentiated cells continue to divide, and a mass of cells (a tumor) grows at the expense of the normal surrounding tissue. If the growth is truly malignant, it invades other tissues and becomes a central mass, with "claws" like those of a crab extending into the normal tissue (*cancer* = crab). So distinctive is the appearance of a malignant tumor that the term cancer was used by both Hippocrates and Galen.

The causes of cancer are at present very poorly understood. It is known that ionizing radiation and certain chemicals, especially mutagenic chemicals, are carcinogenic (cancer-inducing). Chronic irritation may also be a causative factor. At least in experimental animals some types of cancer are caused by viruses, and much current cancer research is directed toward the possibility of viral origin. One theory concerning the cause of cancer is that a somatic mutation may allow a cell to fail to respond to the body's regulatory mechanisms. The role of genetics in the causation of cancer is, in general, relatively minor, though there are exceptions.

A few genetic disorders are characterized by a high risk of cancer. Retinoblastoma and diffuse polyposis of the colon are two autosomal dominant examples. Fanconi's anemia and Bloom's syndrome are autosomal recessives in which leukemia may eventually develop. There is about a sixfold increased risk of leukemia in Down's syndrome. However, except in these and similar conditions, heredity does not play an important role in human cancer.

A noteworthy feature of genetically determined forms of cancer, such as the type associated with polyposis, is that affected persons may have more than one (often many) primary tumor. This is very unusual for other forms of cancer and suggests that each cell of the susceptible tissue has a high probability of undergoing malignant change.

Twin studies show the concordance rate for total incidence of cancer to be similar in monozygotic (MZ) and dizygotic (DZ) twin pairs; but if cancer of a particular organ is studied, MZ twins show a higher concordance rate than do DZ twins. The difference in concordance rate is less for MZ versus like-sexed DZ twins than it is for MZ versus all DZ pairs.

Family studies also indicate that the role of heredity is relatively minor, but where it is a factor the site is usually the same in members of a family group. The incidence of cancer of a specific organ is raised in relatives of propositi, especially in MZ twins, although the increase is relatively slight.

Mitoses in cancer cells are often abnormal. On occasion, one cell can divide into three or even four daughter cells rather than the usual two. With recent advances in cytogenetic technique (page 11), it has been shown that the actual karyotype is altered in many malignant cells, and that as malignancy progresses, more and more bizarre changes may occur in the chromosome complement. Studies of the DNA content of malignant cells confirm the occurrence of abnormal karyotypes.

Only one type of cancer has a consistent, recognizable, cytogenetic abnormality. This abnormality is the Philadelphia chromosome (Ph^1), a deleted chromosome 22 consistently found in cases of chronic granulocytic leukemia (page 155).

CANCER IN THE LABORATORY MOUSE

Much of our knowledge of malignancy in animals comes from studies on mice, which were instituted on a grand scale by the late C. C. Little of the Jackson Laboratory in Bar Harbor, Maine. Many inbred strains show a characteristic high or low incidence of cancer of some particular organ.

Almost 100 per cent of female mice of the C3H/HeJ strain develop mammary cancer by the time they reach one year of age. On the other hand, females of the C57BL/6J strain practically never develop mammary cancer. Actually, the development of mammary cancer in these strains depends primarily upon a virus, or virus-like agent (**mammary tumor agent**), passed from mother to offspring in the milk of the mother. The agent is found in several strains, and strain differences in susceptibility to mammary tumor agent are known. Andervont (1964) has shown that one strain with a very low incidence of mammary cancer (C57BL) is able, within a few generations, to rid itself of the agent; whereas another low-incidence strain (BALB/c) lacks this ability. This may be a true genetic difference affecting the incidence of cancer, but in a rather indirect way.

Other types of tumors have been shown to occur more frequently in certain strains than in others: A/J mice have a high incidence of tumors of the lung, the CE/J strain has a high incidence of ovarian tumors and the CBA/J strain shows some hepatomas (tumors of the liver). In one strain, males often develop testicular teratomas (neoplasms derived from primordial germ cells), whereas other strains rarely show this condition.

AKR/J mice have a very high incidence of leukemia and provided one of the original stimuli for work on viruses as a cause of leukemia. Some murine leukemias may be spontaneous, and others can be produced by chemical agents. Old et al. (1964) have shown that there are several viruses which produce leukemia in mice, and that these, judging by their antigenic properties, are often closely related. The leukemias caused by viruses have antigenic properties quite different from those caused by chemicals or those that appear spontaneously.

GENERAL REFERENCES

Carter, C. O. 1965. The inheritance of common congenital malformations. *Progr. Med. Genet.* 4:59–84.

Carter, C. O. 1969. Genetics of common disorders. *Brit. Med. Bull.* 25:52–57.

Fraser, F. C. 1970. The genetics of cleft lip and cleft palate. *Amer. J. Hum. Genet.* 22: 336–352.

Green, E. L., ed. 1966. *Biology of the Laboratory Mouse.* 2nd ed. McGraw-Hill Book Company, New York.

Markert, C. L., and Ursprung, H. 1971. *Developmental Genetics.* Prentice-Hall, Englewood Cliffs, New Jersey.

Warkany, J. 1971. *Congenital Malformations: Notes and Comments.* Year Book Medical Publishers, Inc., Chicago.

WHO Technical Report Series No. 438. 1970. *Genetic Factors in Congenital Malformations.* Report of a WHO Scientific Group. World Health Organization, Geneva.

11

Elements of Mathematical Genetics

The mathematical aspects of genetics are usually less palatable to medical students than are the biochemical and chromosomal aspects; in fact, medical students actually often seem to recoil from any mathematical treatment of genetic problems. Since some mathematics is unavoidable, we have tried to make it as simple and painless as possible.

Though not all problems in medical genetics require a mathematical approach, there are areas that cannot be handled in any other way. This does not imply that a medical geneticist must himself be an expert mathematician, but he should at least be aware of the pitfalls that await mathematical incompetents and of the importance of obtaining expert assistance with the statistical aspects of his research, beginning at the planning stage and continuing through the analysis of the observations.

Special mathematical techniques are important in human genetics because of the special difficulties in dealing with human material. If human families were large enough for mendelian ratios to be approximated (12 members would be a convenient number), many of the difficulties incurred by bias of ascertainment would disappear. Similarly, if geneticists lived longer than the subjects of their research and could extend their observations over several generations, many other complications of the genetic analysis of human pedigrees would vanish. (Here, pediatricians have an advantage over other clinicians, since they normally see two generations of a family at a time.) Most important of all, the impossibility of arranging test matings makes it necessary to accumulate the results of nature's own experiments, a sometimes laborious process which adds to the statistical difficulties and makes statistical methodology unavoidable.

PROBABILITY

There is a very close analogy between genetic transmission from parent to child and tossing coins. In both cases, there are two alternative possibilities: the coin can turn up heads or tails, and the parent can transmit one or the other of the two alleles at a locus. In either case, the outcome is determined by chance.

Statisticians define probability, the mathematical expression of chance, as the ratio of the number of occurrences of a specified event to the total number of all possible events. The specified event may be that a tossed coin turns up heads. There are two possible events, heads and tails, so, for a single toss:

$$\frac{\text{Possibility of heads}}{\text{Total number of possibilities}} = \frac{1}{2}$$

The probability of tossing heads is $\frac{1}{2}$.

Similarly, the probability that a child whose father is of blood group AB will receive the *A* allele rather than the *B* allele is:

$$\frac{\text{Possibility of } A}{\text{Total number of possibilities}} = \frac{1}{2}$$

The probability of *A* is $\frac{1}{2}$.

PROBABILITY OF COMBINATION OF TWO INDEPENDENT EVENTS

A child receives his genotype from two parents, not just one. The probability of receiving certain combinations of alleles from the two parents can be simulated by tossing two coins at once. The probability of heads is $1/2$ for each coin, and the probability of heads turning up on both coins at once is $1/2 \times 1/2 = 1/4$. Similarly, the chance that a child, both of whose parents are of blood group AB, will receive the *A* allele from each parent is $1/2 \times 1/2 = 1/4$.

There are many genetic applications of this simple but important principle. Perhaps one of the most useful to the physician has to do with sex ratios in human families. The probability of a boy as the first-born child is $1/2$. (Actually it is not exactly $1/2$, but this is close to the precise figure and simplifies the arithmetic.) The next child is an "independent event" since spermatozoa, like dice, have no memory; so the probability that the second-born child will be another boy is again $1/2$. The probability that the first two children in a family will both be boys is $1/2 \times 1/2 = 1/4$. This procedure can be continued to obtain the probability of sons in any number of successive births; for example, the probability of a sequence of six sons is $(1/2)^6$, or $1/64$.

THE BINOMIAL DISTRIBUTION

The seventeenth century mathematician Jakob Bernouilli was a member of a family noted for its many brilliant mathematicians, but his genetic significance as an example of hereditary genius is outweighed by his contribution to genetic theory: the binomial distribution (Bernouilli distribution). Though the binomial formula was originally discovered by Sir Isaac Newton in 1676, it was Bernouilli who gave the first rigorous proof of it.

Suppose that the problem to be solved is not the probability of a boy at each of two births, but the probable distributions of boys and girls in two births. This is an extension of the problem of the probability of coincidence of random events.

If the probability of a boy is $1/2$ and the probability of a girl is also $1/2$, then the distribution of families with no boys, one boy and two boys among all two-child families is given by the expansion of $(1/2 + 1/2)^2$. In general terms: when there are two alternative events, one having a probability of p and the other having a probability of $q = 1 - p$, then in a series of n trials, the frequencies of the different possible combinations of p and q are given by the expansion of the binomial $(p + q)^n$.

Thus, for two-child families:

$$(p + q)^2 = p^2 + 2pq + q^2$$

$$p = q = \frac{1}{2}$$

$$p^2 = \text{families of 2 boys} = \frac{1}{4}$$

$$2pq = \text{families of 1 boy and 1 girl} = \frac{1}{2}$$

$$q^2 = \text{families of 2 girls} = \frac{1}{4}$$

For three-child families:

$$(p + q)^3 = p^3 + 3p^2q + 3pq^2 + q^3$$

$$p^3 = \text{families of 3 boys} = \left(\frac{1}{2}\right)^3 = \frac{1}{8}$$

$$3p^2q = \text{families of 2 boys, 1 girl} = 3\left(\frac{1}{2}\right)^2\left(\frac{1}{2}\right) = \frac{3}{8}$$

$$3pq^2 = \text{families of 1 boy, 2 girls} = 3\left(\frac{1}{2}\right)\left(\frac{1}{2}\right)^2 = \frac{3}{8}$$

$$q^3 = \text{families of 3 girls} = \left(\frac{1}{2}\right)^3 = \frac{1}{8}$$

The difference between a simple multiplication of the separate probabilities and the use of the binomial expansion is that the binomial expansion includes all the possible combinations of the two events. For example, in three-child families there are eight possible sequences:

♂	♂	♂	♀	♂	♂
♂	♂	♀	♀	♂	♀
♂	♀	♂	♀	♀	♂
♂	♀	♀	♀	♀	♀

The probability of each sequence is 1/8, but among families with two sons and a daughter, the daughter might be born first, second or last; the three possible sequences account for the total 3/8 probability of having two sons and a daughter among three children. The expected distribution of males and females in families of one to five children is summarized in Table 11–1.

Pascal's triangle is a convenient way to arrive at the coefficients of the binomial expansion. Each new coefficient in the triangle is made up by adding the two coefficients nearest it in the line above; e.g., on the fourth line, the 3 of $3p^2q$ is obtained by adding the 1 of p^2 and the 2 of $2pq$.

n	***Expansion***
0	1
1	$1p + 1q$
2	$1p^2 + 2pq + 1q^2$
3	$1p^3 + 3p^2q + 3pq^2 + 1q^3$
4	$1p^4 + 4p^3q + 6p^2q^2 + 4pq^3 + 1q^4$
5	$1p^5 + 5p^4q + 10p^3q^2 + 10p^2q^3 + 5pq^4 + 1q^5$

Note that the number of terms in the expansion is always $n + 1$. For families of size n, there can be 0, 1, 2 . . . n sons; hence, there are $n + 1$ classes of family. If there are four children in a family, there can be five different ♂:♀ distributions: 4 : 0, 3 : 1, 2 : 2, 1 : 3, 0 : 4.

TABLE 11–1 DISTRIBUTION OF BOYS AND GIRLS IN FAMILIES

Number of Children in Family	$(p + q)^n$	Distribution
1	$\left(\frac{1}{2}+\frac{1}{2}\right)^1$	$\frac{1}{2}(1♂) + \frac{1}{2}(1♀)$
2	$\left(\frac{1}{2}+\frac{1}{2}\right)^2$	$\frac{1}{4}(2♂) + \frac{1}{2}(1♂:1♀) + \frac{1}{4}(2♀)$
3	$\left(\frac{1}{2}+\frac{1}{2}\right)^3$	$\frac{1}{8}(3♂) + \frac{3}{8}(2♂:1♀) + \frac{3}{8}(1♂:2♀) + \frac{1}{8}(3♀)$
4	$\left(\frac{1}{2}+\frac{1}{2}\right)^4$	$\frac{1}{16}(4♂) + \frac{4}{16}(3♂:1♀) + \frac{6}{16}(2♂:2♀) + \frac{4}{16}(1♂:3♀) + \frac{1}{16}(4♀)$
5	$\left(\frac{1}{2}+\frac{1}{2}\right)^5$	$\frac{1}{32}(5♂) + \frac{5}{32}(4♂:1♀) + \frac{10}{32}(3♂:2♀) + \frac{10}{32}(2♂:3♀) + \frac{5}{32}(1♂:4♀) + \frac{1}{32}(5♀)$

Mathematicians may prefer to use the general term of the binomial expansion:

$$\frac{n!}{m!\,(n-m)!}\,p^m q^{n-m}$$

n = total number in the series
$n!$ ("n factorial") is $n(n-1)\,(n-2)\ldots 1$.
p = probability of a specified event
$q = 1 - p$ = probability of the alternative event
m = number of times p occurs (in other words, the exponent of p).

For example, the probability of having three sons and two daughters in a five-child family is:

$$\frac{5 \times 4 \times 3 \times 2 \times 1}{(3 \times 2 \times 1)(2 \times 1)}\left(\frac{1}{2}\right)^3\left(\frac{1}{2}\right)^2 = \frac{10}{32}$$

In the examples used so far, the values of p and q have been equal ($p = q = 1/2$), but the binomial distribution can be used for other values of p and q. The same method can be applied, for example, to give the distribution of a recessive trait in the progeny of two heterozygous parents, but now the probability of the specified event (that a specific child will be affected) is $1/4$, and the probability that a specific child will be unaffected is $3/4$. We will return to this topic in the following section when discussing tests of genetic ratios. Gene frequencies in populations also depend upon simple binomial distributions (see Hardy-Weinberg law, page 277).

FURTHER COMMENTS ON THE SEX RATIO

The sex ratio is defined as the ratio of the number of male births to the number of female births. Since sex is determined by whether the sperm contributes an X or a Y chromosome to the zygote, and since X-bearing and Y-bearing sperm should be formed in equal numbers at meiosis, the expectation is that the primary sex ratio (ratio at fertilization) should be 1.00. However, in all parts of the world there is an excess of male babies, and currently in the United States and Canada the secondary sex ratio (ratio at birth) is about 1.05 (105 boys per 100 girls). It has been thought that more males than females die before birth, but sex ratio studies in abortuses contradict this idea (Carr, 1971). After birth, males have a higher mortality rate than females. During adult life the sex ratio reaches 1.00, and thereafter the excess of females becomes more and more pronounced. The reason for the higher death rate in males is not known, but it may be due in large part to the fact that males have only one X chromosome, so that harmful X-linked recessive genes are exposed to selection in males but not in females.

What effect does parental preference for one sex or the other have upon the sex distribution in families and in the population? Let us take an extreme example: that in which parents either have no more children after they have had one son, or continue to have children until they have a son. In this case, all one-child families will consist of one boy; all two-child families will be made up of a girl and a boy, in that order; all three-child families will be made up of two girls and a boy, in that order; and so on. The distribution of sex within families will be quite different from the binomial proportions set out in Table 11–1. However, at each birth the total number of boy babies will equal the total number of girl babies, so the overall sex ratio in the population will not be altered.

Population data on sex distribution in sibships of different sizes fit the binomial expectation reasonably well. The literature does include, however, a few possible or real exceptions, such as excesses of unisexual sibships, and a preponderance of males in some pedigrees and of females in others. There are some examples in experimental animals of genetic mechanisms which upset the normal sex ratio (such as a "sex ratio" gene in the fruit fly which leads to almost 100 per cent female offspring, and strikingly different sex ratios in different inbred strains of mice). However, on the whole, the obstetrician can safely tell a prospective mother that the probability that the baby will be a boy is $1/2$.

TESTS OF GENETIC RATIOS

An essential first step in the analysis of any genetic trait is the demonstration of its pattern of inheritance. Sometimes inspection of a collection of pedigrees will make the pattern obvious, but even then more refined tests are required to alert the investigator to discrepancies due to such factors as abnormal segregation ratios or different prenatal viability in the two sexes.

Autosomal dominant inheritance is usually the easiest pattern of transmission to recognize, since an autosomal dominant trait appears in each generation. The pattern most likely to be confused with it is X-linked dominant inheritance, but the two types are readily distinguishable if one looks at the male and female offspring of affected males and females separately. X-linked dominant traits are transmitted by affected males to all of their daughters and none of their sons, so a pedigree in which an affected male has any affected sons or any normal daughters cannot be an X-linked dominant.

If penetrance is reduced or expressivity is highly variable, it is sometimes difficult to distinguish between autosomal dominant and multifactorial inheritance. Here, the chief means of differentiation is by comparison of the incidence of the trait in first-degree, second-degree and third-degree relatives. For autosomal dominants the incidence of the trait drops off by half with each increasingly remote degree of relationship, but for multifactorial traits the reduction in incidence is much more precipitate. Twin studies may also help to distinguish between these two pedigree patterns (page 274).

When calculating the relative proportion of affected and unaffected members in autosomal dominant or X-linked dominant pedigrees, it is often convenient to classify the offspring, but not the sibs, of affected members. As usual in pedigree analysis, the index patient (also known as the propositus or proband), through whom the pedigree comes to the attention of the investigator, must be omitted from the calculations (see below).

BIAS OF ASCERTAINMENT

An important source of error in pedigree analysis is the inclusion of propositi among the affected. As a demonstration of the effect of this kind of bias, consider the sex ratio of the families of a class of medical students. This class, composed of 100 men and 10 women, reported the numbers of males and females in their sibships to be 190 males and 96 females. The sex ratio of 2:1 deviated significantly from the expected 1:1. But when the data were broken down separately for male and female students, it was found that:

The 100 males had	78 brothers and	72 sisters.
The 10 females had	12 brothers and	14 sisters.
Total	90 males	86 females

Thus the sex ratio in the sibs of the propositi was 90 males:86 females or 1.05, which is not significantly different from the expected ratio. The discrepancy in the first results was due to the inclusion of the propositi.

The same kind of bias can arise in studies of genetic disorders if the propositus is included. If, as usually happens, a series of families is studied because there is at least one affected child in each sibship, the expected ratio of normal to abnormal sibs will not be observed unless an appropriate correction is made.

TESTS FOR AUTOSOMAL RECESSIVE INHERITANCE

Pedigree analysis to prove autosomal recessive inheritance has special difficulties not encountered with other pedigree patterns. An autosomal recessive trait appears in one-fourth of the offspring of two carrier parents. This is a simple and well-known rule, but to prove the quarter ratio is not always easy. Since carrier parents are usually identifiable only because they have produced an affected child, collections of data on families with recessive traits usually consist of parent-child sets in which both parents are phenotypically normal and at least one child is affected. Thus, carrier parents go undetected if all their children are normal, and these normal children are not enumerated. As a result, the proportion of affected children in the families that are actually observed is well above the expected one-fourth.

The number of families missed because they include no affected child decreases with family size. Seventy-five per cent of one-child families are not recognizable, but only 3 per cent of the 12-child families are missed. For families of typical sizes, the correction is large enough to be important. (The terminology used here is that of Morton, 1959, and Crow, 1965, but the reader is cautioned that the usage of these two authorities is not universally followed.)

Complete and Incomplete Ascertainment

If every pair of parents heterozygous for the condition under study could be identified, and if all their offspring could be included in the study, ascertainment would be **complete.** This is rarely possible for autosomal recessive inheritance. In actual practice, ascertainment is almost always **incomplete,** i.e., only those sibships which contain at least one affected child are used.

Truncate, Single and Multiple Ascertainment

Given incomplete ascertainment, the next point to be determined is how the sibships for study have been selected from the population. Selection is **truncate** if every affected child in the population is included in the survey (or if the group studied is a true random sample of the total group of affected children). If the population has been fully screened, any one affected child is equally likely to be included regardless of the number of affected children in his sibship. Selection is **single** if there is almost no chance that any one sibship will be ascertained more than once. In this model, the probability that any one child will enter the series is quite low; for example, if an affected child has only a 1 per cent chance of being included and a 99 per cent chance of being missed, then a family in which there are two affected children each of whom has a 1 per cent chance of being included, still has only a 2 per cent ($1 - 0.99^2$) chance of entering the study and a 98 per cent chance of being missed. However, in actual practice, more often than either of these two extremes there is **multiple** ascertainment. Some sibships are ascertained more than once, through more than one of their affected members, and families with more than one affected child have a higher chance than one-child families of being ascertained, though still less than a 100 per cent chance. For example, assume that the probability of ascertaining a particular child is only 80 per cent. Then 20 percent of all sibships with one affected child will be lost to the study, but only 4 per cent (0.20^2) of sibships with two affected members and less than 1 per cent (0.20^3) of sibships with three affected members are missed.

To illustrate possible situations in which truncate, single and multiple selection might apply, consider how a study of the genetics of cystic fibrosis (CF) might be organized.

1. *Truncate Ascertainment.* Every child with CF in a given area might be independently identified, regardless of the number of affected sibs, age, clinical status or other factors that might affect the child's likelihood of being found. Total ascertainment could involve an exhaustive search through clinics, hospital records, school health records, private physicians' practices and so forth. If the search is successful in finding every case of CF in the community, ascertainment is truncate, and in the collected family data, sibships with one, two, three or more affected members are distributed binomially except for the class with no affected members. (Thus the distribution is truncate, which means "shortened by cutting off a part." The cut-off part is the group of families with heterozygous parents but no affected children.)

2. *Single Ascertainment.* Suppose our study of the genetics of CF is limited to a survey of all the children entering Grade 1 in a given year. It is very unlikely that any sibship with affected members would be encountered more than once. However, the more affected children a family contains, the more likely it is that one of them will be entering Grade 1. In this situation, single selection methods are appropriate.

3. *Multiple Ascertainment.* If the case load of a particular CF clinic at a pediatric hospital is the source of the data, a good many affected children in the community might be missed because they do not attend the clinic. Some of these might be treated by private physicians, others might be in a hospital for chronically ill children and still others might be managed at home without regular medical attention. Among the children attending the clinic, there would be some with sibs who also were attending; in other words, some sibships would be ascertained more than once, and the probability that a particular family would be included would depend partly but not entirely upon the number of cases of CF in the sibship. This is the multiple ascertainment situation, which requires special statistical techniques that are beyond the scope of this discussion.

Corrections for Bias of Ascertainment

1. The Proband Method of Weinberg. In 1912, Weinberg, of whom we shall have more to say later, proposed a simple method of correction of data by omitting the proband and examining the family incidence of the disorder in his sibs. The proband is used only as the individual through whom the sibship is ascertained. If there is more than one proband in the sibship, the sibship is counted separately for each. This is the kind of correction applied to the data on sex ratio in sibships mentioned previously. The proband method is often useful for a preliminary analysis, and may be sufficient for all practical purposes.

2. The Apert or *a priori* Method. If it is suspected that a specific disease is inherited as an autosomal recessive, one method of testing is to assume that recessive inheritance is present; in other words, to set up an *a priori* expectation of recessive inheritance, and then to test the data for agreement with this hypothesis. Since the probability that a family with two heterozygous parents will be missed varies with family size, the calculation must be made separately for each size of sibship. The method can be used for either truncate or single ascertainment, but the expected proportion of affected children is different in the two cases, as will be shown.

TRUNCATE SELECTION. As an example, let us begin by considering 16 two-child families, each with two heterozygous parents. One-quarter of the children are expected to be affected. How will these eight children be distributed in their respective sibships?

One-fourth of the families will have an affected child at the first birth, and one-fourth at the second birth:

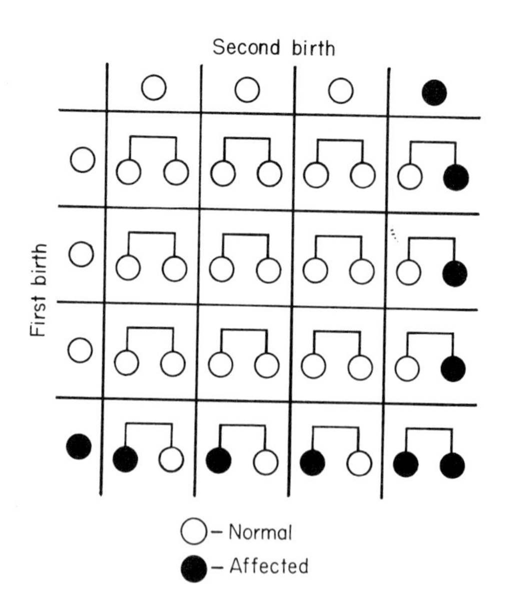

The chance distribution will produce 1 family with 2 affected children, 6 with 1 affected child each and 9 with no affected children (which will not be ascertained because they are not distinguishable from the rest of the population). The incidence of affected children in the observed families is 8/14, or 57.1 per cent.

Note the mathematical relationship: if the probability of a normal child is 3/4 and the probability of an affected one 1/4, then the distribution of the trait in two-child families is given as in the general situation, by the expansion of $(p + q)^n$. Here $p = 3/4$, $q = 1/4$ and $n = 2$.

Now consider a series of sibships of three children each, with heterozygous parents. The proportion of families with 1, 2 and 3 affected children can again be calculated from the expansion $(p + q)^n$, but now $n = 3$.

The population distribution of such families is:

$$\left(\frac{3}{4}\right)^3 + 3\left(\frac{3}{4}\right)^2\left(\frac{1}{4}\right) + 3\left(\frac{3}{4}\right)\left(\frac{1}{4}\right)^2 + \left(\frac{1}{4}\right)^3$$

$\frac{27}{64}$ have no affected child and are not ascertained.

$\frac{27}{64}$ have 1 affected child

$\frac{9}{64}$ have 2 affected children

$\frac{1}{64}$ have 3 affected children

Since 27/64 of all the three-child families of two heterozygous parents are missed, then among the 37/64 that can be studied:

$\frac{27}{37}$ (73.0%) have 1 affected child

$\frac{9}{37}$ (24.3%) have 2 affected children

$\frac{1}{37}$ (2.7%) have 3 affected children

The expected incidence of affected children in the sibships that can be enumerated can now be calculated:

$$\frac{1}{3} \text{ of } 73.0\% = 24.3\%$$

$$\frac{2}{3} \text{ of } 24.3\% = 16.2\%$$

$$\frac{3}{3} \text{ of } 2.7\% = 2.7\%$$

Expected proportion of affected children = 43.2%

This method of correction of the expected proportion of affected children can be extended to sibships of any size. When the expectations in a given sample have been calculated, the expected and observed numbers of affected sibs can be compared by the X^2 method (page 280).

SINGLE ASCERTAINMENT. It has already been pointed out that whereas under truncate selection the assumption is that all sibships which have at least one affected child are included in the series, under single selection the assumption is that only a small proportion of the available sibships have been located. Under single selection, the probability of any one sibship's being included is low and directly proportional to the number of affected persons it contains.

Consider again the distribution of a recessive trait in three-child sibships, which is the expansion of $(p+q)^3 = p^3 + 3p^2q + 3pq^2 + q^3$.

Since the probability that any sibship is included is now proportional to the number of affected sibs, each term of the expansion is multiplied by the number of affected members, and the expression becomes:

$$0(p)^3 + 1(3p^2q) + 2(3pq^2) + 3(q)^3$$

or

$$3p^2q + 6pq^2 + 3q^3$$

If we now divide by the common term $3q$, the distribution becomes

$$p^2 + 2pq + q^2$$

which is $(p + q)^2$. In other words, the distribution in the *ascertained* families is binomial.

Let us now compare the proportion of affected individuals expected under truncate selection with the proportion expected under single selection, again considering the proportion of homozygous recessives in a series of three-child sibships, each of which has two heterozygous parents. See table opposite page.

Note that under truncate selection, 43 per cent of the children in ascertained three-child sibships are expected to be affected, but under single selection the expectation is 50 per cent. The expectation under either type of ascertainment depends upon family size.

Collections of data strictly suitable for analysis by either complete or single ascertainment cannot often be acquired, and all the corrections to allow for the difficulty of collecting unbiased samples are not within the scope of this book. Differences in methods of ascertainment have been discussed here chiefly to alert students to the necessity of paying attention to how cases for analysis have been collected.

Type of Ascer-tainment	Number of Sibs Affected	Relative Probability of Ascer-tainment	Relative Frequency in Data	Proportion of Sibs	
				Affected	*Unaffected*
Truncate					
	0	0	0	0	0
	1	1	$\frac{27}{64}$	27	54
	2	1	$\frac{9}{64}$	18	9
	3	1	$\frac{1}{64}$	3	0
				48	63
		Expectation $= \frac{48}{48 + 63} = 43\%$			
Single					
	0	0	0	0	0
	1	1	$\frac{27}{64}$	27	54
	2	2	$2\left(\frac{9}{64}\right) = \frac{18}{64}$	36	18
	2	3	$3\left(\frac{1}{64}\right) = \frac{3}{64}$	9	0
				72	72
		Expectation $= \frac{72}{72 + 72} = 50\%$			

TESTS FOR MULTIFACTORIAL INHERITANCE

Multifactorial inheritance, also known as **polygenic** or **quantitative** inheritance, is not easy to analyze genetically, but a number of methods have been developed to demonstrate it. The simplest of these is the method of resemblance between relatives; i.e., the demonstration that the more closely related two individuals are, the more closely they resemble one another in the trait in question.

A useful example of multifactorial inheritance in man, and one of the best in any organism, is the total fingerprint ridge count (TRC) (page 324). If a trait is determined purely by heredity, the genes concerned strictly additive in their effects and mating random for the trait under investigation, then the theoretical correlations between relatives for the trait should be proportional to the genes in common, i.e., the number of genes both have inherited from a common ancestor (Fisher, 1918). "Genes in common" are considered to be those genes inherited from a common ancestral source, though actually any two individuals have in common a great many genes for which the whole population is homozygous, and probably share a great many others which are relatively frequent in the population.

A parent transmits to a child one or the other of the two alleles he possesses at any one locus on a pair of homologous chromosomes; thus, except for X-linked or Y-linked genes, the parent and child have *precisely* half their genes in common. There is a 50 per cent chance that a parent will transmit the same allele of the pair to any two of his children; thus, *on the average,* sibs have half their genes in common. Second-degree relatives (grandparents, uncles, aunts, nephews, nieces, grandchildren, half-sibs) have one-quarter of their genes in common. Third-degree relatives (first cousins) have one-eighth of their genes in common. Second cousins, who are fifth-degree relatives, share 1/32 of their genes. Identical twins, of course, have all their genes in common (Table 11–2).

The simplest measure of resemblance between relatives is the correlation coefficient, which is described in any standard textbook of statistics. As mentioned above, Fisher has shown that the correlation coefficient for multifactorial traits is the same as the proportion of genes in common. In Table 11–2, the correlation coefficients found in Holt's study of fingerprint ridge counts are compared with the proportion of genes in common for each class of relative.

Quasicontinuous Variation

Even if the underlying distribution of variability with respect to a given trait is continuous, a threshold effect may make it appear discontinuous (Fig. 11–1). Such a trait is said to be a **quasicontinuous** or **threshold** trait (Gruneberg, 1952; Edwards, 1960). Quasicontinuous

TABLE 11–2 EXPECTED AND OBSERVED CORRELATION BETWEEN RELATIVES FOR TOTAL FINGERPRINT RIDGE COUNTS*

Relationship	***Expected Correlation Coefficient*** *(Proportion of Genes in Common)*	***Observed Correlation Coefficient***
MZ twins	1.00	0.95
DZ twins	0.50	0.49
Sib–sib	0.50	0.50
Parent–child	0.50	0.48
Parent–parent	0	0.05

*After data of Holt, 1961, 1968.

variation is probably quite important in the etiology of developmental defects; for example, it is easy to understand that if the timing of development shows continuous variation, extremely slow growth could lead to failure of normal development at a later stage that depends upon punctual completion of this stage; e.g., if the optic vesicle reaches the ectoderm too late, the lens is not formed; or if the ureteric bud of the mesonephric duct

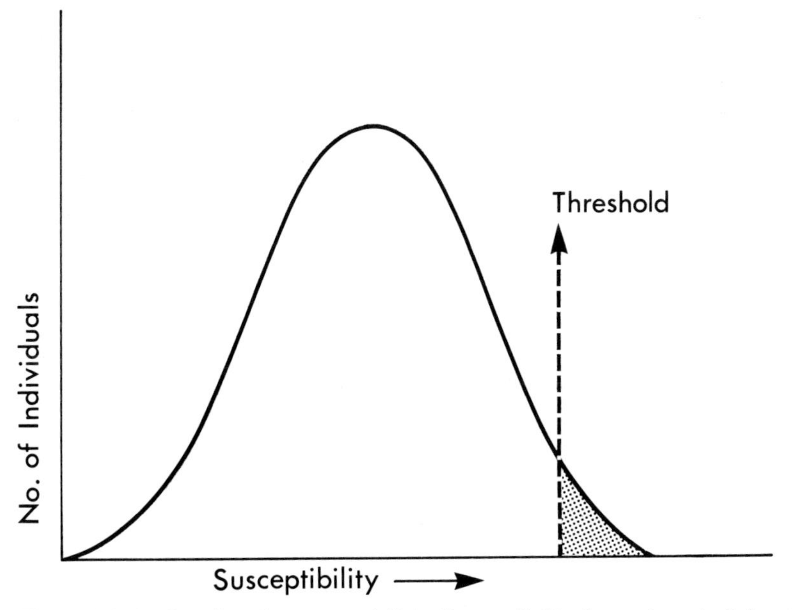

Figure 11–1 *Quasicontinuous variation. Susceptibility to a given trait is normally distributed, but the population is divided into normal and abnormal types by a threshold.*

does not properly complete the development of the collecting tubule system on schedule, the kidney becomes polycystic and nonfunctional.

The peculiar familial incidence of congenital pyloric stenosis can be understood if pyloric stenosis is a threshold trait (Carter, 1964). Before the Ramstedt operation came into general use about 1920, few children born with this condition survived; but since then the mortality has been markedly reduced and the incidence of the disease has now been studied in a large series of the offspring of persons who had pyloric stenosis in infancy. Pyloric stenosis is much more common in boys than in girls; about five of every 1000 boys and only one of every 1000 girls are affected. The proportion of first-degree relatives affected also differs strikingly with sex (Table 11–3). An interpretation of these observations is shown in Figure 11–2.

If the underlying variability with respect to the trait is continuous, but there is a lower threshold for males than for females, then affected females are more extreme with respect to this trait than are affected males, as shown by the relative area of each curve beyond the cutoff line. One characteristic of quantitative inheritance is that the mean of the progeny lies approximately halfway between the mean of the two parents and the population mean. In other words, children of "extreme" parents are on the average less extreme than the parents, but have a mean closer to the extreme than is the mean of the population as a whole. Applying this principle to the pyloric stenosis data, one sees that the mean for the children of these patients would be above the population mean, and since the children are normally distributed around their own mean, more of them fall beyond the cutoff line. Furthermore, since affected females are more extreme than affected males, their children are also more extreme than the children of affected males, and therefore female patients are more likely than male patients to have children who fall beyond the cutoff line.

TABLE 11–3 THE INCIDENCE OF PYLORIC STENOSIS IN FIRST-DEGREE RELATIVES OF PATIENTS, COMPARED WITH THE GENERAL POPULATION*

Male relatives of male patients	×10
Female relatives of male patients	×25
Male relatives of female patients	×30
Female relatives of female patients	×100

*After data of Carter, 1964.

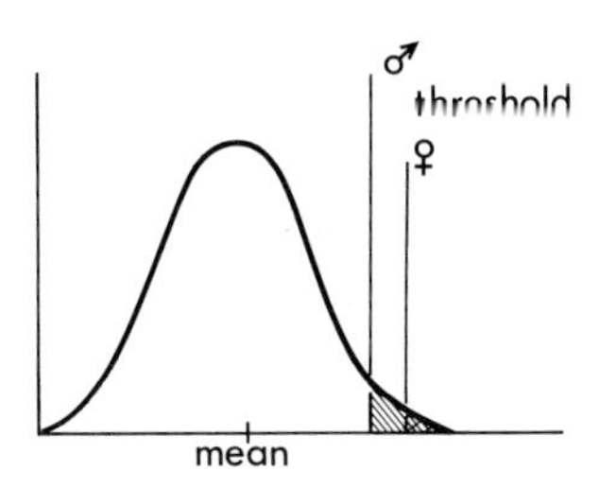

General population

Figure 11–2 *An interpretation of the sex ratio and familial distribution of pyloric stenosis in terms of multifactorial inheritance, with a more extreme threshold in females than in males. In the normal population and among the first-degree relatives of affected individuals, more males than females are affected; but affected females have more affected relatives of both sexes than do affected males. For further discussion, see text. [Based upon Carter 1964. In:* Congenital Malformations: Papers and Discussions Presented at the Second International Conference. *(Fishbein, M., ed.) International Medical Congress, p. 311.*]

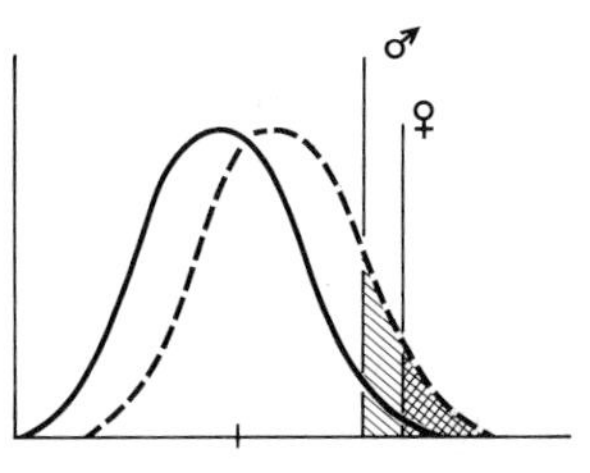

First-degree relatives of males

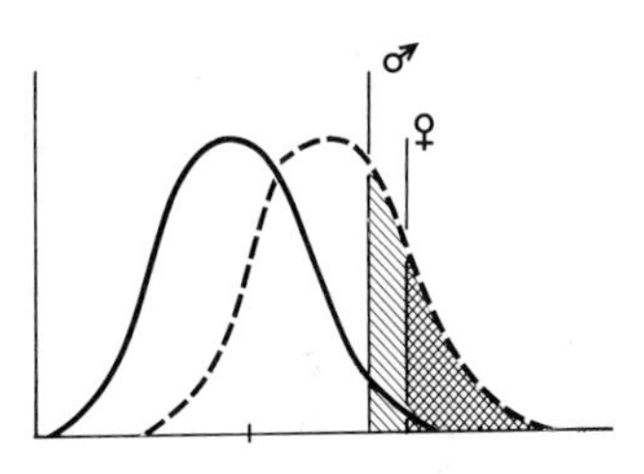

First-degree relatives of females

How to Identify Polygenic Inheritance

Three methods of identifying polygenetic inheritance have been mentioned in the foregoing discussion:

1. For polygenic traits which are not threshold traits (e.g., total fingerprint ridge count, stature), the correlation between relatives is equal to the proportion of genes in common.

2. The mean for the offspring is approximately halfway between the mean for the parents and the mean for the general population.

3. Many common congenital malformations which behave as polygenic threshold traits are much more frequent in one sex than in the other (e.g., pyloric stenosis is five times as common in males as in females, and congenital dislocation of the hip is about seven times as frequent in females as in males). In these, the recurrence risk is higher for relatives of patients of the less susceptible sex. The data for pyloric stenosis given in the foregoing section illustrate this point.

There are other characteristics of polygenic inheritance which can be used to distinguish it from single-gene inheritance. These rules hold only if there is strict additivity of the genes concerned (i.e., no dominance or recessiveness).

4. If the population frequency is p, the risk among sibs of patients is $\sqrt{p}$ (Edwards, 1960). This is not true for traits determined by genes at a single locus. Indeed, if a number of hereditary traits are graphed as to the proportional increase of risk in sibs versus the population incidence, the range for multifactorial traits is quite different from the range for simple dominant and recessive traits (Fig. 11–3). Nora (1968) has compared the incidence of several types of congenital heart defects in the general population and in first-degree relatives of patients, and has found close agreement with this rule (Table 11–4).

5. The recurrence risk is higher when more than one member of a family is affected. For example, for cleft lip and palate the recurrence risk after one affected child is about 4 per cent; but after two affected children it is about 9 per cent (WHO Report, 1970). This is not true of single-gene defects.

6. With polygenic inheritance, the more severe a malformation, the higher the recurrence risk. The risk of cleft lip and palate (CLP) in a subsequent child is about 2.5 per cent if the index patient has unilateral cleft lip, but about 6 per cent if he has bilateral cleft lip and palate (Carter, 1969).

7. The risk to relatives drops off much more rapidly with increasingly remote degrees of relationship for polygenic traits than for autosomal dominants. Some typical figures for common polygenic disorders are shown in Table 11–5.

8. In twin studies, if the concordance rate in monozygotic pairs is

TABLE 11–4 FREQUENCY OF SIX COMMON CONGENITAL HEART DEFECTS IN SIBS OF PROBANDS

Anomaly	*Per Cent Frequency in Sibs**	*Expected Frequency†*
Ventricular septal defect	4.3	4.2
Patent ductus arteriosus	3.2	2.9
Tetralogy of Fallot	2.2	2.6
Atrial septal defect	3.2	2.6
Pulmonary stenosis	2.9	2.6
Aortic stenosis	2.6	2.1

*Data from Nora, 1968.

†$\sqrt{p}$, where p = population frequency of the specific defect.

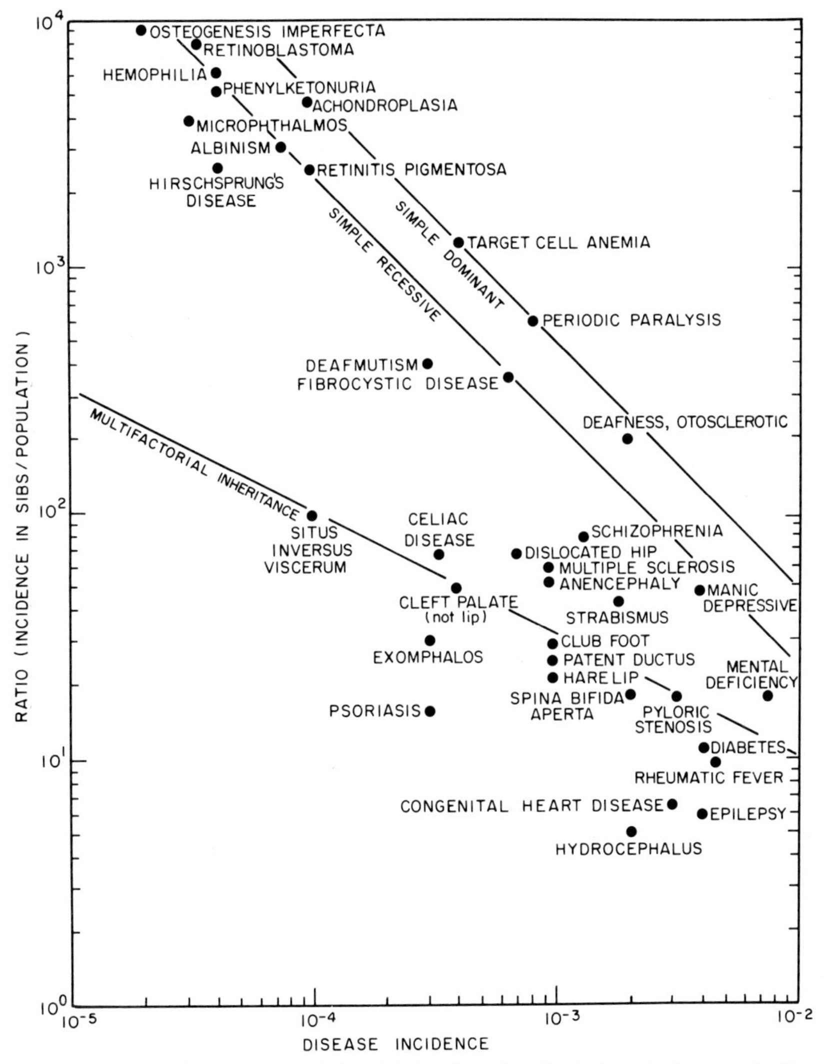

Figure 11–3 *Comparison of the risk in sibs of patients for single-gene traits as compared with multifactorial traits. On a log scale, the incidence of each disease shown is plotted against the ratio of the incidence of the disease in sibs to the incidence in the general population. Single-gene and multifactorial diseases fall into two distinct clusters.* [*From Newcombe 1964. In:* Congenital Malformations: Papers and Discussions Presented at the Second International Conference. *(Fishbein, M., ed.) International Medical Congress, p. 347.*]

TABLE 11–5 FAMILY PATTERNS FOR SOME CONGENITAL MALFORMATIONS*

	Cleft Lip with or without Cleft Palate	*Talipes Equino-varus*	*Congenital Dislocation of the Hip (Males Only)*	*Congenital Pyloric Stenosis (Females Only)*
General population	0.001	0.001	0.002	0.005
First-degree relatives	×40	×25	×25	×10
Second-degree relatives	×7	×5	×3	×5
Third-degree relatives	×3	×2	×2	×1.5

*Data from Carter, 1969.

more than twice as high as the rate in dizygotic pairs, the trait is not a simple dominant; if it is more than four times as high, the trait is not a simple recessive, either. It is then necessary to consider the possibility of polygenic inheritance.

9. Although the demonstration of parental consanguinity is usually taken to imply recessive inheritance, it can also signify that polygenes with additive effects are involved. If recessiveness can be ruled out by demonstration that the parent-offspring correlation is similar to the sib-sib correlation, polygenic inheritance is the only remaining possibility.

10. If a trait is recessively inherited and there is one affected child in a family, the risk for later-born children is the same (25 per cent) whether or not the parents are consanguineous. If the risk is higher when the parents are consanguineous than when they are unrelated, the inheritance is polygenic.

A Note on the Use of Computers for Record Linkage

Useful genetic data for whole populations are gathered routinely by many different agencies for nongenetic purposes. Examples of the kinds of data geneticists would find useful are vital statistics, insurance company records and a wide variety of medical data. Unfortunately, these data are simply too unwieldy to be manageable except by modern methods of information retrieval and data processing. Many questions which concern geneticists could be answered if we could simply extract the information from data already on record. Newcombe and his colleagues have pioneered in this field.

Figure 11–3 is an example of the kind of information that can be obtained by computer methods. At present, a difficulty in the use of these data is that the categories of disorders are broad, some including several separate entities with different frequencies and recurrence risks. More precise records of diagnosis would solve this problem. The problem of confidentiality in the use of vital statistics for such purposes is an area of concern to many people, and may hamper the rapid development of this field.

GENE FREQUENCIES IN POPULATIONS: THE HARDY-WEINBERG LAW

The Hardy-Weinberg law, which is the basis of population genetics, is named for Hardy, an English mathematician, and for Weinberg, a German physician, who independently developed it in 1908.

For any gene locus, in a population in which there is random mating the frequencies of the genotypes are determined by the relative frequencies of the alleles in the population. When we speak of population frequencies, we are thinking about a "gene pool" in which are pooled all the alleles at that particular locus for the whole population. We may think of the gene pool as a beanbag containing beans of two colors, white and black. (Because the beanbag example is so common, population genetics is sometimes called "beanbag genetics.") The chance that with two draws a person will draw any one of the three possible combinations (two white, two black, one of each color) is calculated from the frequency of each color of bean in the bag. If p is the frequency of white beans and $q = 1 - p$ is the frequency of black beans, then the relative proportions of the three combinations are:

$$p^2 \text{ (2 white)} + 2pq \text{ (1 white, 1 black)} + q^2 \text{ (2 black)}$$

Now let the white beans represent the allele T (taster) and the black beans represent the allele t (nontaster). The relative frequencies of the genotypes for arbitrarily selected values of p and q are shown below:

RELATIONSHIP OF GENE FREQUENCY TO GENOTYPE FREQUENCY

Gene Frequencies		***Genotype Frequencies***		
p	q	T/T	T/t	t/t
0.5	0.5	0.25	0.50	0.25
0.6	0.4	0.36	0.48	0.16
0.7	0.3	0.49	0.42	0.09
0.8	0.2	0.64	0.32	0.04
0.9	0.1	0.81	0.18	0.01
0.99	0.01	0.98	0.02	0.0001

If the different genotypes in a population are present in these proportions, the population is said to be in Hardy-Weinberg equilibrium. Note that this is a very simple application of the binomial expansion for $(p + q)^2$.

Hardy-Weinberg equilibrium can be upset by various factors, including nonrandom mating, mutation and selection (see Chapter 12). Nevertheless, it holds true for many genes, and its failure to hold true in a specific situation suggests the possibility of a factor which may be disturbing the equilibrium.

An important consequence of the Hardy-Weinberg law is that the proportions of the genotypes do not change from generation to generation. Table 11–6 shows the next generation following random mating of a population in which the genotypes T/T, T/t, t/t are present in the proportions $p^2{:}2pq{:}q^2$.

TABLE 11–6 FREQUENCIES OF MATING TYPES AND OFFSPRING

Mating Types	*Frequency*	*Offspring* *T/T*	*T/t*	*t/t*
$T/T \times T/T$	p^4	p^4		
$T/T \times T/t$	$4p^3q$	$2p^3q$	$2p^3q$	
$T/T \times t/t$	$2p^2q^2$		$2p^2q^2$	
$T/t \times T/t$	$4p^2q^2$	p^2q^2	$2p^2q^2$	p^2q^2
$T/t \times t/t$	$4pq^3$		$2pq^3$	$2pq^3$
$t/t \times t/t$	q^4			q^4

$$\text{Sum of } T/T \text{ offspring} = p^4 + 2p^3q + p^2q^2 = p^2(p^2 + 2pq + q^2)$$

$$\text{Sum of } T/t \text{ offspring} = 2p^3q + 4p^2q^2 + 2pq^3 = 2pq(p^2 + 2pq + q^2)$$

$$\text{Sum of } t/t \text{ offspring} = p^2q^2 + 2pq^3 + q^4 = q^2(p^2 + 2pq + q^2)$$

Since $p^2 + 2pq + q^2 = 1$, this common factor can be dropped, and the proportions of the three genotypes are then seen to be $p^2{:}2pq{:}q^2$, as in the parental generation.

A widely used application of the Hardy-Weinberg equilibrium is the calculation of gene frequency and heterozygote frequency, if the frequency of the trait is known.

If the frequency of an autosomal trait $(q^2) = 1/10{,}000$

$$\text{The frequency of } q \text{ is } \sqrt{1/10{,}000} = 1/100$$

$$\text{The frequency of } p \text{ is } 1 - \frac{1}{100} = 99/100$$

$$\text{The frequency of heterozygotes is } 2pq = 2\left(\frac{99}{100}\right)\left(\frac{1}{100}\right) \cong \frac{1}{50}$$

Because p (the frequency of the normal allele) is usually close to 1, the heterozygote frequency is usually close to $2q$ (twice the gene frequency).

Frequencies of X-linked Genes

On page 228, we have set out the relative frequencies of the X-linked Xg blood groups in males and females. These data will now be used to show how frequencies of X-linked genes relate to genotype frequencies.

Often the first clue that a trait is X-linked is its different frequency in males and females. For the Xg blood system, 67 per cent of males are Xg(a+) and 33 per cent are Xg(a–), whereas for females 89 per cent are Xg(a+) and only 11 per cent are Xg(a–).

Since males are hemizygous for X-linked genes, there are only two male genotypes:

Genotype	Phenotype	Frequency
Xg^a/Y	Xg(a+)	$p = 0.67$
Xg/Y	Xg(a–)	$q = 0.33$

The frequencies of the two alleles are given directly by the frequencies of the two phenotypes.

Females have two X's, therefore two *Xg* alleles. Three genotypes are possible:

Genotype	Phenotype	Frequency
Xg^a/Xg^a	Xg(a+)	p^2 } $= 0.89$
Xg^a/Xg	Xg(a+)	$2pq$
Xg/Xg	Xg(a–)	$q^2 = 0.11$

From the male data, we see that $q = 0.33$. The square root of the frequency of the Xg(a–) phenotype in females gives the same result. Substituting the male values of p and q, we find that Xg(a+) females should have a frequency of 0.89, which is very close to actual observation.

Note that for deleterious X-linked recessives the ratio of affected males to affected females is q/q^2, or $1/q$. This relationship demonstrates that males are far more frequently affected.

For X-linked dominants, the ratio of affected males to affected females is $q/2pq + q^2$. For rare traits (i.e., when q is very small) this ratio is very close to 1/2, which is to be expected since males have only one X (one "draw" from the gene pool) and females have two. For more common traits the ratio is greater than 1/2.

THE X^2 TEST OF SIGNIFICANCE

Statistical tests of significance are used to determine whether a set of data conforms to a certain hypothesis. Textbooks of statistics and of genetics describe a variety of such tests and the circumstances under which they can appropriately be applied. One of the most useful is the X^2 (Chi square) test, which is applicable to many problems in genetics because it does not require that the data analyzed be more or less normally distributed, but only that the numbers in the different categories be known.

The calculation of X^2 is quite simple. If O is the observed number in each category and E is the number expected in that category on the basis of the hypothesis being tested, then X^2 is the square of the difference between O and E, divided by E, summed over all the categories; in other terms,

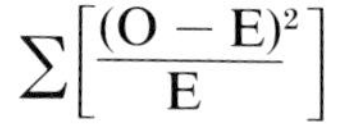

$$\Sigma\left[\frac{(O - E)^2}{E}\right]$$

The probability associated with a given value of X^2 can be obtained from X^2 tables originally prepared by Fisher (1944) and reprinted in many textbooks. (Before using these tables, it is necessary to know the number of "degrees of freedom" available; the examples given subsequently should show how this can be determined.) A value of X^2 associated with a probability of less than 0.05 is considered to be "significant," i.e., to indicate a significant disagreement between the observations and the hypothesis being tested.

The hypothesis being tested may be a "null hypothesis," in the sense that the observed values do not differ from the expected, or that two variables being compared are not in any way associated. If the null hypothesis is disproved, the observed and expected values are not in agreement.

Example 1. A series of patients with congenital pyloric stenosis includes 25 boys and five girls. Does this distribution differ from the normal sex ratio?

	Number Observed	***Number Expected***
Boys	25	15
Girls	5	15
Total	30	30

$$X^2 = \frac{(25 - 15)^2}{15} + \frac{(5 - 15)^2}{15} = 6.7 + 6.7 = 13.4$$

The expected values are calculated on the basis of 1:1 sex ratio. There is one degree of freedom: since the total must remain fixed at 30, only one of the two values can be freely assigned. Consulting a table of X^2, we see that for one degree of freedom, this value of X^2 would occur by chance with a probability of <0.01, i.e., less than once in 100 times. (The acceptable level of significance is $p = 0.05$.) Thus, it can be concluded that in this series the observed excess of boys is statistically significant.

Example 2. In a series of twins, at least one member of each pair having congenital dislocation of the hip, it was found that the defect was present in both members (concordant) in 40 of 100 monozygotic pairs, and five of 100 like-sexed dizygotic pairs. (Since congenital dislocation of the hip is about six times as frequent in females as in males, it is appropriate to compare like-sexed pairs rather than all dizygotic pairs.)

	Concordant	*Discordant*	*Total*
MZ pairs	40	60	100
Like-sexed DZ pairs	10	90	100
Total	50	150	200

In this example, the expected proportion in each category can be calculated on the basis of the null hypothesis (that the degree of concordance is the same for MZ and DZ pairs). In the whole series, the ratio of concordant to discordant pairs is 50/150 = 1/3. The expected numbers then are:

MZ concordant	33.3
discordant	66.7
DZ concordant	33.3
discordant	66.7

$$X^2 = \frac{(40 - 33.3)^2}{33.3} + \frac{(60 - 66.7)^2}{66.7} + \frac{(10 - 33.3)^2}{33.3} + \frac{(90 - 66.7)^2}{66.7} = 26.5$$

Since the marginal totals in the table above must remain fixed, only one of the numbers of twin pairs can be filled in at random; thus there is one degree of freedom. Again, the X^2 table shows the calculated value of

X^2 to be beyond the level of significance ($p < 0.01$). The conclusion is that the observed excess of concordant pairs in MZ as compared with DZ pairs is statistically significant. This suggests that congenital dislocation of the hip is partly genetically determined.

THE RECOGNITION AND MEASUREMENT OF LINKAGE

If two genes have their loci on different chromosomes, or far apart on the same chromosome, they assort at meiosis independently of one another, i.e., they are equally likely to pass to the same or different gametes. If two genes have their loci on the *same* chromosome, especially if the two loci are close together, they do not assort independently—in other words, they are transmitted together to the same gamete more than 50 per cent of the time—and are said to be *linked.* The inheritance of linked genes is an exception to Mendel's law of independent assortment.

Linkage can best be studied in the progeny of a person who is doubly heterozygous (heterozygous at each of two loci), mated to a person who is homozygous for the recessive allele at each locus ($A/a\ B/b \times a/a\ b/b$). This is a "double backcross" mating. A backcross mating is a mating of a heterozygote and a recessive homozygote. Its special advantage is that because the one parent contributes only recessive alleles, the phenotypes of the offspring reveal the genotype of the heterozygous parent. If there is no linkage, in an $A/a\ B/b \times a/a\ b/b$ mating, the progeny would be 25 per cent $A/a\ B/b$, 25 per cent $A/a\ b/b$, 25 per cent $a/a\ B/b$ and 25 per cent a/a, b/b; if wide deviations from these ratios are found, assortment of A and B is nonrandom.

Linkage may be in **coupling** or in **repulsion;** or, in other terms, in *cis* or *trans* configuration. If in an individual whose genotype is $A/a\ B/b$, the genes A and B are on the same chromosome (i.e., A and B came from one parent and a and b from the other), A and B are linked in **coupling;** if they are on opposite chromosomes, A and B are linked in **repulsion.** The linkage phase can be shown by writing the genotypes as follows:

$\frac{AB}{ab}$	$\frac{Ab}{aB}$
Coupling	***Repulsion***

The symbols for these linked genes may also be written AB/ab and Ab/aB, respectively. (The slash represents the chromosome pair on which the genes are located.)

These symbols designate the following distribution of the two genes on the chromosome pair concerned:

The two possible matings are as follows:

$\frac{A\quad a}{B\quad b} \times \frac{a\quad a}{b\quad b}$ or $\frac{A\quad a}{b\quad B} \times \frac{a\quad a}{b\quad b}$

In either case, the doubly recessive parent can transmit only an *a* and a *b* allele to any child, so the children's genotypes will expose the genes each has inherited from the doubly heterozygous parent. If recombination has occurred between the A and B loci, the recombination will be obvious from the phenotype of the child if the linkage phase in the parent is known. In the following pedigree, individual II–5 is a recombinant.

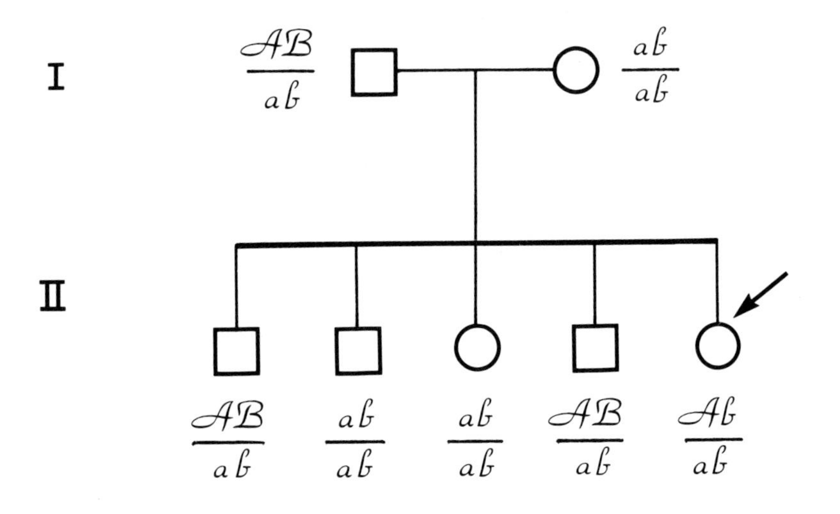

A single family is rarely large enough to allow linkage to be established with certainty, but in a series of families the observations might be as follows:

$$\text{Genotypes of parent:} \quad \frac{AB}{ab} \times \frac{ab}{ab}$$

$$\text{Genotypes of offspring:} \quad 45\% \ \frac{AB}{ab} \quad 5\% \ \frac{Ab}{ab} \quad 5\% \ \frac{aB}{ab} \quad 45\% \ \frac{ab}{ab}$$

These data show that in the offspring of this type of mating, 90 per cent have the parental combinations and 10 per cent have recombinations (are recombinants); if assortment had been independent, the expectation would be 25 per cent for each class, 50 per cent parental and 50 per cent recombinant. In a large enough series, so large a discrepancy from independent assortment of A and B is strong evidence that the gene loci concerned are linked.

In the example given, the genes A and B are linked in coupling. Actually, one cannot tell without family data whether linkage is in coupling or repulsion; that is, if the A allele has been inherited on the chromosome received from the father of the double heterozygote, and the B allele on the homologous chromosome received from his mother, the two genes are in repulsion, not in coupling. Again, the offspring are classified as to whether they carry the two genes in the same chromosomal combination as the doubly heterozygous parent or in the alternative combination. The observations might be as follows:

$$\text{Genotypes of parents:} \quad \frac{Ab}{aB} \times \frac{ab}{ab}$$

$$\text{Genotypes of offspring:} \quad 45\% \ \frac{Ab}{ab} \quad 5\% \ \frac{AB}{ab} \quad 5\% \ \frac{ab}{ab} \quad 45\% \ \frac{aB}{ab}$$

Over sufficient generations, crossing over should take place frequently enough that the coupling and repulsion phases should be equally frequent. Whether the two genes are in coupling or in repulsion, the conclusion from the above data is that their loci are linked, with 10 per cent recombination between them; and the loci are said to be "10 units apart."

Figure 11–4 illustrates the different expectations for the offspring of a doubly heterozygous parent when the two abnormal genes are linked in coupling and when they are linked in repulsion. Each mother is doubly heterozygous, carrying the X-linked genes for a form of color blindness

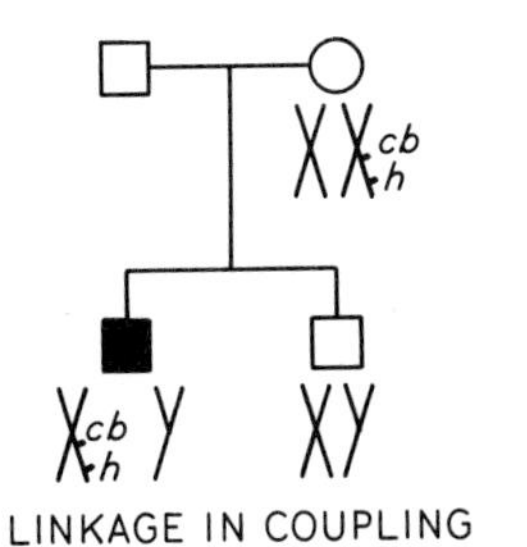

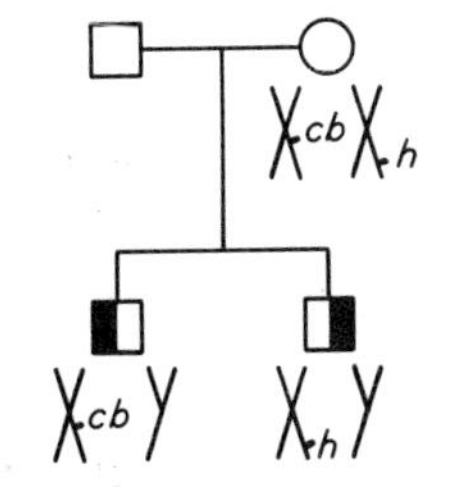

Figure 11–4 *Linkage in coupling and in repulsion. In both pedigrees the mother is heterozygous for color blindness* (cb) *and for hemophilia* (h). *If these genes are linked in coupling, her sons will be either doubly affected or normal. If they are linked in repulsion, her sons will have one condition or the other, but not both!*

(*cb*) and hemophilia (*h*). When the two genes are in coupling, any one son receives both or neither; when they are in repulsion, any one son receives one or the other. Recombination between these two loci by crossing over is known to occur about 3 per cent of the time.

LINKAGE MAPS

In the foregoing section the method of measuring linkage by recombination frequency has been outlined. Recombination frequencies are used to map the positions of genes on chromosomes, and so to construct linkage maps. The lower the percentage of recombination between two loci, the closer together they are on the linkage map.

The physical basis of recombination is the crossing over of chromosomal material that occurs in association with the chiasmata seen during meiotic prophase (Fig. 11–5). If a crossover occurs between the loci of genes *A* and *B*, the genes will then be on different chromosomes of the homologous pair. This is recombination. However, if two crossovers occur, the genes will still be on the same chromosome, and there will have been no recognizable recombination. Since only those crossovers that result in recombinations can be recognized genetically, the recombination frequency is always somewhat lower than the true crossover

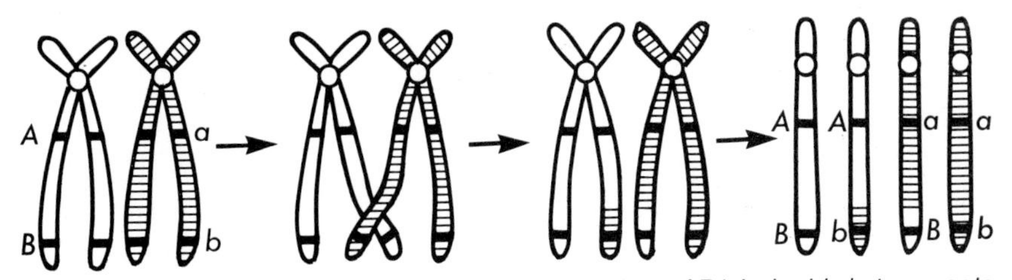

Figure 11–5 *Crossing over and recombination in an* AB/ab *double heterozygote. At the end of meiosis, the gametes are of four types: two* parental types, AB *and* ab, *and two recombinants,* Ab *and* aB.

frequency. The closer together two gene loci are, the more accurately the recombination frequency measures the crossover frequency.

The distances between genes which are far apart on a chromosome cannot be measured directly, but if a gene intermediate between them is known, the distance from each to the intermediate one can be derived. Extensive linkage maps have been worked out for experimental organisms such as *Drosophila* and the mouse. In man, the first autosomal linkage, between the Lutheran blood group locus and the secretor locus, has been known for about 20 years (Mohr, 1951), but the rate of progress in discovering new linkages was discouragingly slow until very recently. Identification of new genetic markers, chromosome variants, cell culture techniques and new chromosome staining techniques have given a great impetus to linkage studies. We now know 10 loci on chromosome no. 1 (Ruddle et al., 1972), and great progress is being made in mapping the rest of the human chromosomes. The genes on the X can readily be identified by their unique pedigree patterns. Because the genes at the Xg locus are both relatively common (Xg^a having a frequency of about 2/3, and Xg about 1/3), many studies of linkage of other loci to Xg have been made. The G6PD locus and the color blindness loci are others which are useful for linkage investigations, but by bad luck they mark almost the same spot on the X. A tentative map of the X is reproduced in Figure 11–6; however, the pace of progress is such that new data change the human chromosome map almost daily.

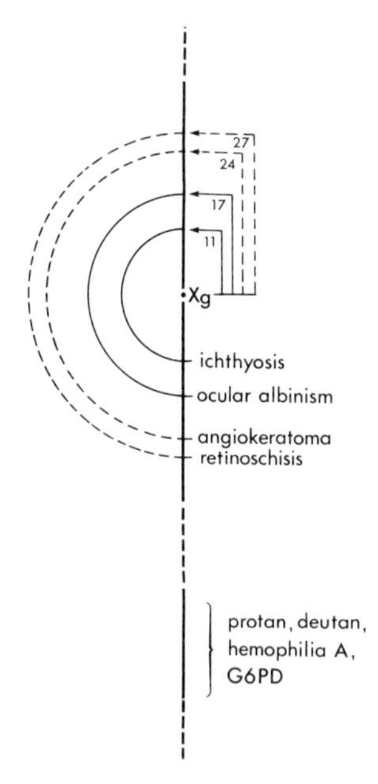

Figure 11–6 *A tentative map of the X chromosome. Four loci are linked to Xg, but though their distances from Xg have been measured, the direction along the chromosome is uncertain. Four other loci (those for the protan and deutan types of color blindness, hemophilia A and glucose-6-phosphate dehydrogenase) are close together at a different part of the chromosome. (Adapted from Race and Sanger, 1968.* Blood Groups in Man. *5th ed. Blackwell Scientific Publications, Oxford, p. 552.)*

OTHER METHODS OF DETECTION OF LINKAGE

In linkage studies in man, the following questions are asked. (1) Are two loci linked? (2) How far apart are they? (3) Can they be assigned to a specific chromosome?

Several different approaches to linkage analysis can now be used:

1. Computer methods have been developed for scanning large bodies of data for linkages. The number of genetic markers available for linkage tests has steadily increased. Consequently, formal linkage studies are going on at an increasingly rapid rate.

2. Chromosome identification techniques have been greatly improved. It is possible to trace variant or aberrant chromosomes through families and to infer what gene loci lie on the unusual chromosome. In this way, the Duffy blood group locus was found to be situated near a heterochromatic region found in a variant chromosome no. 1 (Donahue et al., 1968). (See page 226.)

3. Cell hybridization techniques are a third approach to linkage analysis. Under special conditions, human cells can be made to "hybridize" in culture with cells of other organisms such as the mouse or Chinese hamster. In man–mouse hybrids, the human chromosomes are selectively lost. It is possible to hybridize human and mouse cells and to follow single-cell clones (cultures established from one hybrid cell) to learn which of a battery of biochemical markers disappears at the same time that a particular human chromosome is lost. In this way Migeon and Miller (1968) found that the locus for thymidine kinase is on chromosome no. 17, and Ruddle et al. (1970) found that the loci for two human enzymes, lactate dehydrogenase B and peptidase B, are on the same chromosome (no. 12). These pioneering studies have led to rapid advances in knowledge of the human chromosome map.

For students interested in the theoretical basis and methodology as well as results, Renwick (1971) has recently reviewed human chromosome mapping.

THE MEASUREMENT OF CONSANGUINITY AND INBREEDING

The measurement of consanguinity has some importance in medical genetics because consanguineous marriages have an above-average risk of producing offspring homozygous for some deleterious recessive gene. This risk is proportional to the closeness of the relationship of the parents concerned. For practical purposes, only first-cousin marriages occur often enough and carry a sufficiently increased risk to be of practical importance.

COEFFICIENT OF RELATIONSHIP

The coefficient of relationship (r) is defined as the chance that two persons have a gene in common, *or* the proportion of all their genes which have been inherited from common ancestors. For first cousins, it is one-eighth.

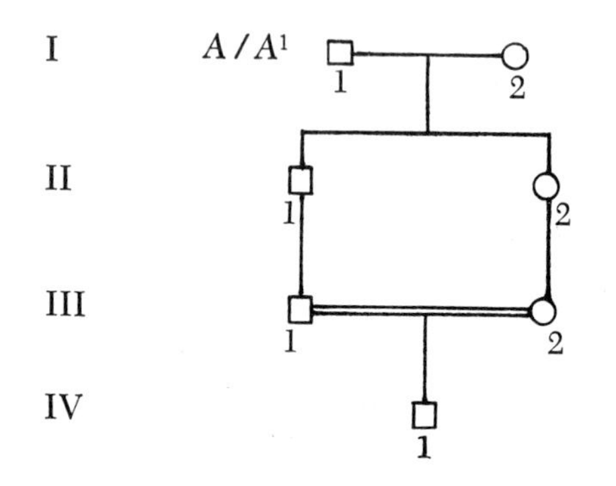

The value of r is calculated as follows. In the above sketch, III–1 and III–2 are first cousins. The chance that III–1 has inherited allele A from his paternal grandfather is $1/2 \times 1/2 = 1/4$. The chance that III–2 has also received it is again $1/2 \times 1/2 = 1/4$. Thus, the chance that both III–1 and III–2 have allele A is $1/4 \times 1/4 = 1/16$. There is the same chance that each has received allele A^1. The total probability that they have each received the *same* allele from the paternal grandfather is $1/16 + 1/16 = 1/8$.

COEFFICIENT OF INBREEDING

The coefficient of inbreeding (F) is the probability that an individual has received both alleles of a pair from an identical ancestral source, or the proportion of loci at which he is homozygous. In the sketch above, IV–1 is the offspring of a first-cousin marriage. It has been shown that for any gene the father has, the chance that the mother also has it is one-eighth. For any gene he gives his child, then the chance that the mother has the same gene and will transmit it is $1/2 \times 1/8 = 1/16$. This is the coefficient of inbreeding for the child of first cousins. It signifies that he has a $1/16$ chance of being homozygous at any one locus, or that he is homozygous at $1/16$ of his loci.

Some values of the coefficient of relationship and coefficient of inbreeding that might be encountered are listed in Table 11–7.

TABLE 11-7 MEASUREMENTS OF CONSANGUINITY

Relationship	**Coefficient of Relationship** *(Genes Possessed in Common)*	**Coefficient of Inbreeding of a Child of Such Parents** *(Proportion of Loci at which the Child is Homozygous)*
Father–daughter	$\frac{1}{2}$	$\frac{1}{4}$
Uncle–niece	$\frac{1}{4}$	$\frac{1}{8}$
First cousins	$\frac{1}{8}$	$\frac{1}{16}$

GENERAL REFERENCES

Carter, C. O. 1965. The inheritance of common congenital malformations. *Progr. Med. Genet.* 4:59–84.

Carter, C. O. 1969. Genetics of common disorders. *Brit. Med. Bull.* 25:52–57.

Crow, J. F. 1965. Problems of ascertainment in the analysis of family data. In: *Genetics and the Epidemiology of Chronic Diseases.* (Neel, J. V., et al., eds.) U. S. Department of Health, Education and Welfare, Public Health Service Publication 1163.

Fisher, R. A. 1944. *Statistical Methods for Research Workers.* Oliver & Boyd Ltd., Edinburgh.

Renwick, J. H. 1971. The mapping of human chromosomes. In: *Annual Review of Genetics* 5:81–120. (Roman, H. L., ed.) Annual Reviews Inc., Palo Alto, California.

PROBLEMS

1. If a coin is flipped six times, what is the probability of obtaining:
   (a) 6 heads
   (b) 3 heads and 3 tails?

2. In a four-child family:
   (a) What is the probability that the fourth child will be a boy?
   (b) What is the probability that all four children will be girls?
   (c) What is the probability that at least one child will be a boy?

3. Two parents are carriers of the same recessive gene for deafmutism.
   (a) What is the probability that their first child will be deaf?
   (b) What is the probability that their five children will all be normal?
   (c) What is the probability that two of their five children will be affected?

4. A man with Huntington's chorea has three children.
   (a) What is the chance that all three will develop Huntington's chorea?
   (b) What is the chance that the first child will be affected and the next two normal?
   (c) What is the chance that one child will be affected and two normal?

5. (a) If the population frequency of brachydactylous individuals is 1/10,000, what is the frequency of the dominant gene responsible for brachydactyly?
   (b) If the population frequency of a certain form of autosomal recessive deafmutism is 1/10,000, what is the frequency of the gene concerned? What is the frequency of carriers?
   (c) If the frequency of hemophilia in males is 1/10,000, what is the frequency of the X-linked recessive gene concerned? What is the frequency of affected females?

6. In a certain population, blood group genes M and N are equal in frequency. What are the proportions of the three possible blood groups in this population?

7. Students in three university classes are asked the number of boys and girls in their sibships, including themselves. The data are pooled for each class. The totals are:

   (Class 1) Males 175, Females 80
   (Class 2) Males 119, Females 137
   (Class 3) Males 73, Females 169

   One of these classes is in the Faculty of Nursing, one in Arts, one in Engineering. Which is which?

8. Given a series of two-child families, in each of which there is one child with cystic fibrosis:
   (a) Under truncate selection, what proportion of all the children in the sibships will be affected?
   (b) Under single selection, what proportion of all the children in the sibships will be affected?

9. In a certain population, three genotypes are present in the following proportions: A/A, 0.81; A/a, 0.18; a/a, 0.01.
   (a) What is the frequency of each gene?
   (b) What will be the frequencies of the genotypes in the next generation?
   (c) What proportion of all the matings within this population are of the $A/a \times A/a$ type?

10. In a certain isolated population of 800 individuals, all members are of blood group O. In another population, all members are of blood group A. Assume that 200 members of the latter population are added to the first group. What are the frequencies of blood groups A and O in the population after a generation of random mating?

11. A color blind man with G6PD deficiency marries a homozygous normal woman. Their daughter marries a normal man. She has four sons: two are color blind and G6PD-deficient, one has normal color vision and G6PD deficiency and one is normal in both respects. Which of the sons are recombinant, and which are nonrecombinant? Are the color blindness and G6PD deficiency genes linked in coupling or in repulsion in the mother?

12. A man of blood group A and with nail-patella syndrome is married to a normal woman of blood group O. Their four children have the following phenotypes:

    2 A, normal
    2 O, nail-patella syndrome.

(a) Are the A and nail-patella syndrome genes of the father in coupling or in repulsion?
(b) Are any of the children recombinants?

13. A woman is heterozygous at three X-linked loci, the Xg blood group locus, a color blindness locus and the Christmas disease (hemophilia B) locus. If she has Xg^{a}, cb and h on one chromosome, and Xg, the normal allele of cb $(+^{cb})$ and the normal allele of h $(+^{h})$ on the other, what are the possible genotypes and phenotypes of her sons? Indicate the most likely and the least likely types.

14. Two sisters marry two brothers. The son of one couple marries the daughter of the other couple. What is the coefficient of inbreeding of their children?

15. Monozygotic twin sisters marry dizygotic twin brothers. What is the coefficient of relationship of the children of one couple to the children of the other couple?

12

Population Genetics

Population genetics is the study of the distribution of genes in populations and of how gene and genotype frequencies are maintained or changed. In Chapter 11 it was pointed out that under certain ideal circumstances, genotypes are distributed in proportion to the gene frequencies in the population and remain constant from generation to generation. If gene and genotype frequencies remained permanently fixed, evolution could not occur; hence, the study of population genetics involves the study of factors concerned in human evolution.

The Hardy-Weinberg law (page 277) is the basis of population genetics. To restate the law briefly: if there are two alleles A and a, with frequencies p and q, then after a generation of random mating the frequencies of the genotypes are $p^2(A/A)$, $2pq(A/a)$, and $q^2(a/a)$. (This assumes, of course, that the frequencies of A and a are the same in the two sexes.) If in a population it is found that Hardy-Weinberg equilibrium is not maintained (that is, if the genotypes are not randomly distributed or if they change from generation to generation) there must be a reason. Factors known to disturb Hardy-Weinberg equilibrium include nonrandom mating, mutation, selection and migration.

SYSTEMS OF MATING

Hardy-Weinberg equilibrium is maintained only if there is **random mating** (also known as panmixis)—i.e., if any one genotype at a locus has a purely random probability of combining with any other genotype at that locus—so that the frequencies of the different kinds of matings are determined only by the relative frequency of the genotypes in the population. The random mating requirement is probably rarely fulfilled in practice. Within any population there are many subgroups differing genetically (e.g., members of different ethnic groups which each have characteristic gene frequencies) and nongenetically (e.g., with respect to religion). Members of such subpopulations are more likely to marry within their own subgroup than outside it. If mating is nonrandom, it is said to be **assortative**, and it is a common observation that within any population mating tends to be positively assortative with respect to intelligence, stature, economic status and many other traits, some of which are genetically, or partly genetically, determined. (Assortative mating may be **positive,** as when there is a tendency for persons who resemble one another to marry more frequently than chance alone would indicate; or **negative**, in the sense that "opposites attract.")

For some traits, it is unlikely that mating is assortative. Probably the blood groups have little primary effect upon random mating, though they may have a secondary effect, since their frequencies vary with ethnic background. The same is true of fingerprint ridge counts and many other traits which have no obvious phenotypic effect, or at least none which is expressed before the mating age.

Consanguineous mating is a special form of assortative mating. It can disturb Hardy-Weinberg equilibrium by increasing the proportion of homozygotes at the expense of heterozygotes. Of course, inbreeding alone does not affect the proportions of alleles in the next generation, but only their assortment into genotypes. However, it may expose recessive genes to selection and loss, and in this way can permanently alter the gene frequencies in the population.

In North America today, the incidence of consanguineous matings is lower than in older countries, probably partly because in this relatively newly settled area there are fewer cousins among whom to find a mate. The lower birth rate also reduces the number of cousins available. In many countries the improvements in communication of the last century have had the effect of breaking down isolates (enclaves within which the members intermarried much more frequently than they married outsiders).

Of course, given true random mating, a certain small proportion of marriages would be consanguineous. For that matter, probably many marriages are at least distantly consanguineous. One can calculate the number of ancestors one had in previous generations (recommended as a better soporific than counting sheep). Estimating about three generations per century, at the time of the Declaration of Independence the young adults of today each had about 64 (2^6) ancestors, and these may well have been 64 different people; but at the time of the Magna Carta the number was 2^{24}, or about 17 million, and it is very unlikely that there were 17 million different people involved. Probably many people are homozygous for genes that have come down in different lines from one heterozygous ancestor. Few people can trace their ancestry in all its ramifications beyond the first few generations. French Canadians who can do this and whose progenitors have been in Canada since the early 1600s can often find themselves to be inbred through 10 or more different lines of descent.

MUTATION

New hereditary variations arise by mutation, and the new gene (or the individual who carries it) is called a **mutant.** Mutations are the only source of the material of evolution, upon which natural selection acts to preserve the fit and to eliminate the less fit.

In a sense, any change in the genetic material may be regarded as a mutation, but here we will follow the usage of Crow (1961) and define a mutation as a change that cannot be shown to depend upon a detectable chromosomal rearrangement or some sort of recombination mechanism. Though small chromosomal aberrations are impossible to distinguish from actual mutations in human material, the following discussion will be restricted as much as possible to "point mutations," i.e., mutations involving changes in base sequence in individual gene loci.

RANDOM AND HARMFUL NATURE OF MUTATIONS

A mutation usually involves loss or change of the function of a gene. The great majority of mutations are harmful. A random change is unlikely to lead to an improvement, the more so since a living organism is complex and highly integrated, and probably is also thrifty with respect to its metabolic processes, not having spare biochemical pathways with which to experiment. Obviously some mutations must confer an advantage, since some do become established in the population by selection. The frequency of a gene in the population represents a balance between the mutation rate of the gene and the effectiveness of selection against it. If either the mutation rate or the effectiveness of selection is altered, the gene frequency changes.

Many mutations are lethal, leading to the death of the persons who receive them. A mutation which leads to sterility is just as lethal in the genetic sense as one which causes an early abortion. Others may be sublethal. Most of those actually observed in man are less severely detrimental, since a mutation cannot be identified unless it is compatible with life.

LOAD OF MUTATIONS

The load of genetic damage in man is not accurately known, but according to one estimate at least 6 per cent of all persons born have some tangible genetic defect. At this rate, some 20,000 Canadian and 200,000 American babies are born each year with some sort of genetic handicap. Some of these defects involve gross chromosomal changes, but the majority involve specific genes. Perhaps a quarter of all genetic defects are caused by single-gene mutations.

It is very difficult to know how many of these mutations are new and how many have been inherited from the previous generation, but Morton et al. (1956) have estimated that each individual in the population carries three to five **lethal equivalents.** A lethal equivalent is defined as one gene which, if homozygous, would be lethal, or two genes which, if homozygous, would be lethal in half the homozygotes and so on. One way of measuring lethal equivalents is by comparison of the mortality of the offspring of consanguineous and random marriages.

Even a slight heterozygote advantage can lead to the preservation and increase in the population of a gene which, when homozygous, is severely detrimental. The classic example is the resistance to malaria afforded by the sickle cell trait. Thalassemia and glucose-6-phosphate dehydrogenase deficiency also provide protection against malaria. It is suspected that a similar but unknown advantage may account for the high frequency of the cystic fibrosis (CF) gene, but there is no reasonable suggestion as to what the advantage might be. Human genes evolved under very different environmental conditions from those that exist today, and we can only speculate about what has made the CF gene so common in Caucasians yet so rare in Orientals.

In experimental organisms, some genes which are detrimental when homozygous are transmitted in some way through the sperm to far more than the usual 50 per cent of the offspring of heterozygotes. A detrimental gene which confers an advantage on the gamete has a greatly enhanced chance of survival and increase.

MUTATION RATE

The spontaneous mutation rate varies for different loci, but only within rather narrow limits, the observed range being between about one in 10,000 (1×10^{-4}) and one in 1,000,000 (1×10^{-6}) per locus per gamete, and the average being about 1×10^{-5}. Measurements of mutation rates in man are usually not highly accurate, but very great differences from these values may mean that mutations at several different loci with similar phenotypic effects are being measured as though all occurred at a single locus.

For rare dominant mutations, the method of measurement is relatively straightforward. The steps required are as follows:

1. Measurement of the number of cases of the disorder that have been born to normal parents in a limited area over a limited period of time, n.

2. Measurement of the total number of births over the same period, N. Then, since the birth of a child with a dominant trait requires a mutation in only one of the two gametes concerned, the mutation rate is $n/2N$.

In practice, it may be necessary to make allowance for unascertained cases, unclassified parents, the possibility of failure of expression in a heterozygous parent and, especially, genetic heterogeneity. If the same phenotype can be produced by mutation at more than one locus or by environmental factors, accurate measurement of the mutation rate is impossible.

For recessive genes the mutation rate is much more difficult to measure because it is impossible to know in a given instance whether the gene in question is a new mutant or has been inherited. Heterozygote advantage, if present, may lead to a gross overestimate of the mutation rate. Similarly, the mutation rate of X-linked genes is difficult to measure because such genes, if recessive, are not expressed in heterozygous females. In a sporadic case it is often impossible to know whether the mother of an affected male is heterozygous or whether she has transmitted a new mutation to her son.

One would expect that the older a parent is at the time a child is conceived, the more likely he is to have accumulated mutations which the child might inherit. For autosomal dominants, this has been demonstrated. It has been shown that the fathers of achondroplastic dwarfs who are new mutants are on the average a few years older than other fathers in the population. The increased risk of mutant offspring with increasing paternal age is an argument for having children while young.

INDUCED MUTATIONS

Mutations can be induced by a variety of mutagenic agents; temperature, chemicals and ionizing radiation can increase the mutation rate. Most of the experimental work on induced mutations has been performed with organisms such as *Drosophila* and mice, but there is no reason to think that the results, especially those from mammals, are not applicable to humans as well.

Temperature

In male mammals, the gonads are normally maintained at a temperature lower than that of the rest of the body. Since temperature affects the rate of any chemical reaction, it is not surprising that temperature affects the mutation rate. In a study in Sweden it was found that the scrotal temperature was about 3 degrees higher in men wearing trousers than in nudists, so it appears that social custom may actually be a factor in raising the mutation rate in males. It is only fair to add that there is no strong evidence for a difference in mutation rate in males and females.

Chemicals

Many chemicals have been shown to be carcinogenic (cancer-inducing), but since most of these never come in contact with germ cells, it is not known how many are also mutagenic. Nitrogen mustards are both carcinogenic and mutagenic in experimental animals. In man, they are used in the therapy of some malignant conditions, but whether they are actually mutagenic is not known. As a general rule, any carcinogenic chemical should be regarded as probably mutagenic, and should not be used prior to the end of the reproductive period unless the indications for its use are very strong.

Caffeine and related substances may be mutagens, since caffeine is a purine and may interfere with the purine metabolism necessary for DNA replication. The importance of its mutagenic effect in mammals is still a matter of speculation.

Ionizing Radiation

Ionizing radiation is by far the most potent mutagen known. The 1946 Nobel prize was awarded to Muller for his discovery that X-rays raised the mutation rate. Since Muller's original work in 1927, concern has developed over the possible effects of diagnostic and therapeutic X-rays, radioisotopes and the fallout from the testing of atomic weapons.

The relationship between the amount of radiation and the number of mutations produced is linear, except possibly at very low intensity. Even at low dosage, there is probably a mutagenic effect for X-rays, radioisotopes and fallout.

From a genetic standpoint, only the gonadal dosage of irradiation is of any importance. Johns (1961) describes the **significant gonadal dosage** as the measured dosage to a particular group of people, adjusted for the average child expectancy in this group. Of course, if the group is past the childbearing age, the significant gonadal dose becomes zero. It is estimated that the significant gonadal dosage per person per year in the United States is 135 ± 100 millirads (mrad). Natural background provides perhaps 95 mrads per year. The accumulated dosage acquired by the average person from conception to the end of the reproductive period is perhaps 3 r. For clarification, the roentgen (r) is the unit of exposure dose, while the rad is the unit of absorbed dose. The absorbed dose is, of course, the significant one from a genetic standpoint.

The following list of estimated gonadal doses in some diagnostic X-ray procedures (Johns, 1961) gives some idea of the importance of X-ray procedures from a genetic standpoint.

Procedure	***Male***	***Female***
X-ray of bony skeleton (pelvis)	6000 mrads	3000 mrads
Upper gastrointestinal series	220 mrads	300 mrads
Upper gastrointestinal fluoroscopy	500 mrads	750 mrads
Chest X-ray	1.2 mrads	0.3 mrads

Gonadal dose depends upon the location of the beam of X-ray in relation to the gonads. Those procedures which bring the gonads directly into the beam of course produce the greatest gonadal exposure. Beams which miss the gonads still subject them to irradiation from scatter, with the effect diminishing the further the beam passes from the gonads. Narrow beams and adequate lead shielding, where possible, help to decrease gonadal exposure, as does increasing the voltage. It should be stressed that X-rays of the pelvis during pregnancy not only strongly irradiate the mother's gonads but subject those of the fetus to a particularly strong blast.

Evidence of the production of mutations by ionizing radiation in humans must necessarily be indirect, but some is available from studies of the survivors of the atomic bombings of Hiroshima and Nagasaki. Schull and Neel (1958) analyzed the records of children born over a period of years to the survivors of the bombings and found that there was an alteration in the expected ratios of males and females.

Theoretically, radiation to the parents could produce changes in the sex ratio, since X-linked lethal mutations might be induced. However, it is necessary to know the radiation history of each parent because:

1. If only the father is irradiated, any radiation-induced X-linked dominant lethal mutations will be passed only to his daughters, and thus he would be expected to have fewer female than male children.

2. If only the mother is irradiated, her mutant X-linked genes would be transmitted to both male and female children, but males, being hemizygous, would succumb to lethals more frequently than would females. Thus, an excess of female children would be expected.

3. If both parents are irradiated, there should be a decrease in the number of male offspring, but the decrease would not be as marked as when only the mother is irradiated.

Analysis of the Japanese data showed changes in the sex ratio of the children in accordance with the above assumptions; this is evidence that the rate of X-linked mutations (and presumably also other mutations) was increased by the atomic bombings of Hiroshima and Nagasaki. No other significant effect of radiation in humans has been detected in the Japanese data.

SOMATIC MUTATIONS

Somatic cells are subjected to the same mutagenic forces as are germ cells. Mutations in somatic cells have no effect upon future generations, but they may have significance for the individual who carries them. As yet, somatic mutations in humans are difficult to identify and evaluate. A somatic mutation will, of course, result in a mosaic individual with two populations of cells, one bearing the mutation and one not.

One theory that has been advanced to explain the phenomena of aging is that somatic mutations of many cells may render them less capable of fulfilling their normal functions. This theory is not universally accepted, but research is currently in progress in an effort to test it.

SELECTION

Darwin postulated natural selection as the factor of importance in evolution. In modern terms, survival of the fittest is interpreted as taking place through the action of selection upon new genotypes which have arisen by mutation or recombination.

In the biological sense, the term *fitness* has no connotation of superior endowment except in a single respect—the ability to contribute to the gene pool of the next generation. The many factors which affect fitness can operate at any stage of the life cycle, at least until the end of the reproductive period. Perhaps selection can also operate after the reproductive period. It has been suggested, not entirely facetiously, that there is some selective value in a life span that lasts no longer than the productive years.

Dominant genes are openly exposed to selection, in contrast to autosomal recessive genes. Consequently, the effects of selection are more obvious and can be more readily measured for dominant genes than for recessive genes.

SELECTION AGAINST DOMINANT MUTATIONS

A harmful dominant mutation, if penetrant, will be expressed in any individual who receives it. Whether or not it is transmitted to the succeeding generation will depend upon how deleterious it is. If it prevents reproduction and accordingly is not represented in the next generation, its relative fitness is zero. If it is just as likely as the normal allele to be represented in the next generation, its fitness is one. Most deleterious dominant mutations have a fitness value between zero and one. Many children with lethal multiple congenital defects may carry new dominant mutations, but it is impossible to be certain since these children do not reproduce.

If the mutation is deleterious but affected persons are fertile, they may contribute fewer than the average number of offspring to the next generation, i.e., their fitness may be reduced. The mutation will be lost through selection at a rate proportional to the loss of fitness of heterozygotes. The coefficient of selection, s, is the measure of the loss of fitness. Fitness is measured as $f = 1 - s$.

Achondroplastic dwarfs have about one-fifth as many children as do normal members of the population. Thus, their fitness is 0.20, and the coefficient of selection is 0.80. In the next generation, the frequency of the achondroplasia mutations passed on from the current generation is 20 per cent. The remaining 80 per cent are added through new mutation. The observed gene frequency in any one generation represents a balance between *loss* of alleles through selection and *gain* through recurrent mutation.

Selection against a dominant genotype can lower the frequency of the dominant gene precipitately. If no heterozygotes for the dominant gene for Huntington's chorea reproduced, the incidence of the disease would fall in one generation to a level determined by the mutation rate, because the only chorea genes remaining in the population would be the new mutations.

SELECTION AGAINST RECESSIVE MUTATIONS

Selection against deleterious recessive genes is less effective than selection against dominants. Even if there is complete selection against homozygous recessives, it takes 10 generations to reduce the frequency from 0.10 to 0.05; and the lower the gene frequency, the slower the decline. Removing selection (for example, by successful medical management of children with cystic fibrosis or sickle cell anemia, so that they could survive and reproduce at a normal rate) raises the gene frequency just as slowly.

SELECTION AGAINST X-LINKED RECESSIVES

Deleterious X-linked recessive genes are exposed to selection in hemizygous males, but not in heterozygous females. Duchenne muscular dystrophy is an example of an X-linked recessive genetic lethal. In each generation, the initial incidence of affected males is three times the mutation rate (3μ) and the initial incidence of heterozygous females is 4μ. Only the genes in the carrier females are passed on to the next generation. An important consequence is that one-third of all cases of Duchenne muscular dystrophy (or of other X-linked lethals such as HGPRT deficiency or mucopolysaccharidosis of the Hunter type) are new mutants, born to genetically normal mothers who have virtually no risk of having any other children with the same disorder. In less severe disorders (e.g., hemophilia), the fitness of affected males is above zero, and the proportion of new mutants is less than one-third.

SELECTION AGAINST HETEROZYGOTES

If selection occurs at the gametic stage, single alleles are selected against. More frequently, selection takes place at the diploid stage, and loss of one allele involves loss of its partner also. If the two alleles are different (i.e., if the individual concerned is a heterozygote), the frequencies of both alleles (or all alleles, if there are more than two possibilities) are altered, and they achieve a new balance at a different level. An example of selection against heterozygotes is provided by the Rh blood group system.

Let us use a simple terminology, in which *R* represents any Rh-positive allele and *r* represents any Rh-negative allele. In a mating of an *r*/*r* female with an *R*/*R* or *R*/*r* male, a proportion of *R*/*r* infants will die or have their fitness impaired because of hemolytic disease of the newborn. In this instance, one *R* and one *r* allele are lost from the gene pool of the next generation. But since the frequency of *R* is about 0.60 and the frequency of *r* is about 0.40, loss of these two alleles will lower the relative frequency of *r*. Over many generations, the rarer allele should be lost. Why, then, is the frequency of *r* not much less than it is?

There has been much speculation on this problem, but there is little definite evidence to answer it. One possibility is that a woman who loses a child because of Rh-incompatibility may "compensate" by having another child. Such a child, if it were *R*/*r*, would again be subject to selection; but if it were *r*/*r*, its chance of survival would be enhanced. Thus, compensation could help to replace the *r* alleles lost through hemolytic disease of the newborn. Compensation should mean that *r*/*r* women have as many or more living children than do *R*/*R* or *R*/*r* women, but there is little evidence that they do. In fact, among North American women who receive adequate medical care, the birth of an infant with hemolytic disease of the newborn often leads to voluntary cessation of reproduction. The use of anti-Rh immune globulin in the treatment of Rh hemolytic disease is still so new that its population effect has not been assessed.

SELECTION FOR HETEROZYGOTES

Sickle cell anemia is the most familiar example (and one of very few well-established examples) of a situation in which the heterozygote is more fit in a particular environment than is either type of homozygote.

Recall that there are three genotypes possible with relation to normal adult hemoglobin and sickle cell hemoglobin. The frequency of the Hb *S* allele in certain parts of Africa and southern Europe and Asia is higher than can be expected on the basis of recurrent mutation of an allele which, when homozygous, has a fitness of zero. A polymorphism develops when the heterozygote is at an advantage as compared with the homozygotes.

The postulated reason for the heterozygote advantage of the Hb *S* allele is the resistance of heterozygotes to the malaria organism *Plasmo-*

dium falciparum. This is a parasitic protozoan which spends a part of its life cycle in the red cells of vertebrates, to which it is introduced by the bite of the vector, the *Anopheles* mosquito. Thus, in a malarial environment, normal homozygotes are relatively unfit because of their susceptibility to malaria; Hb *S* homozygotes are selected against because of their anemia; but Hb *A*/*S* heterozygotes, whose red cells are apparently not well adapted to allow the malaria organism to complete its life cycle and whose hemoglobin is quite adequate to maintain them under normal circumstances, are more fit than either homozygote and therefore reproduce at a rate that has favored an increase in the incidence of the Hb *S* gene.

A selective advantage in one environment may offer no advantage at all in a different environment. For instance, malaria is not a public health problem in Canada nor in most areas of the United States, and therefore Hb *A*/*S* individuals do not have a particular advantage in these countries. Indeed, they may be at some disadvantage since they have blood which shows some sickling at reduced oxygen tension, such as occurs at high altitudes. It is possible that the eradication of malaria, at present a goal of the World Health Organization, might greatly reduce the world frequency of the Hb *S* gene.

STABILIZING SELECTION

Population geneticists distinguish three main types of selection with respect to quantitative traits: stabilizing, directional and disruptive. The kinds of changes each of these types of selection brings about in regard to phenotypic distribution in the population are illustrated in Figure 12–1.

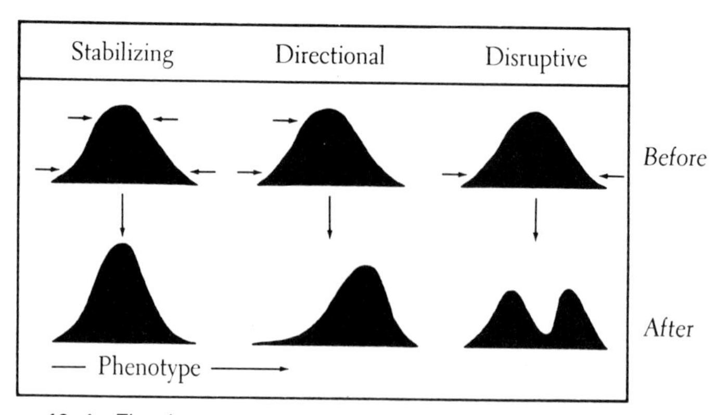

Figure 12–1 *The three main types of selection: stabilizing, directional and disruptive. (After Mather, from Scharloo; reprinted in Wallace 1968.* Topics in Population Genetics. *W. W. Norton & Company, Inc., New York, p. 382.)*

Stabilizing selection favors an intermediate optimum phenotype, and selects against the **phenodeviants** at either extreme. The more extreme the deviant, the less its genetic contribution to the next generation. **Directional selection** is selection directed toward a new optimum, not at the mean of the population, in response to a new environmental challenge. **Disruptive selection**, rather than favoring any one phenotype, favors two quite different forms and selects against the intermediates; this may be viewed as directional selection of two separate subpopulations, in response to two different sets of environmental conditions.

Of the three types of selection, stabilizing selection appears to play the major role. It may be viewed as a tendency to maintain the status quo. For many human characteristics, an intermediate value is the favored optimum. An obvious example is birth weight, and here the effect of stabilizing selection in eliminating the more extreme phenodeviants has been clearly shown: babies much smaller or larger than the average are less likely to survive the perinatal period than those in the intermediate range.

There is not much evidence for directional selection within the human species. For example, except for the last few generations, height appears to have changed little during the last 40,000 generations. Directional selection was basic to Darwin's concept of the mechanism of evolution, and it is much used by agricultural geneticists. Directional selection is a very powerful mechanism; many scientists and writers have speculated about what might happen to mankind if the techniques of cattle breeders were applied to our species.

A striking example of directional selection is provided by the increase in frequency of the melanic (dark) form of the peppered moth *Biston betularia* over the past several decades. This phenomenon is known as industrial melanism; in brief, the melanic form of the moth has rapidly replaced a lighter form as the common type, coincident with the spread of industrial soot over the countryside. The environmental change has given the melanic form better camouflage, with consequent better protection against bird predators.

MIGRATION

Genetic Drift. Changes in gene frequencies may occur when new settlements are formed by members of an older group. There are two possible explanations. Perhaps migrants are themselves a separate subpopulation, differing from the population as a whole in energy, acquisitiveness, curiosity or whatever characteristics stimulate some members of a community to strike out for strange shores. Of more importance, however, is the fact that the gene frequencies of the migrants will probably not be representative of those of the population as a whole; and the smaller the group, the more likely that it will not accurately reflect the gene frequencies of the parent group.

One of the most striking demonstrations of drift is given by McKusick's studies of the Old Order Amish of Pennsylvania. Among the Amish, social custom has provided an excellent situation in which to see the effect of drift. The Amish intermarry, few if any genes being added to the population by marriage with outsiders. As a community grows, it is the custom for a few families to leave to set up a new colony elsewhere. Among one such group studied by McKusick, by chance a founding father was heterozygous for a rare form of dwarfism with polydactyly known as Ellis-van Creveld syndrome. There are about 50 known cases of this rare disease among the 8000 living members of this group, and it is absent in other Amish communities descended from other ancestors. Thus, genetic drift can favor the establishment of genes which are not favorable or even neutral, but actually harmful. Another well-established example of random genetic drift is that of tyrosinemia in a remote area of Quebec (page 61). The socially isolated Amish and the geographically isolated French-Canadian groups involve the "founder principle" of Mayr (1963): if among a small number of founders who form a community there is a member with a rare recessive allele, the frequency of the allele is much higher within the community than without it, and the smallness of the original group of founders allows for a large effect of drift.

Gene Flow. In contrast to the variations in gene frequencies in small populations resulting from genetic drift is the more gradual change in frequency in larger populations resulting from gene flow. A classic example is the steady change (cline) in the frequency of the *B* allele of the ABO blood group system from about 0.30 in Eastern Asia to about 0.06 in Western Europe (Fig. 12–2). The flow of Caucasian genes into American Negroes is another good example; by several different measures (e.g., comparison of the frequency of the strictly African Rh^0 allele in African and American Negroes) it has been shown that some 30 per cent of the alleles now carried by American Negroes are of Caucasian origin. More recently Reed (1969), using data for the "Caucasian" allele Fy^a of the Duffy blood group system and other alleles varying in frequency in Blacks and whites, has found a lower proportion and differences in different geographic areas. He has calculated that the proportion of Caucasian genes in American Blacks is higher in nonsouthern areas of the United States than in the south (22 per cent in Oakland, California and 26 per cent in Detroit, but only 4 per cent in Charleston, South Carolina).

THE ORIGIN OF RACES

New varieties of living things evolve when new genes arise by mutation, are tried out in various combinations with other genes and in various environments, and are preserved or discarded. In man, many different

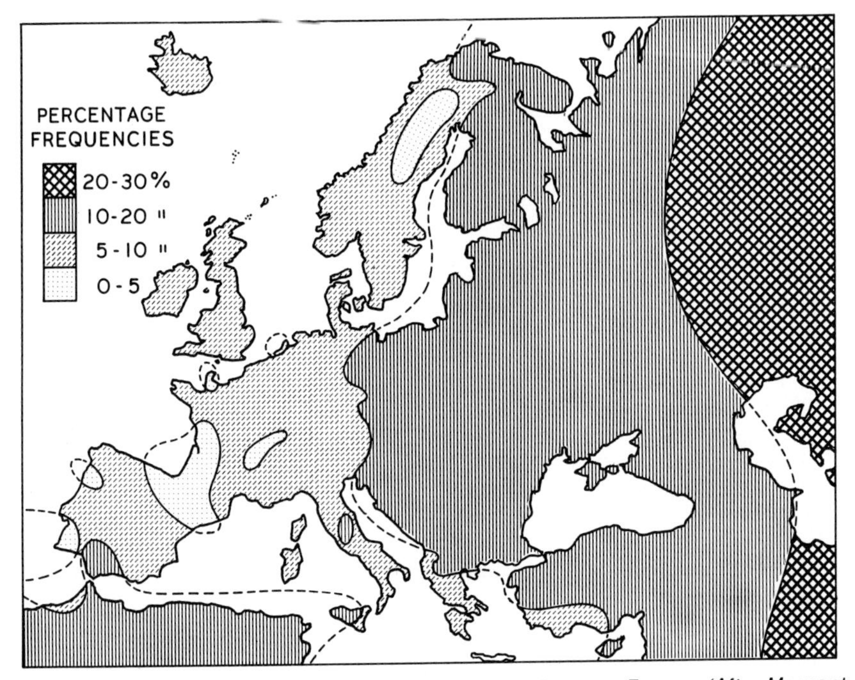

Figure 12–2 *Cline of distribution of blood group B across Europe. (After Mourant 1954.* The Distribution of the Human Blood Groups. *Blackwell Scientific Publications, Oxford.)*

subgroups have thus evolved, though all belong to a single species, *Homo sapiens*. These subgroups are commonly called races. A race is distinguished from other races by a gene pool in which gene frequencies (at a few or many loci) are characteristic of that population and of no other.

One of the most active areas of current research in human genetics is the accumulation of data on the frequency of many genetic markers in different populations. This is a matter of some urgency because, with improvement in communication and the consequent breakdown of isolates, few human populations will retain for many more years the characteristic gene frequencies they have acquired over the long time span of human evolution. The analysis of such data is the only way to obtain information about the evolution of human populations and the origin and maintenance of genetic polymorphisms in man.

GENERAL REFERENCES

Cavalli-Sforza, L. L., and Bodmer, W. F. 1971. *The Genetics of Human Populations*. W. H. Freeman and Company, San Francisco.

Mettler, L. E., and Gregg, T. G. 1969. *Population Genetics and Evolution.* Prentice-Hall Inc., Englewood Cliffs, New Jersey.
Morton, N. E. 1972. The future of human population genetics. *Progr. Med. Genet.* 8:103–124.
Wallace, B. 1968. *Topics in Population Genetics.* W. W. Norton & Company, Inc., New York.

PROBLEMS

1. Among 20,000 successive births in a certain hospital, there were 14 children with brachydactyly (a trait inherited as an autosomal dominant). Ten of these 14 children had a parent with brachydactyly; the remaining four had normal parents. What is the mutation rate of the gene concerned?

2. Many cases of retinoblastoma (a form of cancer affecting the retina of the eye) show autosomal dominant inheritance. This condition was formerly fatal in childhood, but now many affected children are cured by ophthalmectomy and live to reproduce. What will be the long-term effect of this treatment upon the frequency of the retinoblastoma gene?

3. Name three conditions that could be almost eliminated in a generation if no affected persons reproduced.

13

Twins in Medical Genetics

Twins have a special place in human genetics. This is because diseases caused wholly or partly by genetic factors have a higher concordance rate in monozygotic than in dizygotic twins. Even if a condition does not show a simple genetic pattern, comparison of its incidence in monozygotic and dizygotic twin pairs can reveal that heredity is involved; moreover, if monozygotic twins are not fully concordant for a given condition, nongenetic factors must also play a part in its etiology. The importance of twin studies for comparison of the effects of heredity and environment, or "nature and nurture," was pointed out by Galton in 1875.

MONOZYGOTIC AND DIZYGOTIC TWINS

There are two kinds of twins, monozygotic (MZ) and dizygotic (DZ). In common language, these are called identical and fraternal twins. Monozygotic twins arise from a single fertilized ovum, the zygote, which divides into two embryos at an early developmental stage, i.e., within the first 14 days after fertilization. Because the members of an MZ pair normally have identical genotypes, they are like-sexed, and identical with respect to such genetic markers as blood groups. They are less similar in traits readily influenced by environment; for example, they may be quite dissimilar in birth size, presumably because of differences in prenatal nutrition. Phenotypic differences between MZ co-twins may be produced by the same factors that cause differences between the right and left sides of an individual; for example, cleft lip may be bilateral or unilateral in an individual, and may be concordant or discordant in an MZ pair.

Dizygotic twins result when two ova, shed in the same menstrual cycle, are fertilized by two separate sperm. DZ twins are just as similar genetically as ordinary sib pairs, having, on the average, half their genes "in common" (page 270). Phenotypic differences between the members of a DZ pair reflect their genotypic dissimilarities as well as differences arising from nongenetic causes.

Relative Frequency of Monozygotic and Dizygotic Twins

There is a simple way of finding out how many of the twin births in a population are MZ and how many are DZ. MZ twins are always like-sexed, while approximately half the DZ twin pairs are boy-girl sets. Therefore, the total number of DZ pairs is twice the number of unlike-sexed pairs, and the number of MZ pairs can be found by subtracting the number of unlike-sexed pairs from the total number of like-sexed pairs.

$$\frac{\text{All twin pairs} - 2(\text{unlike-sexed pairs})}{\text{All twin pairs}} = \text{Frequency of MZ pairs}$$

For precision, a small correction is required because the sex ratio is not exactly 1:1, but the simple method described gives a good approximation. Among white North Americans, approximately 30 per cent of all twins are MZ, 35 per cent are like-sexed DZ and 35 per cent are unlike-sexed DZ.

Comparison of the ratio of like-sexed to unlike-sexed pairs in populations with varying frequencies of twin births has shown that the proportion of MZ births relative to all births is much the same everywhere, about one in 300 births, but that the proportion of DZ births varies with ethnic group, maternal age and genotype (see below).

FREQUENCY OF TWIN BIRTHS AND HIGHER MULTIPLE BIRTHS

Among white North Americans, about one birth in 87 is a twin birth. Thus, two of every 88 babies—over 2 per cent of the population—are born as twins. Triplets are born about once in $(87)^2$ births, quadruplets about once in $(87)^3$ births and so on, although because of a much higher mortality, this mathematical relation (known as Hellin's law) is less closely approximated for multiple sets of four or more. Since twins have a higher infant mortality than singletons, the incidence of twins in the general population is somewhat lower than 2 per cent.

The frequency of twin births varies with ethnic origin. In the United States the rate is higher in Blacks than in whites. One of the highest frequencies (one in 20 to 30 births) is reported from Nigeria, and one of the lowest (one in 150 births) is reported from Japan.

The frequency of DZ twins rises sharply with maternal age to age 35 and later declines, but the frequency of MZ twins is hardly affected by the mother's age. Scheinfeld and Schachter have calculated, for United States whites, the probability of producing MZ and DZ twins in relation to maternal age and parity (Table 13–1). These probabilities may be used, together with blood group data and other evidence, to help assess the probability that a given pair of twins is DZ.

The tendency for twins to run in families has been noted by many authors. It is questionable whether there is a genetic factor in MZ twinning, but certainly a disposition to DZ twinning is genetically determined. Whether the father plays any part in the occurrence of twinning among his children has been the subject of dispute for many years. The studies of Greulich (1934) seemed to show that the father's genotype was at least as important as the mother's in the production of DZ twins. This observation seemed inexplicable in view of the fact that the multiple ovulation which accounts for DZ twins has been considered to be a purely maternal event, related to the mother's FSH (pituitary follicle-stimulating hormone) level. White and Wyshak (1964) have presented data obtained from the genealogical records of the Mormon church showing that female DZ twins produce twins at a rate of 17.1 sets per 1000 maternities, whereas the wives of male DZ twins produce only 7.19 sets per 1000 maternities. The sisters of DZ twins also produce twins at a high rate, but the brothers do not. These authors conclude that the genotype of the mother affects the frequency of DZ twins among her offspring, but the genotype of the father has no effect. In other words, a disposition to multiple ovulation is an inherited trait and can be expressed only in females. The trait can be transmitted through males as well as through females; i.e., the brothers of DZ twins do not themselves produce a higher than average number of twins, but their daughters do.

TABLE 13–1 PROBABILITY THAT TWINS ARE MONOZYGOTIC IN RELATION TO MATERNAL AGE AND PARITY*

Maternal Age			***Birth Rank***		
	1	*2*	*3*	*4*	*5*
15–19	0.58	0.50	0.41	–	–
20–24	0.46	0.39	0.34	0.32	–
25–29	0.38	0.33	0.30	0.27	0.24
30–34	0.36	0.31	0.27	0.25	0.23
35–39	0.34	0.29	0.27	0.25	0.21
40–44	0.41	0.35	0.34	0.31	0.25

*For each maternal age group and parity, the probability that twins are DZ = 1 − probability that they are MZ; e.g., if a mother aged 17 bears twins at her first pregnancy, the probability that the twins are MZ is 0.58, and the probability that they are DZ is 1 − 0.58 = 0.42. After data of Scheinfeld and Schachter, 1963.

Although it has been previously stated that there is no genetically determined factor in MZ twinning, there are some apparent exceptions, especially in families in which chromosomal abnormalities occur. Many authors have pointed out that twins are more common among individuals with chromosomal abnormalities and their families than in the general population. For example, Nance and Uchida (1964) noted an increased frequency of twins (presumably MZ) in 34 families in which there were XO aneuploids. The twinning rate was eight twin births in a total of 128 births, a fivefold to tenfold increase over the normal rate. Their data and those of other workers also show a slight increase in the frequency of twins in sibships of children with Down's syndrome. Observations such as these suggest a common mechanism in the causation of twinning and chromosomal disorders.

When women are treated for sterility with massive doses of FSH to induce ovulation, they may produce several ova at once. Multiple births (quintuplets, sextuplets and even septuplets) have been reported following FSH treatment.

DETERMINATION OF TWIN ZYGOSITY

There are many good reasons why the zygosity of a twin pair (or of triplets or higher multiple births) should be promptly and accurately determined. If transplantation should be required, a monozygotic co-twin is the most useful donor. It is also essential for MZ and DZ twin pairs to be accurately classified in order to conduct research in genetic and developmental disorders in twins. Zygosity may be investigated by examination of the placenta and fetal membranes, and on the basis of similarities and differences between co-twins.

DETERMINATION OF ZYGOSITY BY PLACENTA AND FETAL MEMBRANES

A developing fetus is invested in two membranes: the inner, delicate amnion and the outer, thicker chorion attached to the placenta. If two fetuses are developing simultaneously in the same uterus, there are several variants of the placenta and fetal membranes. The most common types are shown in Figure 13–1 and summarized in Table 13–2.

A twin placenta is monochorionic if there is a single chorion, and dichorionic if there are two chorions (which may be secondarily fused where they meet). A monochorionic twin placenta may be either monoamniotic (with one amnion), or diamniotic (with two amnions).

Dizygotic twins have separate placentas, chorions, and amnions, but in about 50 per cent of cases the two placentas and chorions are secondarily fused and superficially resemble a monochorionic placenta.

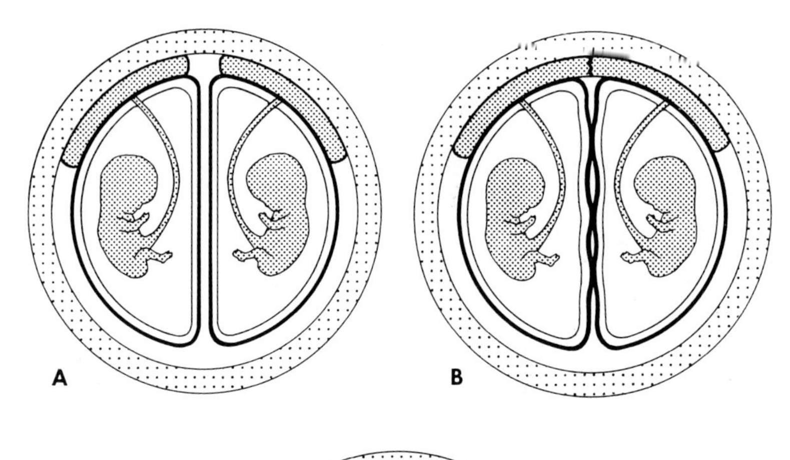

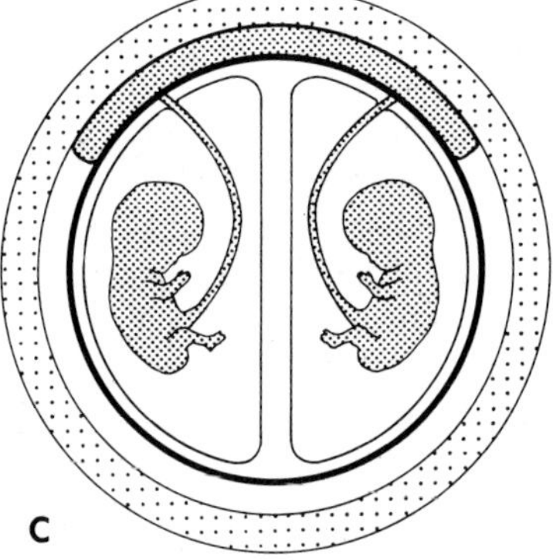

Figure 13–1 *Diagrammatic representation of some common arrangements of the placenta and fetal membranes in twins. Uterine wall is lightly stippled; fetus, umbilical cord and placenta are heavily stippled; chorion is shown by heavy line; and amnion is shown by light line.*

A, *Twins with separate placentas and separate membranes. This arrangement occurs in either DZ or MZ twins, but is unusual in MZ twins.*

B, *Twins with separate but secondarily fused placentas and chorions. The two halves of the placenta have separate circulations. This arrangement occurs in either DZ or MZ twins.*

C, *Twins sharing a single placenta with a common circulation and a single chorion, but having separate amniotic sacs. This arrangement occurs only in MZ twins and is diagnostic of MZ twinning.*

TABLE 13-2 TYPES OF PLACENTAS AND FETAL MEMBRANES IN TWINS

	Monochorionic		***Dichorionic***	
	Monoamniotic	*Diamniotic*	*Single Placenta (by secondary fusion)*	*Two Placentas*
MZ	Rare	75%	25%	Rare (~1%)
DZ	–	–	50%	50%

Monozygotic twins may have either dichorionic or monochorionic placentas, depending on the time in early embryonic development at which twinning occurred. In about 25 per cent of cases, twinning of the embryo occurs before the third day, i.e., before the development of the chorion, so that two separate chorions are formed. There may be two separate placentas or, more commonly, a single secondarily fused placenta. Both these types of placentation also occur in DZ twins. In the majority of MZ twins, however, twinning occurs between the third and eighth day of development, by which time the differentiation of the blastocyst has already proceeded too far to allow duplication of the chorion. A monochorionic placenta (which is usually diamniotic) can be regarded as proof of monozygosity.

There are several varieties of monochorionic placenta, depending again on the stage at which twinning of the embryo occurred. Some of the distinguishable types, arranged in order from earlier to later time of twinning, are the following:

1. Each embryo has a separate amnion, and the two fetal circulations remain separate.

2. Each embryo has a separate amnion, but a common fetal circulation has developed by anastomoses of vessels in the placenta. This is the most common type of monochorionic placenta. The common circulation can be demonstrated by injection of colored latex into one of the two umbilical cords. The fluid passes through the capillary bed and appears in the other cord.

3. Rarely, an even later twinning of the embryo results in a single amniotic sac for both twins. Many monoamniotic twins do not survive.

4. Rarely, still later twinning of the embryo occurs after the formation of the umbilical cord, resulting in the rare conjoined (Siamese) twins. Conjoined twins are monoamniotic and usually have a common, branched umbilical cord.

Though superficially similar, the fused dichorionic placenta and fetal membranes can be clearly distinguished from the monochorionic type if the placenta is washed and floated in running water, and an examination is made of the layer of tissue separating the two amniotic cavities. In a dichorionic placenta, the layer of tissues is opaque, being composed of two layers of chorion and two layers of amnion, as shown in Figure 13–1B. The two chorions can be readily separated, since the connective tissue which joins them is delicate and loose. However, in a monochorionic placenta, the two amniotic cavities are separated only by a semitransparent layer composed of two thin amnions, as demonstrated in Figure 13–1C. (According to an old tale, if twins are of different sexes but share a placenta, a heavy curtain is drawn between them to protect their modesty.)

In summary, the only kind of placenta which is diagnostic of the type of twinning is the monochorionic type, and it is diagnostic of MZ twinning. Separate placentas and membranes, and secondarily fused placentas without common circulation, can be found with either type of twin set. It is helpful to examine and record the nature of the placenta and membranes at like-sexed twin births, since the information about twin zygosity then available may not be easy to obtain in later life, except by extensive tests.

Exceptionally, prenatal mixing of the blood of DZ twins occurs as shown by cases of chimerism in which stem cells of a DZ twin have populated the marrow of the co-twin and continue to proliferate and differentiate there (page 318). Presumably such cases arise either by accidental mixing or more probably through vascular anastomoses established between the two parts of a secondarily fused placenta.

DETERMINATION OF ZYGOSITY BY GENETIC MARKERS

Monozygotic twins are always alike in sex and in genetic markers such as blood groups or serum factors. MZ twins are more alike than are DZ twins in multifactorial traits such as dermatoglyphics and congenital malformations. Dizygotic twins may be alike or may differ in these characteristics. To determine whether a twin pair is MZ or DZ, it is necessary to compare twins with respect to as many traits as possible. A difference in any genetic marker proves twins to be dizygotic; but it is impossible to prove monozygosity, since by chance any two children of the same parents might resemble one another in each of many hereditary traits. However, it may be possible to show that the probability of monozygosity is very high.

The probability that twins are MZ can best be worked out if the genotypes of the parents and sibs are known. The following example shows how to use the method in an oversimplified case. The data are taken from Race and Sanger (1969), where a fuller discussion may be found.

Blood Groups

Father	*Mother*
A_2/O	O/O
Ms/NS	MS/MS
P_1/P_2	P_1/P_2
Twin A	*Twin B*
A_2/O	A_2/O
MS/Ms	MS/Ms
P_2/P_2	P_2/P_2

		DZ	MZ
Relative frequency in the population		0.70	0.30
Chance that twins will be alike	in sex	0.50	1.00
Chance that twins born to these parents will be alike	in ABO groups	0.50	1.00
	in MNSs groups	0.50	1.00
	in P groups	0.25	1.00
Product of chances		0.022	0.30

Probability that twins are MZ $= \frac{0.30}{0.022 + 0.30} = 0.9314$

This probability (93 per cent) is not high enough to make it safe to assume that the twins are MZ, for indeed there is almost one chance in 10 that they are not. But with additional genetic markers it may be possible to increase the probability close to the 100 per cent level, or on the other hand to show that the twins are not MZ. Useful tables of probabilities for situations in which the parental genotypes are unknown are given by Maynard-Smith et al. (1961).

Dermatoglyphics, though not inherited according to a simple single-gene pattern, can often be helpful in the determination of zygosity. Homolateral comparisons of the dermatoglyphic pattern of the right and left sides of co-twins may be made, and the differences expressed as percentages. In identical twins the right hand of one twin is more similar to the right hand of the co-twin than to his own left hand, but in fraternal twins the right and left hands of one twin are more alike than either of his hands is to the homolateral hand of his co-twin. Total fingerprint ridge counts are also useful; for example, when the difference in total fingerprint ridge counts is 33 or more, the relative probability that the twins are DZ is 34:1 (Maynard-Smith et al., 1961). Dermatoglyphic methods can be used if one member of a twin pair has recently had a blood transfusion, in which case the genetic markers of his blood can not be immediately determined. In one such case, a twin had been severely burned and had required several transfusions. It then became necessary to determine the zygosity of the twin pair because the co-twin, if MZ, was available as a source of skin grafts. Dermatoglyphic comparisons helped establish that the twins were indeed MZ.

The most rigorous test for monozygosity, next to the finding of a monochorionic placenta, in the skin graft test. Grafts are normally accepted between MZ pairs, but rejected between DZ pairs.

LIMITATIONS OF THE TWIN METHOD

The chief drawback of the twin method is that, though it tells something about the strength of the genetic predisposition to develop a disorder, it gives no insight into the genes concerned, their mode of action or their pattern of transmission. Perhaps this problem is becoming less acute as methods of assessing multifactorial inheritance are becoming better known. Many of the traits in which twin comparisons are used are multifactorial (e.g., congenital malformations). In a sense, comparing MZ and DZ twin pairs is like comparing the risk that a given trait will occur in an individual of a certain genotype with the risk that it will occur in his sib, who has inherited half the same genes. If a trait is determined by a single autosomal dominant gene, it should be half as common in a sib or in a DZ co-twin of the propositus as in an MZ co-twin; if it is an autosomal recessive, it should affect one-quarter as many sibs or DZ co-twins as MZ co-twins. If the proportions vary by much from these ratios, multiple factors (genetic or environmental) are probably involved. In most of the common malformations, the MZ concordance rate is well below 50 per cent, indicating that environmental factors operative before birth are important in causing them (Carter, 1964).

Although the twin method assumes that postnatal environmental differences are constant for both types of twins, this assumption is unwarranted in many cases. MZ twins, because they are more alike, may seek the same environment and develop along much the same paths. DZ twins, who may even be of different sexes, probably have more different environments than do MZ twins. It is noteworthy that DZ twins become less and less alike as they grow older, whereas MZ twins remain remarkably similar throughout life, aging in much the same way and being subject to the same geriatric disorders. In many studies a comparison can be made more fairly between MZ twins and like-sexed DZ twins than between all MZ pairs and all DZ pairs.

Still another limitation of the twin method is related to bias of ascertainment. MZ pairs who are concordant are much more likely to be reported than any other combination. If a twin series is compiled from the literature, it is likely to include a preponderance of concordant MZ pairs.

SOME EXAMPLES OF TWIN STUDIES IN MEDICAL GENETICS

A few examples will indicate ways in which twins are used in genetic research.

Chromosomal Aberrations

If one member of an MZ twin pair has Down's syndrome, the other twin is always affected; if one DZ twin is affected, the other is nearly always normal. This is one of the observations that led to the hypothesis that a chromosomal defect might be the cause of mongolism, and is an example of the way in which a twin study can point to an unusual causative mechanism of disease.

Mutation in Sporadic Cases of Disease

Many cases of muscular dystrophy indistinguishable from the X-linked recessive Duchenne type arise as sporadic cases presumably caused by new mutations. The observation of concordance for muscular dystrophy in an MZ twin pair in an otherwise normal family helps to verify the genetic causation of sporadic cases of the disease (Fig. 13–2).

Congenital Malformations

The concordance rates for congenital malformations in MZ and DZ twins allow a comparison of the strength of genetic predisposition. Some typical figures follow:

	*MZ Pairs**	*DZ Pairs**
Congenital dislocation of the hip	40%	3%
Cleft lip	40%	5%
Club foot (talipes equinovarus)	35%	3%
Congenital heart defects	5%	5%

*After the data of Carter, 1964.

These few examples show that in some congenital malformations, the concordance rate is rather high in MZ pairs as compared with DZ pairs. For congenital heart defects the risk is low, about the same in both types of twins, so that in this type of malformation genetic factors appear to be of little significance.

Diabetes Mellitus

Concordance rates for diabetes mellitus are complicated by the variability of onset age, which is more marked in DZ than in MZ pairs. The concordance rate is about 80 per cent in MZ pairs and only about 10

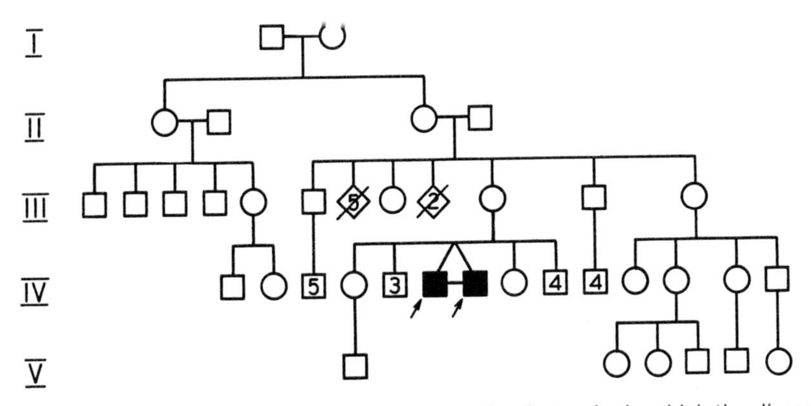

Figure 13–2 *A pedigree of Duchenne muscular dystrophy in which the disease appeared only in a pair of MZ twins. The large number of normal males related through females to the propositi makes it unlikely that the mutant gene had been present previously in the kindred. Its occurrence in both members of an MZ twin pair is evidence for its genetic basis even though it is sporadic. (Redrawn from Stephens and Tyler 1951.* Amer. J. Hum. Genet. 3*:111–125, Kindred 31.)*

per cent in DZ pairs. Follow-up investigations of cases of diabetes mellitus in one or both twins should provide useful information about the genetics, onset age, prognosis and therapy of the disease in genetically predisposed individuals.

Cerebral Palsy

Cerebral palsy is mentioned here as an example of a disorder in which twin concordance rates can be misleading. Cerebral palsy is usually not strongly genetically determined, but is produced by difficulties at the time of birth, especially prematurity or anoxia. These factors are more prominent in twin births than in single births, and for this reason a high frequency of twins, whether MZ or DZ, is found in a series of children ascertained through a cerebral palsy clinic. Nevertheless, if a sufficiently large series of twins with cerebral palsy could be obtained, comparison of the concordance rates in MZ and DZ could be informative.

UNUSUAL TYPES OF TWINS

Monozygotic Twins with Different Karyotypes

By 1970, about 12 pairs of twins were known in which the karyotypes of the twins differed, though evidence from genetic markers, placenta and physical appearance pointed to their MZ origin. Among the 12 pairs, four comprised a normal child and a 21-trisomic Down's syndrome member; six comprised a normal child and a 45, X Turner's syndrome; the others were a Turner's with a Turner's mosaic twin, and an X/XXX "normal" child with an X/XXX Turner's twin. In almost all these cases, both twins were mosaic, each with a minor population of cells like the co-twin's major population.

Twins such as these are very rare, and of great theoretical interest. Presumably they originate in postzygotic nondisjunction (as do chromosomal mosaics) followed by twinning.

Conjoined Twins

Conjoined twins are believed to be MZ twins produced by an incomplete split of the original embryo, occurring relatively late in development (i.e., after the eighth day). Their frequency is approximately one in 400 MZ twin births, or one in 120,000 births.

Chimeras

In Greek mythology, a chimera was a fire-breathing monster with the head of a lion, the body of a goat and the tail of a serpent. More prosaically, the term is used in human genetics to denote those rare individuals who are composed of a mixture of cells from two separate zygotes. There are two types, both exceedingly rare: blood group chimeras and whole-body chimeras.

Blood group chimeras are produced when DZ twins exchange hematopoietic stem cells in utero. The grafted cells are not recognized as foreign, and are retained. Each twin then has two populations of blood cells, one with his own genetic markers, the other with those of the twin.

Whole-body chimeras are thought to develop from fusion of fraternal twin zygotes in very early development. Separate fertilization of the egg and a polar body is one of several other suggested mechanisms. If a whole-body chimera has developed from fused XX and XY zygotes, true hermaphroditism usually results. Indeed, the very first case to be described was found after a deliberate search for a true hermaphrodite with different-colored eyes (Gartler, Waxman and Giblett, 1962).

GENERAL REFERENCES

Benirschke, K. 1972. Origin and clinical significance of twinning. *Clin. Obstet. Gynec.* 15:220–235.

Bulmer, M. G. 1970. *The Biology of Twinning in Man.* Clarendon Press, Oxford.

Maynard-Smith, S., Penrose, L. S., and Smith, C. A. B. 1961. *Mathematical Tables for Research Workers in Human Genetics.* Churchill, London.

PROBLEMS

1. A pair of like-sexed dichorionic twins each have the following blood group genotypes:

Twin A	Twin B
A/B	A/B
MS/MS	MS/MS
CDe/cde	CDe/cde
P_1/P_2	P_1/P_2
Xg^a	Xg^a

   The father's genotype is A/O, MS/Ns, cde/cde, P_1/P_1, Xg.
   The mother's genotype is B/O, MS/MS, CDe/cDE, P_1/P_2, Xg^a/Xg.
   What is the probability that the twins are MZ?
   If it is found that Twin A has haptoglobin type HP 2-1 and Twin B has haptoglobin type Hp 1-1, what is the conclusion as to the zygosity of the twins?

2. In a certain population, 33 per cent of all twins are unlike-sexed. What is the proportion of MZ twins in this population?

14

Dermatoglyphics in Medical Genetics

Dermatoglyphics are the patterns of the ridged skin of the palms, fingers, soles and toes. Though these patterns show great diversity in detail and in the combinations found in any one person, they can be systematically classified into a reasonable number of different types. Dermatoglyphics are important in medical genetics chiefly because of the characteristic combinations of pattern types found in Down's syndrome and, to a lesser extent, in other chromosomal disorders. Moreover, the dermatoglyphic traits themselves are genetically determined, at least in part; for example, the fingerprint ridge counts offer a classic example of multifactorial inheritance (page 270). Dermatoglyphics therefore provide a useful means of measuring resemblance between twins in the determination of twin zygosity.

The flexion creases—the heart, head and life lines of palmistry—are not, strictly speaking, dermal ridges, but they are formed at the same time, during the third fetal month, and affect the course of the dermal ridges. Indeed, flexion creases may themselves be determined in part by the same forces that affect ridge alignment. A single transverse crease (simian crease) in place of the usual two creases is found in 1 per cent of Caucasians and in a larger percentage of Asiatics. Simian creases are not unusual in abnormal individuals, e.g., in children with congenital malformations, even when the dermal patterns themselves are not obviously disturbed. In Down's syndrome and other chromosomal disorders, single flexion creases are much more common than in controls; about half the palms of patients with Down's syndrome have a single crease (Fig. 14–1).

The history of dermal pattern studies goes back to antiquity, but the scientific classification of patterns used today was proposed by Galton, who was the first to study dermal patterns in families and racial groups. Cummins, who coined the term dermatoglyphics ("writing on

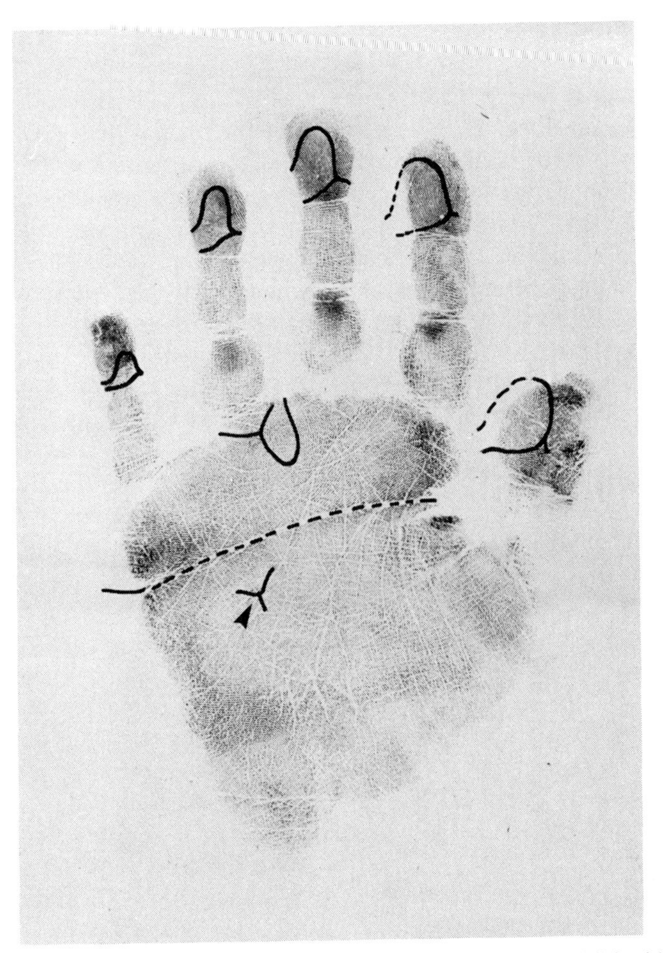

Figure 14–1 *Single flexion crease (simian line) on palm of child with Down's syndrome. Note also the distal axial triradius (arrow) and ulnar loops on all digits.*

the skin"), a modification of the term hieroglyphics ("symbol writing"), has made great contributions to the methodology and scientific basis of the field. It was Cummins who first noted that the dermal patterns in Down's syndrome show certain distinct trends that differentiate them from those of normal persons. Walker (1958) developed the first dermatoglyphic index for detection of Down's syndrome, showing that some 70 per cent of Down's patients could be distinguished from normal patients on the basis of the dermal patterns alone. Following the discovery of the chromosomal basis of Down's syndrome, interest in dermal patterns intensified, and dermal patterns of patients with other chromosomal disorders were found to be distinctive, though usually not distinctive enough to be used as diagnostic aids. Dermal pattern analysis is now recognized as a useful technique for screening patients for chromosome study, and more generally as a research tool for disorders of development.

EMBRYOLOGY

Because of the growing importance of dermal pattern analysis in clinical medicine, some understanding of the factors that influence their development is required. Differentiation of the dermal ridges begins in the third fetal month, and is complete in all but minor details by the end of the fourth month. The initial phase, in which the primary ridges differentiate and are aligned, is the most critical, since the dermal patterns as then established remain essentially unchanged throughout the remainder of the developmental period, after which they undergo no further alteration throughout life.

The alignment of the ridges is strongly influenced by the growth of the whole hand or foot. The hand of a 10-week human fetus bears conspicuous volar pads, which are comparable in relative size to a cherry on an adult fingertip. At about the thirteenth week these pads begin to regress, and during the period of regression the dermal ridges differentiate in the thickening skin. The development of the dermal patterns of the foot takes place slightly later, but the sequence of events is the same. Thus, the alignments of ridges which constitute the patterns of the fully differentiated palm and sole are determined by the relative growth rates of the various structures involved, including the supporting skeletal and muscular elements. Developmental disturbances occurring during ridge differentiation can therefore produce abnormalities of the dermal patterns.

Little is known about the developmental conditions that determine dermatoglyphics in either normal or abnormal fetuses. Ridge alignments give the impression that a system of parallel lines has been drawn as economically as possible over an irregular terrain. A defect which disturbs limb growth before the dermal patterns have developed will be reflected in the dermal ridges. Probably the origin of the pattern abnormalities is a direct result of mechanical rather than biochemical disturbances. In Turner's syndrome, for example, there is generalized edema in the early fetus. Edematous sacs in the neck region of the early fetus account for the redundant folds of skin (webbing) seen postnatally. The feet are often still edematous at birth. If in early prenatal life the fingertips are also swollen with excessive fluid, the area to be covered by the differentiating dermal ridges will be greater. This would account for the high ridge count and large finger patterns seen in Turner's syndrome.

CLASSIFICATION OF DERMAL PATTERNS

Prints of the digits, palms and soles can be made by one of several standard methods. In our experience, a technique used in many hospitals for identification of newborns produces excellent results. In very young infants, who have soft skin and fine ridges which are usually obscured by dry, scaling epithelium, clear prints may be extremely difficult to obtain. It is much easier to take a print of a child at least a month old than it is of a newborn.

The dermatoglyphics of importance in medical genetics are fingerprints (rolled prints of the ridged skin of the distal phalanx of each finger), palmar prints and prints of the hallucal area of the sole. Rules for the formulation of dermal patterns in these areas are to be found in Cummins and Midlo (1943, 1961), Penrose (1968) and Holt (1968).

Pattern combinations and frequencies are more significant than pattern types alone as indicators of abnormal development. In Down's syndrome, for example, there is no single characteristic of the dermal patterns that does not also occur in controls; but the combination of a number of patterns, most of which are more common in Down's syndrome than in normal persons, permits definite recognition of the majority of affected children.

FINGERPRINTS

Fingerprints are classified, according to Galton's system, as whorls, loops or arches. Examples of each type are shown in Figure 14–2. The classification is made on the basis of the number of triradii: two in a whorl, one in a loop and none in an arch. (A triradius is a point from which three ridge systems course in three different directions, at angles of 120°; see Fig. 14–1.) Loops are subclassified as radial or ulnar, depending upon whether they open to the radial or ulnar side of the finger. In a fingerprint formula, the patterns are conventionally listed in the following order: left 5, 4, 3, 2, 1; right 1, 2, 3, 4, 5.

The frequency of the different patterns varies greatly from finger to finger; for example, in northern Europeans the frequency of radial loops

Figure 14–2 *Examples of three basic fingerprint types:* A, *arch;* B, *loop; and* C, *whorl. Ridge counts are obtained by counting along the indicated lines, excluding the triradius and the pattern core. The ridge count of the loop is 16 (expressed as 16-0, since here the triradius is on the left). The ridge count of the whorl is 14-10. By definition, an arch has no triradius and therefore no ridge count; this is expressed as 0-0.*

is about 20 per cent on the second digit but less than 0.3 per cent on the fifth digit.

The size of a finger pattern is expressed as the ridge count, i.e., the number of ridges that touch a line drawn from the triradial point to the pattern core. An arch has a count of zero, since it has no triradius. The line of count for a loop and the two lines for a whorl are shown in Figure 14–2. Pattern size is especially important in assessing conditions characterized by a high frequency of arches (e.g., trisomy 18), since small loop patterns with low ridge counts merge into arches. The total ridge count (TRC) of the 10 digits is a useful dermatoglyphic parameter. In a normal British series, the mean TRC is 145 (standard deviation 50) for males, and 126 (standard deviation 52) for females (Holt, 1968).

PALMAR PATTERNS

Palmar patterns are defined chiefly by five triradii: four digital triradii, near the distal border of the palm, and an **axial triradius,** which is commonly near the base of the palm but sometimes displaced distally, especially in Down's syndrome and other chromosomal disorders. Its position is usually somewhere along the fourth metacarpal. Interdigital patterns (loops or whorls) may be formed by the recurving of ridges between the digital triradii. Hypothenar or thenar patterns may be present. Commonly, in a normal palm the ridges course obliquely toward the proximal portion of the ulnar side. Palm prints also show whether a single transverse palmar crease is present. Figure 14–3 indicates the chief landmarks of the palm.

The position of the axial triradius is perhaps the single most im-

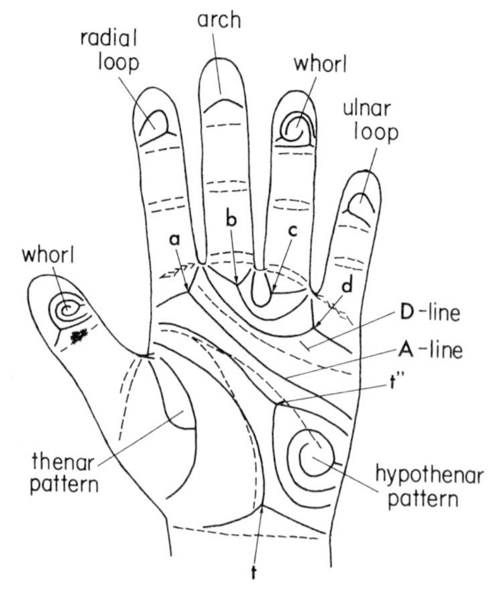

Figure 14–3 *The nomenclature of the dermatoglyphics of the palm. Note the four digital triradii (a, b, c and d) and the axial triradius t (here designated as t'' to indicate that its location is distal); two main lines, A and D, traced from the corresponding digital triradii; and the patterns in the thenar and hypothenar areas. Digital patterns are also shown. Broken lines are flexion creases. (From Penrose and Smith 1966.* Down's Anomaly, *The Williams & Wilkins Co., Baltimore, p. 63, by permission.)*

portant feature, because it is distally displaced in many abnormal conditions. Its location may be measured either in terms of the total length of the palm, or as the "*atd* angle," as shown in Figure 14–4. An axial triradius in a position of 0.40 or more, or an *atd* angle greater than 57 degrees, is much more common in patients with Down's syndrome and several other chromosomal syndromes than in the general population.

PLANTAR PATTERNS

Plantar patterns have been studied less extensively than palmar patterns, chiefly because they are less readily obtained. Only in the hallucal area have distinctive patterns been described in clinical syndromes. The unusual "arch tibial" (A^t) pattern, which is found in nearly 50 per cent of all cases of Down's syndrome and very rarely (0.3 per cent) in controls, is probably the single most useful dermal pattern in Down's syndrome (Fig. 14–5).

CLINICAL APPLICATIONS OF DERMATOGLYPHICS

Some of the more outstanding dermatoglyphic peculiarities seen in chromosomal disorders are listed in Table 14–1. It should be noted that,

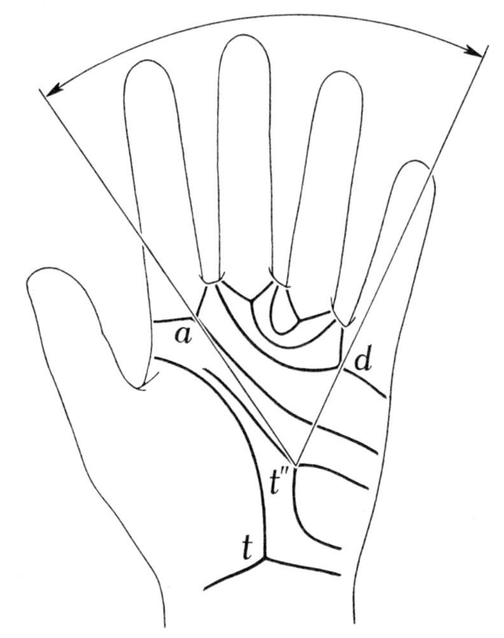

Figure 14–4 *Measurement of the* atd *angle of a palm. Here the angle measures 60°. Two digital triradii (a and d) and two axial triradii (t and t'', defined by their position in the palm) are shown. (From Holt 1968.* The Genetics of Dermal Ridges. *Charles C Thomas, Publisher, Springfield, Illinois, p. 86, by permission.)*

TABLE 14–1 CHARACTERISTIC DERMATOGLYPHICS IN CHROMOSOMAL ABERRATION SYNDROMES*

	Fingers	*Palms*	*Hallucal Area of Sole*
Down's syndrome (trisomy 21 or translocation type)	Ten ulnar loops (60%) Radial loop on 4th and/or 5th digits	Distal axial triradius (85%) Single flexion crease (50%)	Arch tibial (50%) Small loop distal (35%)
D_1 trisomy	Increased number of arches Low TRC†	Very distal axial triradius Single flexion crease Thenar pattern	Large pattern, loop tibial or arch fibular
18 trisomy	6–10 arches (also on toes) Single flexion crease on 5th digit Very low TRC	Distal axial triradius Single flexion crease	
Cri du chat syndrome	Increased number of arches Low TRC	Distal axial triradius "Bridged" flexion crease	Open field
Turner's syndrome	Variable; patterns usually large loops or whorls High TRC	Axial triradius slightly more distal than average	Very large pattern, usually loop or whorl
Klinefelter's syndrome	Increased number of arches TRC below average	Axial triradius more proximal than average	
XYY syndrome	Normal	Normal	Normal
Other syndromes with extra X and Y chromosomes	Increased number of arches Reduced TRC; the more X's and Y's present, the greater the reduction	–	–

*Based chiefly upon the data of Walker (1958), Penrose (1963), Uchida and Soltan (1963) and Holt (1964).
†TRC = total (digital) ridge count.

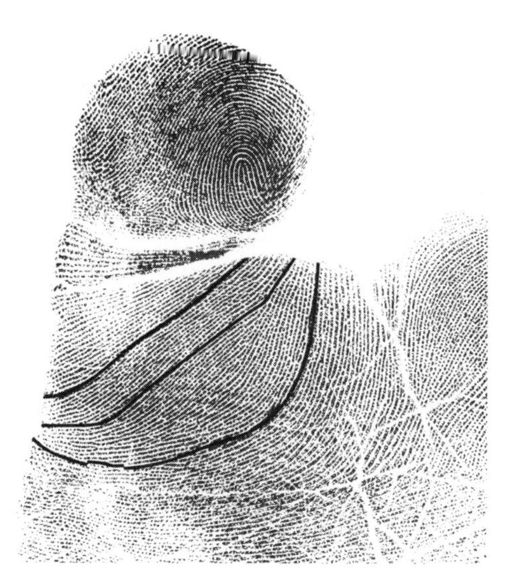

Figure 14–5 *The hallucal arch tibial pattern which is seen in the feet of 50 per cent of patients with Down's syndrome but which is otherwise very rare.*

except for Down's syndrome, no large series of any one type have been described, and that larger series may permit other typical features to be recognized.

DOWN'S SYNDROME

In Figure 14–6, the dermal patterns of a normal individual and a Down's syndrome patient are compared. Though no single feature of the dermal patterns of Down's syndrome is unique, the frequency with which it appears in controls is often quite different from its frequency in Down's patients. For example, a radial loop is rarely found on the second digit in Down's patients (2 per cent), whereas it is common in controls (20 per cent); again, a radial loop on the fifth finger occurs in about 4 per cent of Down's syndrome patients but in only three per 1000 controls. The most common digital pattern combination found in Down's syndrome is that of 10 ulnar loops.

Typically, the palm in Down's syndrome bears an axial triradius which is displaced distally and lies near the middle of the palm. The distal axial triradius is often associated with a pattern in the hypothenar area. There is an increased frequency of patterns in the third interdigital area. The dermal ridges tend in general to course transversely across the palm, rather than obliquely as in controls. A single flexion crease is present in about half the patients.

As previously noted, the pattern of the hallucal area of the sole may be very characteristic. The feet of over 80 per cent of persons with

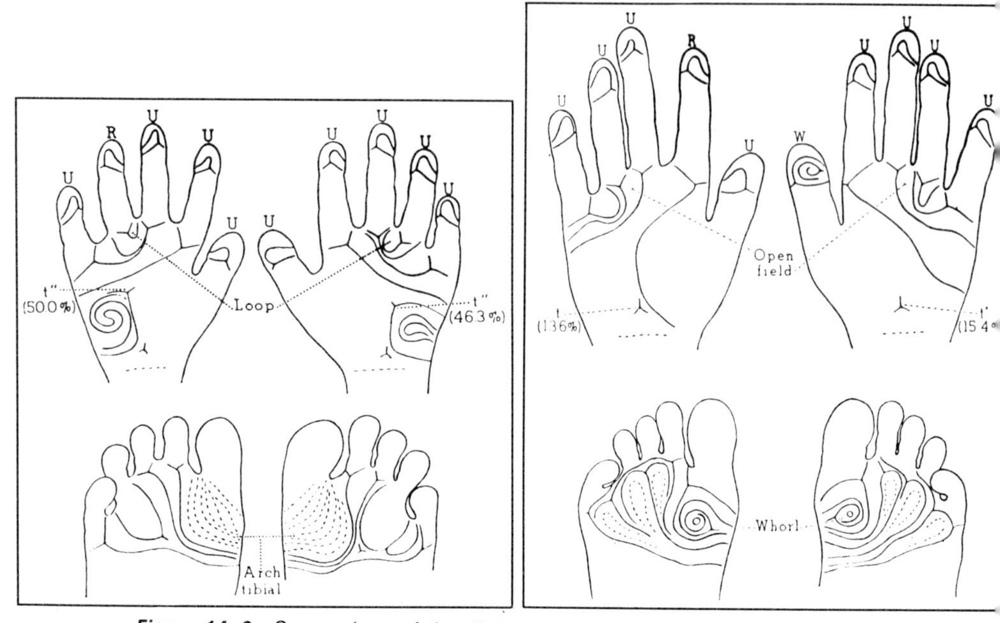

Figure 14–6 *Comparison of the digital, palmar and plantar patterns of a patient with Down's syndrome (left) and a control (right). For discussion, see text. (From Walker 1958.* Pediat. Clin. N. Amer. *5:531–543.)*

Down's syndrome have either an "arch tibial" (as shown in Fig. 14–5) or a very small "loop distal," which is a transitional type of pattern between arches and the more common large loops.

The "Walker index" (Walker, 1958) is a useful aid to the diagnosis of Down's syndrome. In mathematical terms, it is the log of the probability that any given set of patterns will be found in a Down's patient rather than in a control. A histogram of the index values for Walker's series of Down's patients and controls is shown in Figure 14–7. Note that approximately 70 per cent of each group can be distinguished, but that an overlap group remains in which the index is consistent with Down's syndrome, but not diagnostic. Though dermal pattern analysis cannot be used in place of karyotyping for the diagnosis of chromosome abnormalities, it has its place as a screening technique or as an objective means of assessment which can be used in conjunction with other morphological criteria when an immediate diagnosis is required. Since dermal patterns show ethnic variation, patients and controls should be matched for ethnic background. The Walker index is appropriate for whites of Northern European origin.

A useful dermatoglyphic nomogram for the diagnosis of Down's syndrome has been described (Reed et al., 1970). The nomogram (Fig. 14–8) uses only four variables: the patterns on the left and right index

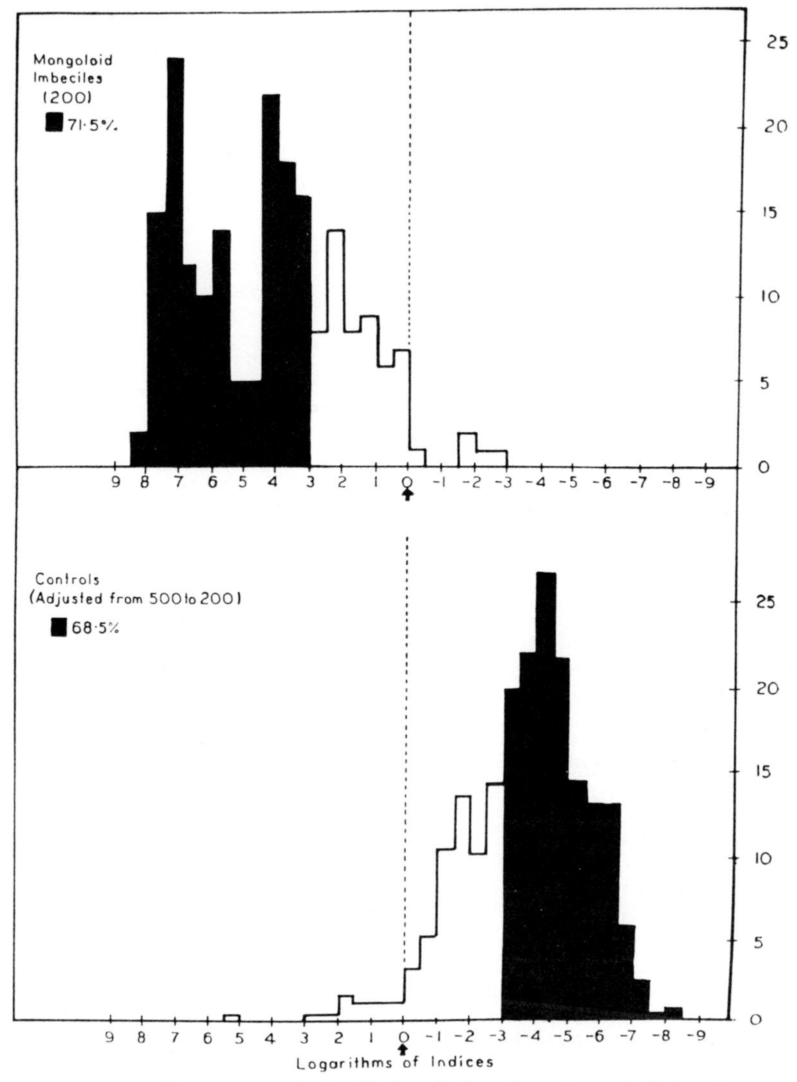

Figure 14–7 *Histograms of the Walker index, based on digital, palmar and plantar patterns in 200 patients with Down's syndrome and 500 controls. The method of calculation of the index is given in the original reference. (From Walker 1958.* Pediat. Clin. N. Amer. *5:531–543.)*

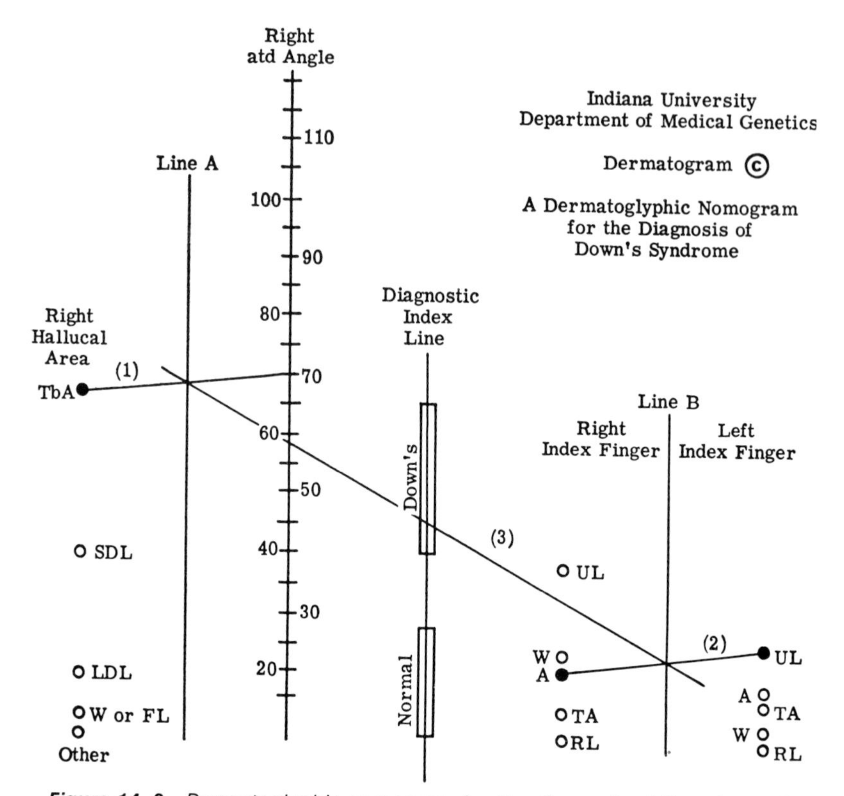

Figure 14–8 Dermatoglyphic nomogram for the diagnosis of Down's syndrome. Three lines are constructed. Line 1 is drawn connecting the appropriate point on the atd axis to the circle corresponding to the right hallucal pattern type. Line 2 connects the corresponding circles for the pattern types of the right and left index fingers. Line 3 connects the points of intersection of lines 1 and 2 with lines A and B, respectively. The point at which line 3 crosses the diagnostic index line determines whether the individual has dermatoglyphics within the Down's syndrome range, normal range or overlap area. In the example shown, the patient's dermatoglyphics are concordant with Down's syndrome. Abbreviations: TbA, tibial arch; SDL, small distal loop; LDL, large distal loop; FL, fibular loop; W, whorl; UL, ulnar loop; A, arch; TA, tented arch; RL, radial loop. (Copyright © 1970, Indiana University Foundation. All rights reserved. Reprinted from Reed et al. 1970, J. Pediat. 77:1024–1032.)

fingers, the right hallucal pattern and the right *atd* angle. A mathematical method of discrimination which separates Down's patients from controls, with less overlap than in the Walker method, has been developed by Borgoankar et al. (1971).

DERMATOGLYPHICS IN OTHER CHROMOSOMAL ABERRATIONS

As Table 14–1 shows, on the whole the dermal patterns associated with most chromosomal syndromes are not sufficiently distinctive to be useful in diagnosis. Distal axial triradii and single flexion creases are more frequent in nearly all chromosomal syndromes than in controls. It is not yet possible to diagnose chromosomal syndromes other than Down's by means of characteristic dermal pattern combinations.

One of the most consistent findings in any chromosomal syndrome is the high frequency of arch patterns seen in trisomy 18. Most infants with trisomy 18 have six or more digital arches. However, this is a nonspecific finding which is seen in other syndromes also, e.g., in D trisomy and in individuals with extra X chromosomes. It is also found in some single-gene disorders (e.g., brachydactyly Type A) and in many children with multiple congenital malformations but with normal karyotypes. About 3 per cent of normal individuals have six or more arches.

DERMATOGLYPHICS IN NONCHROMOSOMAL DISORDERS

Now that chromosomal disorders have aroused clinical interest in dermal patterns, dermatoglyphic aberrations have been described in a number of nonchromosomal conditions as well. Among the wide variety of disorders in which unusual dermal patterns have been reported are Wilson's disease, various congenital heart disorders, rubella syndrome (following maternal rubella during early pregnancy) and childhood leukemia. Many of these reports remain unconfirmed or have been challenged. The differences described may in some cases reflect small sample size and inappropriate controls.

Gross distortions of the dermal patterns can occur in association with malformations of the limbs, whether genetic or nongenetic in origin. This is not surprising, because the differentiation of the limbs precedes the differentiation of the dermatoglyphics, which must then conform to the shape of the underlying structures. Examples of dermatoglyphic distortions are found in the various autosomal dominant types of brachydactyly, and in thalidomide embryopathy, a malformation syndrome which in the early 1960's occurred in several hundred children whose mothers were given the drug thalidomide to control vomiting in early pregnancy.

GENERAL REFERENCES

Borgaonkar, D. S., Davis, M., Bolling, D. R., and Herr, H. M. 1971. Evaluation of dermal patterns by predictive discrimination. I. Preliminary analysis based on frequencies of patterns. *Johns Hopkins Med. J.* 28:141–152.

Cummins, H., and Midlo, C. 1961. *Fingerprints, Palms and Soles; An Introduction to Dermatoglyphics.* Dover Publications, Inc., New York.

Holt, S. B. 1968. *The Genetics of Dermal Ridges.* Charles C Thomas, Publisher, Springfield, Illinois.

Penrose, L. S. 1963. Fingerprints, palms and chromosomes. *Nature* 197:933–938.

Penrose, L. S. 1968. *Memorandum on Dermatoglyphic Nomenclature.* Birth Defects Original Article Series Vol. IV, No. 3, The National Foundation, New York.

Penrose, L. S. 1969. Dermatoglyphics. *Sci. Amer.* 221:72–84.

Reed, T. E., Borgaonkar, D. S., Conneally, P. M., Yu, P. L., Nance, W. E., and Christian, J. C. 1970. Dermatoglyphic nomogram for the diagnosis of Down's syndrome. *J. Pediat.* 77:1024–1032.

Uchida, I. A., and Soltan, H. C. 1963. Evaluation of dermatoglyphics in medical genetics. *Pediat. Clin. N. Amer.* 10:409–422.

Walker, N. F. 1958. The use of dermal configurations in the diagnosis of mongolism. *Pediat. Clin. N. Amer.* 5:531–543.

15

Applications of Medical Genetics

Not all disorders that affect more than one member of a family are genetic. On the contrary, occasionally a clearly definable environmental cause (for example, an infection or teratogen) may affect more than one member of a family at a time. Since it is not always easy to judge whether a particular problem is genetically determined, Neel and Schull (1954) have provided a useful guide to genetic etiology:

1. The occurrence of the disease *in definite proportions* among persons related by descent, when environmental causes can be ruled out.

2. The failure of the disease to appear in unrelated lines (e.g., spouses or in-laws).

3. A characteristic onset age and course, in the absence of known precipitating factors.

4. A greater concordance of incidence in monozygotic than in dizygotic twins.

The foregoing list was prepared some years before the role of chromosomal disorders was known. Now it is possible to add the following criterion:

5. The presence in the propositus of a characteristic phenotype (usually including mental retardation) and a demonstrable chromosomal abnormality, with or without a family history of the same or related disorders.

THE FAMILY HISTORY

Taking an adequate family history is an essential part of the assessment of a patient with a genetic disorder; it is also helpful in eliminating the possibility that a condition has a genetic basis (Fraser, 1963). Because few of us know our pedigrees in any detail, family history details are often difficult and time-consuming to obtain and verify.

A genetic history should always deal with data pertinent to the patient's condition. (Many hospital histories record the family history only with respect to a few defects such as diabetes, asthma and "mental retardation," regardless of their applicability to the patient's problem.) Negative information, i.e., the absence of a disorder in any relative, may be as important as a positive finding. To record "family history negative" after one or two brief questions to the patient or his nearest available relative may give a completely erroneous impression. It is necessary to ask specifically about age, sex and health (present and past) of parents, sibs and near relatives, and to ask about each person separately. Miscarriages and stillbirths should be listed. The age at death of deceased relatives, and the cause of their deaths (if known) should be recorded. If the cause was established by autopsy, this fact should also be noted. Usually the pedigree need not extend over many generations; the more remote the relative, the less accurate the information. It is especially important to check whether there is any consanguinity in the pedigree, especially in the parents of the propositus, and to determine whether both parents are from the same geographic area or ethnic isolate.

The exact relationship of the relatives to the propositus and to one another should be established, and for this purpose it is useful to construct a pedigree chart, which will show at a glance the relationship of affected relatives to the propositus and to one another.

GENETICS IN DIAGNOSIS

The knowledge that a genetic disorder is present in a patient's family may indicate the diagnosis and suggest diagnostic tests to confirm it. For example, a male child was found to have a history of polydipsia and polyuria. His urine had very low specific gravity. A variety of disorders can lead to inability to concentrate the urine normally, but in this case the child was accompanied to the hospital by a 10-generation pedigree linking him to a well-known kindred—the "water-drinkers of Nova Scotia"—in which nephrogenic diabetes insipidus is inherited as an X-linked recessive trait (Bode and Crawford, 1969). Thus, the confirmatory test—resistance to the antidiuretic effect of vasopressin—was immediately indicated.

Not many patients come for diagnosis accompanied by detailed pedigrees, but far less extensive family histories may also be useful. If a child is the first member of his kindred to have Duchenne muscular dystrophy, his mother has a two-thirds chance of being a carrier, and a one-third chance of being normal, in which case the child is a new mutant. Carrier testing in this disorder can identify only 70 per cent of the heterozygous mothers. If in this case the mother has a second boy, the child can be tested for the disorder several months before it would have been clinically apparent. Though in this disorder early diagnosis does not affect the ultimate prognosis, diagnostic tests can confirm the presence or absence of the disorder and thus allay uncertainty. Moreover, if the second son is found to be affected, the mother's carrier status is confirmed (see later).

There are numerous examples of rare and obscure disorders that can be diagnosed if it is known that the patient has a relative with a particular disease, or belongs to an ethnic group in which a specific disorder has a high frequency.

DIAGNOSIS OF CHROMOSOMAL ABERRATIONS

The diagnosis of chromosomal anomalies depends on cytogenetic analysis. Because autosomal anomalies are frequently classic in their signs and symptoms, clinical assessment may often indicate the diagnosis but karyotyping is needed to confirm it. In anomalies such as rare translocations, for which the phenotype may not have been described previously, a combination of developmental defects, mental retardation and failure to thrive may suggest that a chromosomal defect is present; in these cases, application of the new banding techniques for chromosome analysis is often helpful, because banding may make evident abnormalities not revealed by conventional staining.

Sex chromosome anomalies may not be diagnosed until puberty, unless the affected child is found in a population screening program. Chromosome analysis is indicated in patients with abnormal sexual development, ambiguous genitalia or any of the classic signs of Turner's syndrome, though of course many conditions which are not associated with any chromosomal anomaly may also involve abnormal sexual or mental development.

Chromosome anomalies are slightly more common in women who undergo multiple abortions, or in their husbands, than in the general population. Hence, chromosome analysis of both partners is a necessary part of the management of women with histories of multiple abortions.

MASS SCREENING

Population screening for treatable genetic disorders of metabolism, particularly in newborns, is a public health measure used to discover patients with remediable disorders and make prompt treatment possible. About 90 per cent of North American newborns are screened for PKU. A number of other diseases, such as galactosemia, tyrosinemia and cystinuria, can also be identified in newborns by screening urine for abnormal metabolites or blood for amino acids. The use of mass screening is certain to become more widespread as additional tests are devised and laboratory facilities for doing the tests become generally available. Since the purpose of these programs is to identify disorders that can be successfully treated, the patients that are found require prompt verification of the diagnosis and long-term treatment; they and their families also need genetic counseling.

A different kind of mass screening with a different objective is the screening of high-risk populations to find heterozygotes who might have affected children (see later). The prototype programs are screening for carriers of Tay-Sachs disease in Ashkenazi Jews, for carriers of sickle cell anemia in American Blacks and for carriers of thalassemia in certain areas of Italy. The practicality and acceptability of Tay-Sachs screening have been well demonstrated. Carriers of sickle cell anemia among American Blacks are even more common than carriers of Tay-Sachs disease among Ashkenzai Jews (8 to 14 per cent of American Blacks have sickle cell trait, as compared with an average of about 4 per cent for carriers of the Tay-Sachs allele among Ashkenazi Jews), but in contrast to Tay-Sachs disease there is still no routinely applicable prenatal test for sickle cell anemia. Until prenatal diagnosis is possible, the value of routine testing for carriers is doubtful, since the anxiety caused by carrier identification has no compensatory benefit. However, sickle cell trait itself is not entirely benign, and it is probably useful to identify carriers for their own sakes. Thalassemia carriers in high-frequency parts of Italy are actively discouraged from marrying other carriers, but no further preventive measures are taken.

Many other genetic disorders can be identified, actually or potentially, in homozygotes, heterozygotes or both, by population screening. For example, in type II hyperlipoproteinemia, the most common of the hyperlipoproteinemias, the serum cholesterol is very high and xanthomata composed chiefly of cholesterol are formed in tendons and in skin. Inheritance is autosomal dominant and heterozygotes have a relatively normal life span, but homozygotes have a severe form of the disease which is usually fatal in childhood. Screening for identification of heterozygotes for this defect would have the double benefit of allowing treatment of their high serum cholesterol and identifying couples at high risk of producing homozygous children.

Chromosome disorders can also be identified by population screening, but in this as in other cases the benefits of a screening program must be weighed against the cost.

GENETICS IN PARENTAGE PROBLEMS

Parentage problems that can be resolved genetically are usually questions of disputed paternity involving illegitimate children. Occasionally, other types of parentage problems arise involving interchange of newborns in hospitals, return of abducted children or even the concern of an engaged couple that they might be half-sibs. The genetic markers, especially the blood groups, of the parents and child can sometimes resolve these problems.

In cases of disputed paternity, there are two general types of exclusion: the putative father is excluded if the child has an antigen which he and the mother lack (e.g., if he and the mother are group O and the child is group A), or if the child lacks an antigen which he must transmit (e.g., if he is AB and the child is O). Caution must be used in interpreting the findings, because peculiar genetic situations, such as the Bombay phenotype or a silent allele (described in Chapter 9), may lead to false conclusions.

If a laboratory is well equipped and expertly staffed, it can exonerate about two-thirds of white men falsely accused of paternity. Conversely, it may be able to affirm with very high probability (though not with absolute certainty) that a particular man is the true father. If the man and child are both group B, NS, $C^{w}D^{u}e/cde$, Lu(a+) and K+, and the mother has only common blood groups, it is virtually certain that the man is the father of the child, since this combination of rare blood groups is found in only one in 10 million whites. (Of course it may also be present in other members of the family of the accused man.) Other rare dominant traits, such as brachydactyly or an abnormal hemoglobin in man and child, have also been used to provide positive evidence of paternity.

GENETIC COUNSELING

The field of genetic counseling has developed in response to the special needs for information and support of families with genetic disorders. Typically, those seeking counsel are parents of a child with a genetic defect, who must make a decision about having more children but first need help to understand the recurrence risk of the condition and the burden it is likely to impose on the patient, the family and society as a whole. The genetic counselor's role is to determine the recurrence risk, interpret it to the family and help the parents make a responsible adjustment. The parents' decision may be to have no more children; if so, they may need to be referred for sterilization or contraceptive advice, or to an adoption agency. If they want more children, prenatal diagnosis or artificial insemination are options they may want to consider. Referral to parent groups or to agencies concerned with handicapped children may also be appropriate in some cases.

Much genetic counseling is done by clinicians as an integral part of the overall management of genetic disorders. Most medical schools and teaching hospitals now have medical genetics units for service, teaching and research; genetic counseling is an important part of the service component.

Though the most common kind of case seen by genetic counselors is that of parents concerned about the recurrence risk of a genetic defect, several other genetic problems are also encountered. Examples are late-onset disorders, multiple abortions, late maternal age, consanguineous marriage and preadoption counseling. The scientific background of the advice given in each of these situations has been outlined in the foregoing chapters.

RECURRENCE RISK

When the specific diagnosis and pattern of transmission of a trait are known, the recurrence risk is usually obvious; for example, cystic fibrosis is an autosomal recessive with a 1-in-4 risk for subsequent children, and classic hemophilia is an X-linked recessive in which the risk that any son of a carrier woman will be affected is 1/2 and the risk that a daughter will be a carrier is also 1/2. For multifactorial disorders, empiric risk figures based on previous experience can be provided (see later). For chromosomal aberrations, recurrence risks depend on the patient's karyotype and on whether the abnormality is sporadic or familial. Examples have been given in previous chapters.

There are many situations in which accurate recurrence risk information cannot be provided. For example:

1. The specific diagnosis may be in doubt. Because of genetic heterogeneity, it is almost always risky to attempt to give genetic counseling when the diagnosis is not precisely known. Sometimes, however, the family history strongly suggests a specific pattern of inheritance even when the diagnosis is uncertain.

2. The diagnosis may be known, but the trait may be so rare that whether it has a genetic basis is unknown.

3. The trait may be one in which both genetic forms and nongenetic forms (phenocopies) exist or in which there is more than one genetic form.

4. Variable expressivity or reduced penetrance may blur the pedigree pattern.

Carter and colleagues (1971) have compared the subsequent reproductive behavior of parents of children with disorders having a high recurrence risk (10 per cent or more) and parents of children with disorders having a low recurrence risk (less than 10 per cent). They found that high-risk parents are much less likely than low-risk parents to take the chance of having another child. This shows that on the whole most parents take a responsible attitude toward further reproduction when the risk of losing a child or having an ill or deformed child is high.

BURDEN

The recurrence risk of a disorder is only one of the factors to be considered in genetic counseling. The counselor must also take into account the burden it imposes on affected people, their families and society.

Though any genetic disorder by definition carries some burden, the burden is much greater in some conditions than in others (Murphy, 1969). The recurrence risk of color blindness is quite high, but the burden is slight, whereas hemophilia has the same risk but is a much greater burden. Tay-Sachs disease is a severe burden for a relatively short time, whereas Duchenne muscular dystrophy and cystic fibrosis are severe burdens for years. Spina bifida and Down's syndrome have a low risk, but the burden in terms of long-range problems may be great. Anencephaly has a low risk and only a moderate burden, since affected children die within hours of birth; but the counselor must also consider the risk that a subsequent child might have spina bifida. Parents who could accept the risk of a second anencephalic might not be prepared to take even a slight risk of a child with spina bifida.

Burden cannot be assessed quantitatively, but for the parents it is an important component of the decision about whether to have another child.

CARRIER DETECTION

Tests for the identification of heterozygotes can add to the accuracy of genetic counseling. Carrier tests can be offered to relatives of affected individuals, or used for large-scale surveys of certain high-risk populations. Unfortunately, reliable carrier tests are not yet available for some of the most common genetic disorders (cystic fibrosis is an outstanding example), or are not sufficiently discriminatory (as in classic hemophilia or Duchenne muscular dystrophy).

The practical value of a carrier test is less for autosomal recessives than for dominants or X-linked recessives. To illustrate: Mrs. A. has a sister whose son has cystic fibrosis. The probability that Mrs. A. is a carrier of cystic fibrosis is 1/2, but she will have an affected child *only* if her husband is also a carrier; the probability that her husband is a carrier (if he is unrelated) is approximately 1/20, i.e., the population frequency of heterozygotes; and if both parents are carriers, the risk that the first child will be affected is 1/4. Thus, the total probability that Mrs. A.'s first child will have cystic fibrosis is only $1/2 \times 1/20 \times 1/4$ or 1 in 160. In contrast, for X-linked disorders, which are handed down by females, the risk that the maternal aunt of an affected boy will herself have an affected son is high. If we ignore the possibility that the disorder has arisen as a new mutation, the chance that Mrs. A. is a carrier is 1/2 and the chance that her first child will be affected is 1/8.

Because autosomal dominants are expressed in heterozygotes, "carrier" detection tests for autosomal dominants are tests to identify genetically susceptible individuals before they are old enough to manifest the trait, or when the expression of the trait is very mild.

Tests for heterozygotes for autosomal dominants of late onset, such as Huntington's chorea, would be very useful from the eugenic standpoint, because those individuals identified as genetically susceptible would be in a position to forego reproduction in order to eradicate the defect. But does a potential choreic really want to know in advance whether or not he is likely to develop the disease at age 40 or so? Among physically unaffected but possibly genetically susceptible people, there seems to be no consensus that a detection test would be welcome.

Autosomal dominants with reduced penetrance or variable expressivity present difficulties in counseling because it may be very hard to distinguish between sporadic cases (new mutants) and inherited cases (subjects who have inherited the gene from a minimally affected parent). Crouzon's craniofacial dysostosis is such a condition. In typical cases this disorder is readily diagnosed by the characteristic facies (Fig. 15–1). The defect is burdensome because it is disfiguring, and extensive plastic surgery is needed to correct the hypertelorism and exophthalmos of the affected child in order to make his appearance more or less normal. Some cases (perhaps one-fourth of the total) are new mutants, and in these of course the only member of the family capable of passing on the

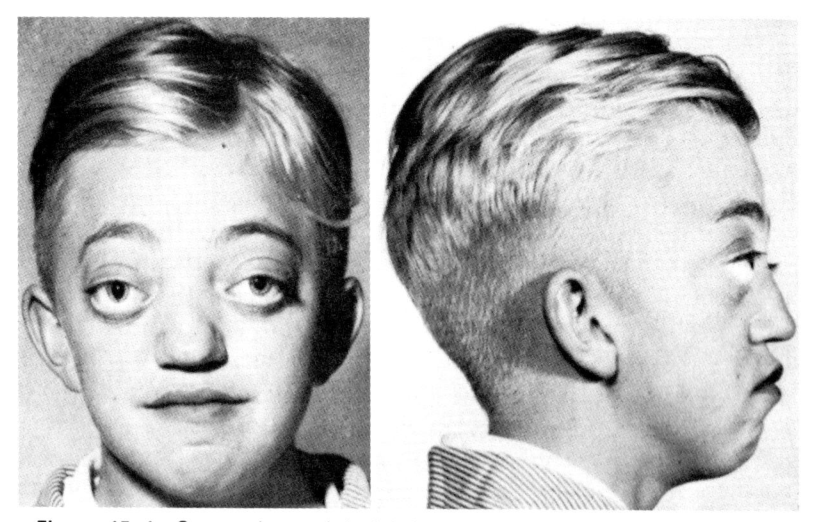

Figure 15–1 *Crouzon's craniofacial dysostosis, an autosomal dominant defect. Not exophthalmos and hypertelorism. For discussion, see text. (From Gorlin and Pindborg, 1964.* Syndromes of the Head and Neck. *McGraw-Hill Book Company, New York, Toronto and London, p. 172, by permission.)*

defect is the affected person. Other cases are inherited but may be expressed so mildly that it is hard to determine whether either parent (or any other family member) has the gene, yet in these instances the risk for subsequent children is 1/2.

Carriers of chromosomal aberrations form a special group of high-risk individuals who can often be identified by karyotyping, especially with the use of the new banding techniques. Prenatal diagnosis can often help such people to have only normal children (see later).

BAYESIAN METHODS IN GENETIC COUNSELING

Bayes' theorem (first published in 1763) gives a method of assessment of the relative probability of each of two alternative possibilities. This theorem can be usefully applied to certain problems in genetic counseling (Murphy and Mutalik, 1969). To illustrate, we will consider some pedigrees of Duchenne muscular dystrophy.

In Figure 15–2*A*, II-3 is the daughter of a definite carrier of the Duchenne gene. The *prior* probability that she is a carrier is 1/2, and the prior probability that she is a noncarrier is also 1/2.

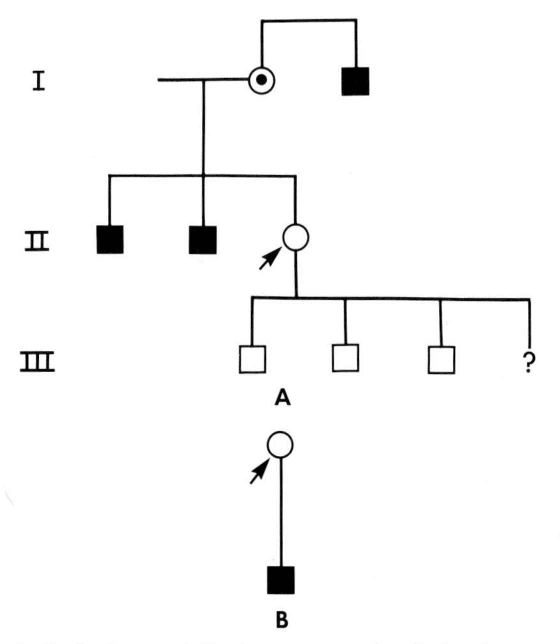

Figure 15–2 A, *Pedigree of Duchenne muscular dystrophy, in which the probability that the mother II-3 is a carrier can be determined by Bayesian analysis (see text).* B, *A sporadic case of Duchenne muscular dystrophy, in which the probability that the mother is a carrier can be determined by Bayesian analysis (see text).*

This woman has three normal sons. Now if she is a carrier, the *conditional* probability that all three sons would be normal is 1/8; and if she is normal, the *conditional* probability that all three boys would be normal is 1 (or very close to 1, since she might have a new mutant child).

We can now consider the *joint* probability, which is the product of the prior and conditional probabilities; the joint probability that she is a carrier is $1/2 \times 1/8 = 1/16$, and the joint probability that she is not a carrier is $1/2 \times 1 = 1/2$.

The posterior probability that she is a carrier is therefore:

$$\frac{1/16}{1/16 + 1/2} = 1/9$$

and the posterior probability that she is not a carrier is:

$$\frac{1/2}{1/16 + 1/2} = 8/9$$

These calculations are summarized in Table 15–1.

Now the probability that the woman is a carrier (1/9) can be applied to genetic counseling. The risk that her next child will be an affected male is $1/9 \times 1/4 = 1/36$. This is appreciably below the prior probability of 1/8 estimated when the genetic evidence provided by her children is not taken into consideration.

A common situation in genetic counseling in Duchenne muscular dystrophy is that of the mother of a sporadic case who may or may not be a new mutant (Fig. 15–2*B*). Here the prior probabilities are quite different. The *prior* probability that the woman is a carrier is 4μ (μ = mutation rate). (This probability is calculated as follows: the chance that she received a new mutation from one of her parents is μ for each parent, and the chance that she received the gene from her mother who is a carrier is 2μ. The prior probability that she is not a carrier is $1 - 4\mu$, which is very close to 1. The *conditional* probability of an affected son is 1/2 if she is a carrier, but only μ if she is not a carrier. The *joint*

TABLE 15–1 PROBABILITY THAT II–3 OF FIGURE 15–1*A* IS A CARRIER

	II–3 a Carrier	*II–3 Not a Carrier*
Prior probability	1/2	1/2
Conditional probability	1/8	1
Joint probability	1/16	1/2
Posterior probability	$\frac{1/16}{1/16 + 1/2} = 1/9$	$\frac{1/2}{1/16 + 1/2} = 8/9$

probabilities are $4\mu \times 1/2 = 2\mu$ and $1 \times \mu = \mu$, respectively. The posterior probability that she is a carrier is therefore $\frac{2\mu}{2\mu + \mu} = 2/3$.

Additional information may be provided by other family members (normal sons or normal brothers, for example), or by the use of a carrier detection test. Two-thirds of carriers of Duchenne muscular dystrophy have a level of serum creatine phosphokinase (CPK) activity higher than that of noncarriers; the remaining one-third are indistinguishable from normals. If the woman in question has normal CPK, this information can be included and the probability that she is a carrier becomes 2/5. The calculations are shown in Table 15–2.

The calculations used on page 309 to estimate the relative risk that a twin pair is DZ also make use of Bayesian probability.

GENETIC COUNSELING FOR MULTIFACTORIAL TRAITS

When a trait is inherited multifactorially, genetic counseling can usually be provided only in terms of the **empiric risk** that it will recur, i.e., on the basis of past experience with the behavior of the trait within families. In earlier chapters we have given examples of empiric risks for a number of common congenital malformations. Though empiric risk figures are useful guides, it must be remembered that they are general averages based on experience and do not necessarily accurately estimate the risk for a specific family.

Diabetes Mellitus. Because of its variable onset age, different frequency in males and females and genetic heterogeneity, diabetes has been termed a "geneticist's nightmare." Nevertheless, when the few cases which are caused by single mutant genes and/or which are parts

TABLE 15–2 PROBABILITY THAT WOMAN OF FIGURE 15–2*B* IS A CARRIER

	Carrier	*Not a Carrier*
Prior probability	4μ	$1 - 4\mu \cong 1$
Conditional probability		
Genetic evidence	1/2	μ
CPK test normal	1/3	1
Joint probability	$2/3\ \mu$	μ
Posterior probability	$\frac{2/3\ \mu}{2/3\ \mu + \mu} = 2/5$	$\frac{\mu}{2/3\ \mu + \mu} = 3/5$

TABLE 15–3 RISKS OF DIABETES IN CHILDREN OF DIABETICS*

	Onset Age of Diabetes in Proband		
	0–19	20–39	40+
Male child of male diabetic	×40	×13	×3
Female child of male diabetic	×41	×7	×2
Male child of female diabetic	×29	×6	×3
Female child of female diabetic	×18	×9	×1

*Risks stated as multiples of risk of diabetes in the general population, by sex and age. After Simpson, 1968.

of syndromes are excluded, the great majority of cases of diabetes appears to have multifactorial inheritance (Simpson, 1964).

Recurrence risk figures for diabetes are available only for first-degree relatives (Simpson, 1968). Tables 15–3 and 15–4 give estimates of the recurrence risk for children and sibs of diabetics. The estimates are stated in terms of multiples of the risk of diabetes in members of the general population of the same age and sex. Table 15–5 summarizes the estimated risks for first-degree relatives, giving separate estimates for the risk of diabetes at various onset ages. Note that the earlier the onset in the proband, the greater the risk for first-degree relatives.

To use these tables to calculate the empiric risk of diabetes in a first-degree relative of a diabetic, we need to know the general population risk of diabetes. Data for the risk of developing diabetes by specific ages would be useful, but there are no recent data on large populations.

TABLE 15–4 RISKS OF DIABETES IN SIBS OF DIABETICS*

	Onset Age of Diabetes in Proband		
	0–19	20–39	40+
Male sib of male diabetic	×14	×5	×4
Female sib of male diabetic	×10	×4	×2
Male sib of female diabetic	×11	×4	×3
Female sib of female diabetic	×13	×4	×2

*Risks stated as multiples of risk of diabetes in the general population, by sex and age. After Simpson, 1968.

TABLE 16–5 SUMMARY OF RISKS OF DIABETES AT VARIOUS ONSET AGES IN FIRST-DEGREE RELATIVES OF DIABETICS*

Onset Age in Diabetic	Relative	Risk for Relative	
		Onset Age <20	Onset Age ≥20
<20	Parent	×5	×2
	Sib	×15	×8
	Child	×22	–
≥20	Parent	–	×2
	Sib	×7	×3
	Child	×5	×2

*Stated as multiple of risk of diabetes in the general population, by sex and age. After Simpson, 1968.

Prevalence data (data on the number of living cases of diabetes in the population in certain age groups) from the United States National Health Survey of 1961 to 1963 were estimated to be as follows (expressed as per cent):

Age	Male	Female	Total
0–24	0.11	0.11	0.11
25–44	0.50	0.52	0.51
45–64	2.34	2.53	2.44
65–74	4.11	5.19	4.70
75+	4.26	5.29	4.83

TREATMENT OF GENETIC DISORDERS

Although there is a persistent impression that labeling a disease as genetic is labeling it as incurable, actually many genetic disorders can be treated with a reasonable degree of success. For genetic disorders of metabolism, the possible forms of therapy are:

1. Restriction of a substance which the patient cannot metabolize (phenylalanine in PKU, galactose in galactosemia).
2. Product replacement (hormones in hereditary hormone deficiencies).
3. Vitamin supplementation, to enhance enzyme activity by increasing ingestion of its coenzyme (vitamin D in vitamin D–dependent rickets).
4. Enzyme replacement (still in the experimental stage).

In some genetic disorders, patients may be at risk only under certain environmental conditions; e.g., patients with G6PD deficiency develop hemolysis only when exposed to primaquine or certain other drugs. Here, "treatment" is merely a matter of keeping the patient away from the precipitating agent.

A model program for diagnosis, counseling and treatment of genetic diseases has been in operation in Quebec province since 1971 (Clow, Fraser, Laberge and Scriver, 1973). In addition to the traditional type of genetic counseling supported by a network of university-based genetics centers, this program has pioneered in the delivery of "continuous" counseling and management to patients with genetic metabolic disorders treatable by special diets. More programs of this kind are needed to bridge the gap between theory and practice in medical genetics.

PRENATAL DIAGNOSIS

The development of the technique of prenatal diagnosis by amniocentesis has added a new dimension to genetic counseling; now, instead of counseling in terms of probabilities, for some conditions it is possible to say definitely that a particular child is or is not affected, and to do so at a stage early enough to allow selective abortion of abnormal offspring.

The developing fetus is enclosed in an amniotic sac wrapped in two membranes, the inner amnion and the outer chorion. The sac is filled with amniotic fluid, which is formed by the fetus and consists chiefly of urine and of secretions from the respiratory tract. This fluid contains cells sloughed off from the fetal skin and epithelium, which are known as amniotic fluid cells (amniocytes).

In amniocentesis, the amnion is tapped transabdominally, and amniotic fluid is withdrawn. The procedure is usually performed at about the sixteenth week of pregnancy. The amniotic fluid cells are spun down and grown in culture. After about two weeks, chromosome preparations can be made. After a somewhat longer period in culture, there are enough cells to allow testing for any of 40 or so different biochemical disorders (Table 15–6).

For certain tests the cells can also be analyzed directly without culture, though with less reliability. For example, uncultured cells can be examined for sex chromatin for the X chromosome and by quinacrine fluorescence for the Y, but the full chromosome set cannot be visualized without culture.

The amniotic fluid itself can also be analyzed for metabolites or infective agents. For example, α-fetoprotein (AFP) is a protein formed in the fetal liver as early as the sixth week of pregnancy and is identifiable in amniotic fluid. It is greatly increased in amount in the amniotic fluid of fetuses with anencephaly, probably because of leakage of cerebrospinal fluid. For women who have a high risk of producing an anen-

TABLE 15–6 SOME GENETIC DISORDERS SUITABLE FOR PRENATAL DIAGNOSIS*

Prenatal Diagnosis Now Possible or Already Made	***Prenatal Diagnosis Potentially Available in the Near Future***
Disorders of Amino Acid Metabolism	
Argininosuccinicaciduria	Citrullinemia
Cystinosis	Hyperlysinemia
Homocystinuria	Others
Hypervalinemia	
Maple syrup urine disease	
Methylmalonic aciduria	
Ornithine-α-ketoacid transaminase deficiency	
Disorders of Carbohydrate Metabolism	
Glycogen storage disease (types II, III, IV)	Fucosidosis
Galactosemia	Pyruvate decarboxylase deficiency
Mannosidosis	
G6PD deficiency	
Disorders of Lipid Metabolism	
Fabry's disease	Mucopolysaccharidoses
Gaucher's disease	Maroteaux-Lamy syndrome
Gangliosidoses	Morquio's syndrome
GM_1 (generalized)	Sanfilippo's syndrome
GM_2 (Tay-Sachs disease)	
Metachromatic leukodystrophy	
Niemann-Pick disease	
Refsum's disease	
Mucopolysaccharidoses	
Hunter's syndrome	
Hurler's syndrome	
Scheie's syndrome	
Miscellaneous Single Gene Disorders	
Adrenogenital syndrome	Acatalasemia
I-cell disease	Chédiak-Higashi syndrome
Lesch-Nyhan syndrome	Congenital erythropoietic porphyria
Lysosomal acid phosphatase deficiency	Cystic fibrosis
Xeroderma pigmentosum	Orotic aciduria
	Hemoglobinopathies
	Others
Chromosomal Disorders	
Those detectable by routine staining or by banding	
Congenital Malformations	
Those recognizable by radiography	Those recognizable by fetoscopy or ultrasound

*Based upon Milunsky et al. (1970), Clow et al. (1973) and other sources. Descriptions of the biochemical defects listed appear in McKusick (1971), Bondy and Rosenberg (1974) and elsewhere.

cephalic, prenatal detection of anencephaly by measurement of AFP in amniotic fluid appears to be feasible.

If the fetus is found to be abnormal, in most provinces and states it is now possible to obtain an abortion, which usually must be done by the twentieth week of pregnancy. Since amniocentesis is difficult before the fourteenth week, the length of time available for culturing and examining the fetal cells is limited.

CHROMOSOMAL INDICATIONS FOR AMNIOCENTESIS

Women who are candidates for amniocentesis because of a high risk of a cytogenetically abnormal child are of three main types:

1. Women who are (or whose husbands are) carriers of a chromosomal anomaly that leads to a high risk of an abnormal baby; e.g., carriers of DqGq translocations.
2. Women who are in the high-risk age group for Down's syndrome. If all pregnancies of women aged 34 and over were monitored by amniocentesis, half of all cases of Down's syndrome could be identified. Even if only women over 40 were tested, about one-sixth of the Down's cases could be found.
3. Women who have had a previous chromosomally abnormal child, e.g., a previous child with 21-trisomic Down's syndrome.

At the time of writing, we have only very preliminary figures about the number of Down's patients identified in these three groups: seven of 22 offspring of carriers, two of 82 fetuses of women aged 40 and over and one of 28 fetuses of women who had had a previous trisomic Down's, were found to be affected.

If a parent has a translocation, the risk of an abnormal child with an unbalanced karyotype is high. Fetuses of such parents are the only really high-risk group among the candidates for prenatal diagnosis of chromosomal aberrations. Down's syndrome is only one of the many kinds of abnormality that can result from translocation. Probably 5 per cent of cases of *cri du chat* syndrome are unbalanced offspring of reciprocal translocation carriers, and many rare kinds have been found. Of course, most translocation carriers are found only after the birth of a chromosomally unbalanced child. The birth of such a child signals the need for karyotyping of the parents. If either parent is a carrier, chromosome studies are indicated in his or her relatives. Any pregnancies in which either mother or father is a carrier can then be monitored by amniocentesis.

When amniocentesis leads to the prenatal diagnosis of a chromosomal or biochemical defect, it is assumed that the parents will elect to abort the defective fetus, though in a minority of cases the parents may decide to let the pregnancy run its course. Though economic factors should not be the chief considerations in prenatal diagnosis, an argument can be made that screening of all pregnancies of older mothers by amniocentesis, followed by selective abortion of defective fetuses, can lead to a reduction in the total costs of health care.

In Table 15–7, data are shown for the distribution of Down's syndrome in a typical population, by maternal age. Assume amniocentesis becomes available to all pregnant women of 40 or over in this population, who number 2450 in a given year. A total of 26 Down's fetuses or about 18 per cent of all Down's syndrome conceptions would be identified. Selective abortion of these would reduce the total Down's births in the population from 146 to 120. On the assumption that the cost of amniocentesis is $250 and the cost of rearing a Down's child is $6000 per year for an average life span of 20 years, the amniocentesis program would cost $500,000 and would save $3.1 million, effecting a net saving of about $2.6 million per year.

The cost of amniocentesis in relation to its benefit depends on the probability of identifying the defect concerned. For women in the 35 to 39 year age group in the population shown, only about one pregnancy in 300 produces a Down's child. It is generally agreed that for women over 40 the risk of producing a Down's syndrome child is high enough to make prenatal diagnosis desirable. There is less general agreement that the risk for women in the 35 to 39 age group is high enough to justify the procedure.

Of course the economic argument is not the only argument in favor of restricting amniocentesis to high-risk mothers. A second and more telling case can be made on the basis of the risks of the procedure itself.

TABLE 15–7 DISTRIBUTION OF DOWN'S SYNDROME IN 134,724 BIRTHS BY MATERNAL AGE

	Total Births[1]		***Down's Syndrome Births***[2]			
Maternal Age	*Number*	*% of All Births*	*Number*	*% of Births At This Maternal Age*	*% of All Down's Syndrome Births*	***"Enrichment"***[3]
Under 20	17,001	12.7	7	0.043	4.8	× 1/3
20–24	47,513	35.3	29	0.062 (~1 in 1500)	19.9	× 1/2
25–29	40,263	29.9	33	0.083	22.6	× 1
30–34	19,189	14.2	22	0.115 (~1 in 700)	15.1	× 1
35–39	8308	6.2	29	0.350 (~1 in 300)	19.9	× 3
40–44	2301	1.7	23	0.993 (~1 in 100)	15.8	
Over 44	149	0.1	3	2.200 (~1 in 50)	2.1	× 10
Total	134,274	100.2	146	0.109 (1 in 910)	100.2	

[1]Ontario Vital Statistics, 1970.
[2]Estimated from incidence data of Collmann and Stoller, 1963.
[3]"Enrichment" is the proportion of all Down's births which occur in a specific age group as compared to the proportion of all births which occur in that age group.

The risks of amniocentesis for mother and fetus appear to be very low, certainly acceptably low if the risk of a defective child is high. However, when the probability of a defect is low, it must be balanced against the slight but inevitable associated risks. The risks are of two kinds: risk to mother and risk to fetus.

Risk to Mother. As with any surgical procedure, there is a slight probability of infection. There is also some risk that a spontaneous abortion will follow, though we do not yet know what this risk is as compared with the overall risk of a second-trimester abortion.

Risk to Fetus. It is possible for the fetus to be injured by the needle itself, though this has been very rare. There is a possibility of causing hemorrhage by puncturing blood vessels of the placenta, but this risk is reduced if the placenta is first localized by ultrasound techniques. Loss of amniotic fluid through the punctured amniotic membrane is a present but slight hazard.

Follow-up studies of "survivors" of amniocentesis now under way in the United States, Canada and other countries will soon allow a more accurate assessment of the risks.

Not everyone agrees that mass screening of older pregnant women to identify trisomy 21 fetuses is desirable or even ethical. Those who disagree point to such factors as the hazards of the test itself, the possibility of laboratory error, the possibility that a prenatal detection and abortion program might be considered to relieve public health officials of all responsibility for the Down's syndrome patients who were not aborted. Nevertheless, it is likely that a screening program would be welcomed by most parents and supported as a public health measure. The task now is to make sure the procedure is done only by competent people and that data are routinely collected on a nationwide basis to ensure that the benefits and hazards can be assessed.

A special category of high-risk mothers is the group who carry deleterious X-linked recessive genes. If the defect is not identifiable in utero, a procedure in current use is the determination of fetal sex, followed by abortion of all male fetuses. This procedure has adverse aspects, since half the number of aborted males are genetically normal whereas half the surviving females are carriers who can in turn transmit the gene to the next generation.

BIOCHEMICAL INDICATIONS FOR AMNIOCENTESIS

Much of what is known about enzyme defects in cultured fetal cells has been learned by studies of cultured cells of other types, especially skin fibroblasts, derived from people homozygous or heterozygous for known defects. If a defect is expressed in fibroblasts, it is usually expressed in amniotic fluid cells also.

There are many advantages to studying a biochemical defect in culture rather than in living patients. Experiments that cannot ethically be performed with patients present no such problem in cell culture. Patients do not remain available, but cell cultures can be kept frozen and brought out again for analysis at any time. The cells can be manipulated by microbiological techniques; for example, clones derived from a single cell can be obtained. "Hybrids" of somatic cells from two different species, such as man and mouse, can be made and the hybrid cells can be used for linkage analysis and other genetic studies. One of the many advances in somatic cell genetics is its application to the detection in utero of biochemical defects, which have previously been studied in fibroblasts.

The biochemical defects actually or potentially identifiable in cultured amniotic fluid cells number about 40 so far (Table 15–6). Nearly all such defects are autosomal recessive, and one-quarter of the monitored pregnancies involve homozygous affected fetuses. Pregnancies likely to involve fetuses homozygous for such defects involve women with two main kinds of problems:

1. Women who in an earlier pregnancy have produced a child homozygous (or hemizygous) for an identifiable single-gene defect.
2. Women who are known carriers of an identifiable defect and married to known carriers. (At present, Tay-Sachs disease is virtually alone in this category.)

If a disorder is recessive and can only be recognized retrospectively (after the parents have revealed themselves as carriers by having a defective child), the proportion of affected fetuses actually identified is much less than the 100 per cent theoretically possible for Tay-Sachs disease.

Assuming the reproductive goal is two normal children per family, only one-fourth of all the carrier × carrier couples would be identified by having an affected child as the firstborn. The next pregnancies of these couples could be monitored. Of the remaining three-fourths, one-fourth would be found at the second pregnancy. (Nine-sixteenths of all the couples at risk would have two normal children and remain in happy ignorance of the risk they have run.)

Assume 64 carrier × carrier couples. At the first birth, there are 16 affected and 48 normal children. At the second birth, 16 amniocenteses yield 4 affected fetuses that are aborted; 48 normal and 12 affected children are born. Thirty-six families are now complete. At the third birth, 28 amniocenteses yield 21 normal children and 7 affected abortuses. Only 10 families have not yet achieved the goal of two normal children. At this point, 28 affected children have been born and only 11 affected fetuses (30 per cent of the total affected homozygotes conceived) have been aborted as a consequence of 44 amniocenteses. If the parents had been identified as heterozygotes in advance, there would have been 156 amniocenteses, but no affected children.

LONG-RANGE GENETIC EFFECTS OF AMNIOCENTESIS

The long-range effect of prenatal diagnosis and selective abortion on the gene pool are difficult to predict, since they depend very much upon the subsequent reproduction of the parents concerned, i.e., upon whether parents who have lost a defective child through abortion eventually have fewer children than they otherwise would have had, or whether, now that they are assured that defective fetuses can be identified and aborted, they will compensate by having as many or perhaps even more children than others in the population.

Selective abortion of fetuses affected with autosomal recessive disorders that are genetic lethals, such as Tay-Sachs disease, will not affect the total gene pool unless the parents replace the aborted fetus by an unaffected child, which has a 2/3 chance of being heterozygous. Even so, the long-range effect on the gene pool will be slight, because far more recessive genes are transmitted by heterozygotes mated to normals than by heterozygotes mated to other heterozygotes.

Only a few X-linked recessive disorders can be diagnosed prenatally. Lesch-Nyhan syndrome (hyperuricemia, choreoathetosis and self-mutilation) is one such disorder. Like Duchenne muscular dystrophy, it is a genetic lethal in which one-third of the affected boys represent new mutations. Thus, if all the affected male fetuses of carrier mothers are aborted, the incidence of affected males will remain at one-third the initial frequency.

If all male fetuses, whether affected or not, are aborted and all female fetuses survive, the number of female carriers rises in the next few generations. The increase in female carriers is most rapid if carriers are diagnosed retrospectively after the birth of an affected son, and if there is full compensation for affected children. (For details, see Motulsky, Fraser and Felsenstein, 1971.) In general, the increase in carrier frequency is a modest price to pay for the decrease in births of affected males; but abortion of *affected* males only would be preferable both for the family and for the population.

OTHER METHODS FOR PRENATAL DIAGNOSIS

Radiography. Some conditions can be diagnosed in utero by X-ray, early enough to allow abortion. Some limb reduction anomalies are suitable for this procedure, which must be performed with great caution because of the risk of radiation damage to the fetus.

Ultrasonography. At present, ultrasound examination can detect only gross malformations such as anencephaly, and only late in pregnancy. Ultrasound scanning to localize the placenta is a routine preliminary to amniocentesis. The possible risk to fetal development has not yet been assessed.

Fetoscopy. Instrumentation for direct visualization of the fetus is now being developed. This technique is highly promising because it will allow detection of external malformations, many of which are common. It will enable obstetricians to take samples of fetal or placental blood, thereby opening the prospect of prenatal detection of a large group of disorders, such as hemophilia and Duchenne muscular dystrophy, which are not detectable in cell culture but might be expressed in blood even in the 16-week fetus.

MANIPULATION OF HUMAN GERM CELLS AND EMBRYOS *IN VITRO*

The objective of attempts to observe and manipulate the immediate prefertilization and postfertilization stages of the human life cycle, in addition to the basic scientific knowledge gained, is to find ways of circumventing certain causes of infertility and genetic abnormality in order to allow childless couples to have children of their own. Human oocytes obtained by surgery have already been brought to maturation, fertilized and developed to the multicellular blastocyst stage, *in vitro*. Much of this work has been done by Edwards, Steptoe and their collaborators (Edwards and Fowler, 1970). In experimental and agricultural animals, work has gone further; implantation of blastocysts fertilized in vitro has led to the birth of normal offspring.

The ethical and social implications of experimental work with human germ cells *in vitro* are causing some controversy. Reimplantation of fertilized ova to complete their development seems very near at hand. The risk of abnormal development, the possibility of selecting the sex and perhaps even the genetic constitution of fertilized ova and the questionable wisdom of attempting to improve the fertility of a single couple when the species as a whole is already reproducing far too rapidly are among the debatable aspects of this work. On the other hand, it is a promising long-range approach to the prevention of many birth defects, both genetic and environmental.

CONCLUDING COMMENTS

The remarkable successes of molecular genetics have opened the prospect of spectacular developments in the modification of human genes. **"Genetic engineering,"** the synthesis of artificial genes and their incorporation into human cells, could theoretically provide a cure for enzyme-deficiency disease. Germ cells or early embryos might be more amenable to experimental engineering than somatic cells, if only because it would be easier to correct one or a few cells than many. **Predetermination of sex** by separation of X and Y sperm should soon be possible. **Cloning** (inducing a single somatic cell to develop into a complete individual, and even into numerous identical individuals) has already been done in frogs. The technique begins with replacing the nucleus of an egg cell with the nucleus of an intestinal cell by micromanipulation. The resulting frog has the exact genotype of its single parent. Any number of clonal individuals, exact genetic replicas of one another, can be produced. Finally, there is **selective reproduction,** a powerful means of genetic manipulation that has long been used by plant and animal breeders but has hardly been applied to human reproduction; in fact, comparison of family size with parents' intelligence led some geneticists of the 1920s to conclude that differential fertility was leading to deterioration of the gene pool rather than to its improvement.

There is much public concern about the dangers of applying these genetic techniques to man. To what extent is this concern justified? Though we can speculate about genetic warfare by gene manipulation, the possibility is so remote as to be in the realm of science fiction. Predetermination of sex could upset the sex ratio we have achieved by millennia of natural selection, and could have unanticipated consequences. Cloning could be misused either to produce squads of specialized workers or to produce numerous copies of exceptionally gifted people; though the prospect of living in a group of one's identical copies is not appealing, it is hard to see that the human genetic endowment would be impaired as long as enough variety is maintained in the species to allow it to adjust to future environmental changes. In any case, scientific knowledge is bound to move forward.

Even while the broader ethical and social issues of the applications of genetics to man are being debated, there is much that can be done to bridge the gap between what is known about medical genetics and how it is applied to genetic disease. Prenatal diagnosis and selective abortion will be an important part of the practice of genetic medicine, but by no means the whole of it. Because genes interact with environment, a large part of the task is to find ways of manipulating the environment in order to allow "harmful" genes to be viable. Will this lead to "genetic pollution"? This is a hard question and one that will have to be resolved, but it is not a new question; medical ethics has always been on the side of the welfare of the individual if it conflicts with the welfare of society as a whole. The application of genetic knowledge to improve human health is the ultimate goal of genetics in medicine.

GENERAL REFERENCES

Clow, C. L., Fraser, F. C., Laberge, C., and Scriver, C. R. 1973. On the application of knowledge to the patient with genetic disease. *Progr. Med. Genet.* 9:159–214.

Davis, B. D. 1970. Prospects for genetic intervention in man. *Science* 170:1279–1283.

Emery, A. E. H., ed. 1973. *Antenatal Diagnosis of Genetic Disease.* Churchill Livingstone, Edinburgh and London.

Friedmann, T. 1971. Prenatal diagnosis of genetic disease. *Sci. Amer.* 225:34–42.

Milunsky, A., Littlefield, J. W., Kanfer, J. N., Kolodny, E. H., Shih, V. E., and Atkins, L. 1970. Prenatal genetic diagnosis. *New Eng. J. Med.* 283:Part I, 1370–1381; Part II, 1441–1447; Part III, 1498–1504.

Motulsky, A. G., Fraser, G. R., and Felsenstein, J. 1971. Public health and long-term genetic implications of intrauterine diagnosis and selective abortion. *Birth Defects: Original Articles Series* 7:22–32.

Nadler, H. L. 1969. Prenatal detection of genetic defects. *J. Pediat.* 74:132–143.

GLOSSARY

Allele Alleles (allelomorphs) are alternative forms of a gene occupying the same locus on homologous chromosomes. Alleles segregate at meiosis, and a child normally receives only one of each pair of alleles from each parent. See also *Multiple alleles* and *Isoalleles*.

Allograft A tissue graft from a donor of one genotype to a host of another genotype, host and donor being members of the same species.

Allotypes Genetically determined differences in antigens, e.g., the Gm and Inv variants, which are detected by their antigenic properties.

Amino acids The building blocks of protein, for which DNA forms the genetic code. Abbreviations for the amino acids are listed on page 31.

Amniocentesis Needle puncture of the uterus and amniotic cavity through the abdominal wall to allow amniotic fluid to be withdrawn by syringe. The term is often applied to the whole procedure of prenatal diagnosis by culture and analysis of amniotic fluid cells.

Amorph A gene that has no detectable product; an apparently inactive gene.

Anaphase The phase of mitosis or meiosis at which the chromosomes leave the equatorial plate and pass to the poles of the cell.

Aneuploid A chromosome number which is not an exact multiple of the haploid number, or an individual with an aneuploid chromosome number.

Antibody An immunoglobulin molecule formed by immune-competent cells in response to an antigenic stimulus, and reacting specifically with this antigen.

Anticipation The term used to describe the apparent tendency of certain diseases to appear at earlier onset ages and with increasing severity in successive generations.

Antigen Any substance which can elicit antibody formation by immune-competent cells and react specifically with the antibody so produced.

Ascertainment The method of selection of families for inclusion in a genetic study.

Association The presence together of two or more phenotypic characteristics in members of a kindred more often than expected by chance. To be distinguished from *Linkage*.

Assortative mating Selection of a mate with preference for a particular genotype, i.e., nonrandom mating. Preference for a mate of the same genotype is *positive* assortative mating; preference for a spouse of a different genotype is *negative* assortative mating.

Assortment The random distribution to the gametes of different combinations of the chromosomes. At anaphase of the first meiotic division, one member of each pair passes to each pole, and the gametes thus contain one chromosome of each type but *this chromosome may be of either paternal or maternal origin.* Thus, nonallelic genes assort independently to the gametes. Exception: linked genes.

Autoimmunity Immunity to self, as exhibited in certain conditions in which an individual forms antibodies against one or more of his own antigens.

Autoradiography A technique in which a living cell incorporates a radioactive label, and a radiogram of it is then made. Used particularly in cytogenetics to study DNA synthesis in chromosomes labeled with tritiated thymidine.

Autosome Any chromosome other than the sex chromosomes. Man has 22 pairs of autosomes.

Backcross A mating of a heterozygote to a recessive homozygote ($A/a \times a/a$) in which the progeny ($\frac{1}{2}$ A/a, $\frac{1}{2}$ a/a) reveal the genotype of the heterozygous parent. The double backcross mating (A/aB/b × a/ab/b) is the most useful mating for linkage analysis.

Banding The techniques of staining chromosomes in a characteristic pattern of cross bands, thus allowing individual identification of each chromosome pair. Giemsa banding (G banding) and quinacrine fluorescence banding (Q banding) are the best known banding techniques.

Barr body The sex chromatin as seen in somatic cells of the female, named for M. L. Barr, who first described sexual dimorphism in somatic cells. See *Sex chromatin.*

Base pairing In the DNA double helix, adenine always pairs with thymine, and guanine with cytosine; in RNA, adenine pairs with uracil. The specificity of base pairing is fundamental to DNA replication, and to its translation into RNA for protein synthesis.

Carrier An individual who is heterozygous for a normal gene and an abnormal gene which is not expressed phenotypically, though it may be detectable by appropriate laboratory tests.

Centric fusion Fusion of the long arms of two acrocentric chromosomes at the centromere; Robertsonian translocation.

Centriole One of the pair of organelles which form the points of focus of the spindle during cell division. The centrioles lie together outside the nuclear membrane at prophase, and migrate during cell division to opposite poles of the cell.

Centromere The small mass of heterochromatin within a chromosome by which the chromatids are held together, and by which the chromosome is attached to the spindle. Also called the *primary constriction.*

Chiasma Literally, a chiasma is a cross. The term refers to the crossing of chromatid strands of homologous chromosomes, seen at diplotene of the first meiotic division. Chiasmata either result in or are evidence of interchanges of chromosomal material between members of a chromosome pair. Such an exchange is called a *crossover.*

Chimera An individual composed of cells derived from different zygotes. In human genetics the term is used with reference to blood group chimerism, a phenomenon in which dizygotic twins exchange hematopoietic stem cells in utero and continue to form blood cells of both types, or to "whole-body" chimerism, in which two separate zygotes are fused into one individual.

Chromatid A chromosome at prophase and metaphase can be seen to consist of two parallel strands held together by the centromere. Each strand is a chromatid. A single-stranded chromosome replicates during the DNA synthesis stage of the cell cycle, and is then composed of two chromatids until the next mitotic division, at which each chromatid becomes a chromosome of a daughter cell.

Chromosomal aberration An abnormality of chromosome number or structure.

Clone A cell line derived by mitosis from a single ancestral diploid cell.

Codominant If both alleles of a pair are expressed in the heterozygote, the traits determined by them are codominant.

Codon A triplet of three bases in a DNA or RNA molecule specifying a single amino acid.

Coefficient of inbreeding The probability that an individual has received both alleles of a pair from an identical ancestral source; or the proportion of loci at which he is homozygous.

Coefficient of relationship The probability that two persons have inherited a certain gene from a common ancestor; or the proportion of all their genes that have been inherited from common ancestors.

Colinearity Term used to describe the parallel relationship between the base sequence of the DNA of a gene (or the RNA transcribed from it) and the amino acid sequence of the polypeptide determined by that gene.

Compound A genotype in which the two alleles are different mutations from the wild type, or an individual who carries two different mutant alleles at a locus.

Concordant A term often used in twin studies to describe a twin pair in which both members exhibit a certain trait.

Congenital trait Trait present at birth, not necessarily genetic.

Consanguinity Relationship by descent from a common ancestor.

Coupling Two genes at different loci on the same chromosome are linked in *coupling* or in *cis* configuration. Antonym: *repulsion.*

Crossover or crossing over Exhange of genetic material between members of a chromosome pair. The chiasmata seen at diplotene are the physical evidence of crossing over.

Cytogenetics The study of the relationship of the microscopic appearance of chromosomes and their behavior during cell division to the genotype and phenotype of the individual.

Degeneracy of the code The term used to indicate that more than one codon may specify the same amino acid.

Deletion A form of chromosomal aberration in which a portion of a chromosome is lost.

Dermatoglyphics The patterns of the ridged skin of the palms, fingers, soles and toes.

Dictyotene The interphase-like stage of the first meiotic division in which the human oocyte remains from late fetal life until ovulation.

Diploid The number of chromosomes in most somatic cells, which is double the number found in the gametes. In man, the diploid chromosome number is 46.

Discordant A term often used in twin studies to describe a twin pair in which one member shows a certain trait and the other does not.

Dispermy Fertilization of a duplicated egg nucleus, or egg and polar body, by two sperm. This event can produce a rare type of chimera in which two separate zygotes fuse to form a single individual.

Dizygotic Type of fraternal twins produced by two separate ova, separately fertilized.

DNA (Deoxyribonucleic acid) The nucleic acid of the chromosomes, which carries the genetic code.

Dominant A trait is dominant if it is expressed in the heterozygote when the allele of the gene that determines this trait is not expressed. Modifications of this definition are discussed in the text.

Duplication Presence of part of a chromosome in duplicate. Duplication may involve whole genes or series of genes, or only part of a gene (see *nonhomologous crossing over* in text).

Empiric risk An estimate of the risk that a trait will recur in a family, based on past experience rather than on knowledge of the mechanism by which the trait is produced.

Epistasis Interaction between the products of two genes at different loci, in which the one gene prevents the phenotypic expression of the other.

Euchromatin Most of the chromosomal material, which stains uniformly. See *Heterochromatin.*

Expressivity The extent to which a gene is expressed. A trait with variable expressivity may range in expression from mild to severe.

F_1 ("F one") The first-generation progeny of a mating.

Fibroblast A type of spindle-shaped cell grown in cell culture, especially from skin biopsies or amniotic fluid cells, often used in genetic studies.

Fingerprint 1. The pattern of the ridged skin on the distal phalanx of a finger. 2. A method of combining electrophoresis and chromatography to separate the components of a protein such as hemoglobin.

Fitness The ability to transmit one's genes to the next generation and to have them survive in that generation to be passed on to the next

Forme fruste An expression of a genetic trait so mild as to be of no clinical significance.

Gene A segment of a DNA molecule coded for the synthesis of a single polypeptide. For further discussion, see text.

Gene(s) in common Those genes inherited by two individuals from a common ancestral source.

Gene flow Gradual diffusion of genes from one population to another.

Genetic code The base triplets that specify the 20 different amino acids. See Table 3–1, page 31.

Genetic drift Random fluctuation of gene frequencies in small populations.

Genetic lethal A genetically determined trait in which affected individuals do not reproduce.

Genetic marker A trait can be used as a genetic marker in studies of cell lines, individuals, families and populations if it is genetically determined, can be accurately classified, has a simple unequivocal pattern of inheritance and has heritable variations common enough to allow it to be classified as a genetic polymorphism.

Genetic trait Trait determined by genes, not necessarily congenital.

Genome The full set of genes.

Genotype The genetic constitution (genome), or more specifically the alleles present at one locus.

Haploid The chromosome number of a normal gamete which contains only one chromsome of each type. In man, the haploid number (n) is 23.

Hardy-Weinberg law The law which relates gene frequency to genotype frequency. See text.

Hemizygous A term which applies to the genes on the X chromosome in a male. Since males have only one X, they are hemizygous (not homozygous or heterozygous) with respect to X-linked genes.

Heterochromatin Chromosomal material staining differently from the rest of the chromosomal material in the cell, e.g., the Barr body. See *Euchromatin.*

Heterogametic The sex which produces gametes of two types; in humans, this is the male, who produces X-bearing and Y-bearing sperm.

Heterogeneity If a certain phenotype (or closely similar phenotypes) can be produced by different genetic mechanisms, the phenotype is genetically heterogeneous.

Heteroploid Any chromosome number other than the normal.

Heterozygote An individual who has two different alleles, one of which is the normal allele, at a given locus on a pair of homologous chromosomes. Adjective: *heterozygous.*

Holandric The pattern of inheritance of genes on the Y chromosome. A trait caused by a Y-linked gene is transmitted by an affected male to all his sons but to none of his daughters.

Homogametic The sex which produces gametes of only one type; in humans, this is the female, since each of her ova bears an X chromosome.

Homograft A tissue graft between two members of the same species. Normally a homograft is rejected, but if the donor and host are *isogenic* it is accepted.

Homologous chromosomes A "matched pair" of chromosomes, one from each parent, having the same gene loci in the same order.

Homozygote An individual possessing a pair of identical alleles at a given locus on a pair of homologous chromosomes. Adjective: *homozygous*.

Iatrogenic Literally, caused by a physician. The term refers to any condition which results from medical treatment.

Immune competent Cells capable of producing antibody in response to an antigenic stimulus are immune competent.

Immune reaction The specific reaction between antigen and antibody.

Immunological homeostasis The characteristic condition of a normal adult, who has certain antigens and the ability to react to antigens by producing antibodies, but who does not produce antibodies to his own antigens.

Immunological tolerance The inability to respond to a specific antigen resulting from previous exposure to that antigen, especially during embryonic life.

Inborn error A genetically determined biochemical disorder in which a specific enzyme defect produces a metabolic block which may have pathological consequences.

Incompatibility The donor and host are incompatible if because of a genetic difference the host rejects cells transplanted from the donor. In maternal-fetal incompatibility the mother forms antibodies against fetal cells which have entered her circulation.

Inversion A chromosomal aberration in which a segment of a chromosome is reversed end-to-end. See *paracentric inversion* and *pericentric inversion* in text.

Isoalleles Allelic genes which are "normal" and can be distinguished from one another only by their differing phenotypic expression when in combination with a dominant mutant allele.

Isochromosome An abnormal chromosome with two arms of equal length and bearing the same loci in reverse sequence, formed by transverse rather than longitudinal division of the centromere.

Isograft A tissue graft between two individuals who have identical genotypes.

Isolate A subpopulation in which matings take place exclusively with other members of the same subpopulation.

Isozymes (Isoenzymes) Multiple molecular forms of an enzyme within a single individual.

Karyotype The chromosome set. The term is often used for photomicrographs of chromosomes arranged in a standard classification.

Kindred An extended family group.

Lethal equivalent A gene carried in the heterozygous state which, if homozygous, would be lethal; or a combination of two genes in the heterozygous state each of which, if homozygous, would cause the death of 50 per cent of homozygotes; or any equivalent combination.

Linkage Linked genes have their loci within measurable distance of one another on the same chromosome. See *Synteny*.

Locus The position of a gene on a chromosome. Different forms of the gene (alleles) are always found at the same position on the chromosome. A *complex locus* is a locus within which mutation and recombination can occur at more than one site.

Lyon hypothesis (Lyonization) Random and fixed inactivation of one X chromosome in the somatic cells of female mammals during early embryonic life, which leads to dosage compensation, heterozygote variability and female mosaicism.

Manifesting heterozygote A female heterozygous for an X-linked disorder, in whom, because of Lyonization, the trait is expressed clinically with approximately the same degree of severity as in hemizygous affected males.

Mapping a chromosome Determining the position and order of gene loci on a chromosome, especially by analyzing the frequency of recombination between the loci.

Meiosis The special type of cell division occurring in the germ cells by which gametes containing the haploid chromosome number are produced from diploid cells. Two meiotic divisions occur, the *first* and *second* (meiosis I and meiosis II). Reduction in number takes place during meiosis I. To be distinguished from mitosis.

Messenger RNA RNA which has a base sequence complementary to that of a DNA strand, and acts as a template for synthesis of a polypeptide.

Mitosis Somatic cell division resulting in the formation of two cells, each with the same chromosome complement as the parent cell. To be distinguished from meiosis.

Mitotic cycle The cycle of a cell between two successive mitoses, in which four periods are distinguished: G_1, S (DNA synthesis), G_2 and mitosis.

Mixoploid Synonym of *mosaic*.

Monosomy A condition in which one chromosome of one pair is missing, as in 45,X Turner's syndrome. Partial monosomy may occur.

Monozygotic Twins derived from a single fertilized ovum. Identical twins.

Mosaic An individual or tissue with at least two cell lines differing in genotype or karyotype, derived from a single zygote.

Multifactorial Determined by multiple factors, genetic and nongenetic, each with only a minor effect. See also *Polygenic*.

Multiple alleles Though only two alleles at a locus can be present in any one normal individual, there may be more than two different alleles (multiple alleles) at that locus in the population as a whole.

Mutant A gene in which a mutation has occurred, or an individual carrying such a gene.

Mutation A permanent heritable change in the genetic material. Usually defined as a change in a single gene (point mutation), although the term is sometimes used more broadly for a structural chromosomal change.

Mutation rate The rate at which mutations occur at a given locus, expressed as mutations per gamete per locus per generation.

Nondisjunction The failure of two members of a chromosome pair to disjoin during anaphase of cell division, so that both pass to the same daughter cell.

Nucleotide A monomer consisting of a base, a pentose sugar, and a phosphate group. A nucleic acid molecule consists of many such nucleotides, held together by sugar-phosphate bonds.

Oogenesis The process of formation of the female gametes, from primordial germ cell to mature ovum.

Operon A postulated unit of gene action, consisting of an operator and the closely linked structural gene(s) whose action it controls.

p 1. The short arm of a chromosome (from the French *petit*). 2. Often used to indicate the frequency of the more common allele of a pair.

Penetrance When the frequency of expression of a genotype is less than 100 per cent, the trait is said to exhibit reduced penetrance. In an individual who has a genotype which characteristically produces an abnormal phenotype but who is phenotypically normal, the trait is said to be nonpenetrant.

Pharmacogenetics The area of biochemical genetics concerned with drug responses and their genetically controlled variations.

Phenocopy A copy of a phenotype which is usually determined by one specific genotype, produced instead by the interaction of some environmental factor with a different genotype.

Phenotype The entire physical, biochemical and physiological nature of an individual, as determined by his genotype and the environment in which he develops; or, in a more limited sense, the expression of some particular gene or genes, as classified in some specific way.

Philadelphia chromosome The deleted chromosome no. 22 typically occurring in a proportion of bone marrow cells in patients with chronic myelogenous leukemia.

Pleiotropy If a single gene or gene pair produces multiple effects, it is said to exhibit pleiotropy (or to have pleiotropic effects).

Polygenic Determined by many genes at different loci, with small additive effects. Also termed *quantitative.* To be distinguished from *multifactorial,* in which environmental as well as genetic factors may be involved.

Polymorphism The occurrence of two or more genetically determined alternative phenotypes in a population, in frequencies too great to be maintained by mutation alone, e.g., the rarer of two alleles having a population frequency of at least 0.01.

Polypeptide A chain of amino acids, held together by peptide bonds between the amino group of one and the carboxyl group of an adjoining one. A protein molecule may be composed of a single polypeptide chain, or of two or more identical or different polypeptides.

Polyploid Any multiple of the basic haploid chromosome number, other than the diploid number; thus, $3n$, $4n$ and so forth.

Prenatal diagnosis Determination of the sex, karyotype or phenotype of a fetus, usually prior to the 20th week of gestation. A variety of techniques, especially amniocentesis and cell culture, is employed.

Proband See *Propositus.*

Prophase The first stage of cell division, during which the chromosomes become visible as discrete structures and subsequently thicken and shorten. Prophase of the first *meiotic* division is further characterized by pairing (synapsis) of homologous chromosomes.

Propositus The family member who first draws attention to a pedigree of a given trait. Also called *index case* or *proband.*

Pseudoalleles Alleles at two loci concerned with the same function and so closely linked that crossing over between the loci rarely occurs. Though known in experimental organisms, they have not been demonstrated in man.

q 1. The long arm of a chromosome. 2. Often used to indicate the frequency of the rarer allele of a pair.

Quasicontinuous variation The type of variation shown by a multifactorial trait which has a threshold effect and thus appears to have a discontinuous distribution.

Quasidominant The pattern of inheritance produced by the mating of a recessive homozygote with a heterozygote, so that recessively affected members appear in two successive generations and the frequency of affected persons in the second generation is ½.

Random mating Selection of a mate without regard to the genotype of the mate. In a randomly mating population, the frequencies of the various matings are determined solely by the frequencies of the genes concerned.

Recessive Strictly, a trait which is expressed only in homozygotes. Less precisely, a *gene* which is expressed only when homozygous. For further discussion, see text.

Recombinant An individual who has a new combination of genes not found together in either parent. Usually applied to linked genes.

Recombination The formation of new combinations of linked genes by crossing over between their loci.

Reduction division The first meiotic division, so called because at this stage the chromosome number per cell is reduced from diploid to haploid.

Regulator genes Genes which control the rate of production of the product of other genes by synthesis of a substance which inhibits the action of an operator.

RNA (Ribonucleic acid) A nucleic acid formed upon a DNA template and taking part in the synthesis of polypeptides. Three forms are recognized: messenger RNA (mRNA), which is the template upon which polypeptides are synthesized; transfer RNA, which in cooperation with the ribosomes brings activated amino acids into position along the mRNA template; and ribosomal RNA, a component of the ribosomes, which functions as a nonspecific site of polypeptide synthesis.

Satellite 1. Chromosomal satellite; a small mass of condensed chromatin attached to the short arm of each chromatid of a human acrocentric chromosome by a relatively uncondensed stalk (secondary constriction). 2. Nucleolar satellite: a minute mass of chromatin found near the nucleolus in the nerve cells of female mammals of many species, representing the sex chromatin.

Segregation In genetics, the separation of allelic genes at meiosis. Since allelic genes occupy the same locus on homologous chromosomes, they pass to different gametes, i.e., they segregate.

Selection 1. In population genetics, the operation of forces that determines the relative fitness of a genotype in the population. 2. The manner in which kindreds are chosen for study, i.e., ascertainment.

Sex chromatin A chromatin mass in the nucleus of interphase cells of females of most mammalian species, including man. It represents a single X chromosome which is inactive in the metabolism of the cell. Normal females have sex chromatin, thus are *chromatin positive*; normal males lack it, thus are *chromatin negative*. Synonym: Barr body.

Sex chromosomes Chromosomes responsible for sex determination. (In man: XX in female, XY in male.)

Sex-influenced A trait which is not X-linked in its pattern of inheritance but which is expressed differently (either in degree or in frequency) in males and females.

Sex-limited A trait which is expressed in only one sex though the gene determining it is not X-linked.

Sex linkage X linkage.

Sib, sibling Brother or sister.

Silent allele An allele which has no detectable product.

Synteny Presence together on the same chromosome of two or more gene loci, whether or not they are close enough together for linkage to be demonstrated. Adjective: *syntenic*.

Telophase The stage of cell division which begins when the daughter chromosomes reach the poles of the dividing cell and lasts until the two daughter cells take on the appearance of interphase cells.

Teratogen An agent that produces or raises the incidence of congenital malformations.

Transformation A form of recombination of genetic material in bacteria in which a bacterium incorporates DNA extracted from other bacteria into its own genetic material.

Translocation The transfer of a piece of one chromosome to a nonhomologous chromosome. If two nonhomologous chromosomes exchange pieces, the translocation is *reciprocal*. See also *Centric fusion*.

Triplet In molecular genetics, a unit of three successive bases in DNA or RNA, coding for a specific amino acid.

Triploid A cell with three times the normal haploid chromosome number, or an individual made up of such cells.

Triradius In dermatoglyphics, a point from which the dermal ridges course in three directions at angles of approximately 120 degrees.

Trisomy The state of having three of a given chromosome instead of the usual pair, as in trisomy 21 (Down's syndrome).

Wild type Term used especially in experimental genetics to indicate the normal allele (often symbolized as +) or the normal homozygote (+/+).

X linkage Genes on the X chromosome, or traits determined by such genes, are X-linked.

Zygote The fertilized ovum. Less precisely, the fertilized ovum and the organism developing from it.

REFERENCES

Aebi, H., and Suter, H. 1971. Acatalasemia. In: *Advances in Human Genetics* 2:143–199. (Harris, H., and Hirschhorn, K., eds.) Plenum Publishing Corporation, New York.

Aird, I., Bentall, H. H., and Roberts, J. A. F. 1953. A relationship between cancer of the stomach and the ABO blood groups. *Brit. Med. J.* 1:799–801.

Alberman, E., Polani, P. E., Roberts, J. A. F., and Spicer, C. C. 1972. Parental exposure to X-irradiation and Down's syndrome. *Ann. Hum. Genet.* 6:195–208.

Allen, F. H., Diamond, L. K., and Niedziela, B. 1951. A new blood-group antigen. *Nature* 167:482.

Anderson, J. W. 1971. Transplantation—Nature's success. *Lancet* 2:1077–1082.

Andervont, H. B. 1964. Fate of the C3H mammary tumor agent in mice of strains C57BL, I and BALB/c. *J. Nat. Cancer Inst.* 32:1189–1198.

Avery, O. T., MacLeod, C. M., and McCarty, M. 1944. Studies on chemical nature of substance inducing transformation of pneumococcal types: induction of transformation by desoxyribonucleic acid fraction isolated from pneumococcus type III. *J. Exp. Med.* 79:137–158.

Bain, B., and Lowenstein, L. 1964. Genetic studies on the mixed leucocyte reaction. *Science* 145:1315–1316.

Bannerman, R. M. 1972. The thalassemias: Recent advances in study and management. *Hematol. Rev.* 3:297–385. (Ambrus, J. L., ed.) Marcel Dekker, New York.

Barr, M. L. 1957. Cytologic tests of chromosomal sex. *Prog. Gynec.* 3:131–141. (Meigs, J. V., and Sturgis, S. H., eds.) Grune and Stratton, New York.

Barr, M. L. 1960. Sexual dimorphism in interphase nuclei. *Amer. J. Hum. Genet.* 12:118–127.

Barr, M. L., and Bertram, E. G. 1949. A morphological distinction between neurones of the male and female, and the behaviour of the nucleolar satellite during accelerated nucleoprotein synthesis. *Nature* 163:676–677.

Batchelor, J. R., and Hackett, M. 1970. HL-A matching in treatment of burned patients with skin allografts. *Lancet* 2:581–583.

Bateson, W., and Punnett, R. C. 1905–1908. Experimental studies in the physiology of heredity. Reports 2, 3, 4 to the Evolution Committee of the Royal Society. Reprinted in: Peter, J. A., ed. *Classic Papers in Genetics*. Prentice-Hall, Englewood Cliffs, New Jersey.

Beadle, G. W., and Tatum, E. L. 1941. Genetic control of biochemical reactions in *Neurospora*. *Proc. Nat. Acad. Sci. USA* 27:499–506.

Beerman, W. 1963. Cytological aspects of information in cellular differentiation. *Amer. Zool.* 3:23–32.

Benirschke, K. 1972. Origin and clinical significance of twinning. *Clin. Obstet. Gynec.* 15:220–235.

Bergan, J. 1971. Structure and function of organ transplant registry of the American College of Surgeons. *Transplant. Proc.* 3:298–302.

Beutler, E., Baluda, M. C., Sturgeon, P., and Day, R. W. 1965. A new genetic abnormality resulting in galactose-1-phosphate uridyl transferase deficiency. *Lancet* 1:353–354.

Black, J. A., and Dixon, G. H. 1968. Amino acid sequence of the alpha chains of human haptoglobins and their possible relation to the immunoglobulin light chains. *Nature* 218:736–741.

Bloom, A. D., and Tjio, J. H. 1964. In vivo effects of diagnostic X-irradiation on human chromosomes. *New Eng. J. Med.* 270:1341–1344.

Bode, H. H., and Crawford, J. D. 1969. Nephrogenic diabetes insipidus in North America—the Hopewell hypothesis. *New Eng. J. Med.* 208:750–754.

Bondy, P. K., and Rosenberg, L. E., eds. 1974. *Duncan's Diseases of Metabolism.* 7th ed. W. B. Saunders Company, Philadelphia.
Borgaonkar, D. S., Davis, M., Bolling, D. P., and Herr, H. M. 1971. Evaluation of dermal patterns by predictive discrimination. I. Preliminary analysis based on frequencies of patterns. *Johns Hopkins Med. J.* 28:141–152.
Boyer, S. H., ed. 1963. *Papers on Human Genetics.* Prentice-Hall, Englewood Cliffs, New Jersey.
Brady, R. O., and Kolodny, E. H. 1972. Disorders of ganglioside metabolism. *Progr. Med. Genet.* 8:225–241.
Buchanan, D. I., and McIntyre, J. 1955. Consanguinity and two rare matings. *–D– / –D–* × *CDe/–D–* and *CDe/–D–* × *cDe/–D–. Brit. J. Haemat.* 1:304–307.
Bulmer, M. G. 1970. *The Biology of Twinning in Man.* Clarendon Press, Oxford.
Callender, S. T., Race, R. R., and Paykoc, Z. V. 1945. Hypersensitivity to transfused blood. *Brit. Med. J.* 2:83–84.
Carr, D. H. 1971. Chromosomes and abortion. *Advances in Human Genetics* 2:201–258. (Harris, H., and Hirschhorn, K., eds.) Plenum Publishing Corporation, New York.
Carter, C. O. 1964. The genetics of common malformations. In: *Congenital Malformations: Papers and Discussions Presented at the Second International Conference on Congenital Malformations.* pp. 306–313. (Fishbein, M., ed.) International Medical Congress, New York.
Carter, C. O. 1965. The inheritance of common congenital malformations. *Progr. Med. Genet.* 4:55–84.
Carter, C. O. 1967. Risk to offspring of incest. *Lancet* 1:436.
Carter, C. O. 1969. *An ABC of Medical Genetics.* Little, Brown and Company, Boston.
Carter, C. O. 1969. Genetics of common disorders. *Brit. Med. Bull.* 25:52–57.
Carter, C., and MacCarthy, D. 1951. Incidence of mongolism and its diagnosis in the newborn. *Brit. J. Soc. Med.* 5:83–90.
Carter, C. O., Roberts, J. A. F., Evans, K. A., and Buck, A. R. 1971. Genetic clinic: a follow-up. *Lancet* 1:281–285.
Caspersson, T., Zech, L., Johansson, C., and Modest, E. J. 1970a. Identification of human chromosomes by DNA-binding fluorescent agents. *Chromosoma* 30:215–227.
Caspersson, T., Gahrton, G., Lindsten, J., and Zech, L. 1970b. Identification of the Philadelphia chromosome as a number 22 by quinacrine mustard fluorescence analysis. *Exp. Cell Res.* 63:238–240.
Cavalli-Sforza, L. L., and Bodmer, W. F. 1971. *The Genetics of Human Populations.* W. H. Freeman and Company, San Francisco.
Chang, T. M. S., and Poznansky, M. J. 1968. Semipermeable microcapsules containing catalase for enzyme replacement in acatalasemic mice. *Nature* 218:243–245.
Clarke, C. A. 1961. Blood groups and disease. *Progr. Med. Genet.* 1:81–119.
Cleaver, J. E. 1968. Defective repair replication of DNA in xeroderma pigmentosum. *Nature* 218:652–656.
Clermont, Y., and Leblond, C. P. 1959. Differentiation and renewal of spermatogoma in the monkey, *Macacus rhesus. Amer. J. Anat.* 104:237–273.
Clever, U. 1964. Actinomycin and puromycin: Effects on sequential gene activation by ecdysone. *Science* 146:794–795.
Clow, C. L., Fraser, F. C., Laberge, C., and Scriver, C. R. 1973. On the application of knowledge to the patient with genetic disease. *Progr. Med. Genet.* 9:159–214.
Collman, R. D., and Stoller, A. 1963. Comparison of age distributions for mothers of mongols born in high and in low birth incidence areas and years in Victoria, 1942–57. *J. Ment. Defic. Res.* 7:60–68.
Coombs, R. R. A., Mourant, A. E., and Race, R. R. 1945. Detection of weak and "incomplete" Rh agglutinins: a new test. *Lancet* 2:15.
Coombs, R. R. A., Mourant, A. E., and Race, R. R. 1946. In-vivo isosensitization of red cells in babies with haemolytic disease. *Lancet* 1:264–266.
Cori, G. T., and Cori, C. F. 1952. Glucose-6-phosphatase of liver in glycogen storage disease. *J. Biol. Chem.* 199:661–667.
Cox, D. W. 1964. An investigation of possible genetic damage in the offspring of women receiving multiple diagnostic pelvic x-rays. *Amer. J. Hum. Genet.* 16:214–230.
Crow, J. F. 1961. Mutation in man. *Progr. Med. Genet.* 1:1–26.
Crow, J. F. 1965. Problems of ascertainment in the analysis of family data. In: Neel, J. V., Shaw, M. W., and Schull, W. J., eds.: *Genetics and the Epidemiology of Chronic Dis-*

eases. U.S. Department of Health, Education and Welfare, Public Health Service Publication 1163.
Cummins, H., and Midlo, C. 1943. *Fingerprints, Palms and Soles: An Introduction to Dermatoglyphics.* Blakiston, Philadelphia.
Cummins, H., and Midlo, C. 1961. *Fingerprints, Palms and Soles: An Introduction to Dermatoglyphics.* Dover, New York. (Paperback reprint.)
Cutbush, M., Mollison, P. L., and Parkin, D. M. 1950. A new human blood group. *Nature* 165:188.
Dallaire, L., and Flynn, D. 1967. Frequency of antibodies to thyroglobulin in relation to gravidity and to Down's syndrome. *Canad. Med. Assoc. J.* 97:209–212.
Dausset, J. 1971. The genetics of transplantation antigens. *Transplant. Proc.* 3:8–14.
Davidson, R. G., Nitowsky, H. M., and Childs, B. 1963. Demonstration of two populations of cells in the human female heterozygous for glucose-6-phosphate dehydrogenase variants. *Proc. Nat. Acad. Sci. USA*, 50:481–485.
Davis, B. D. 1970. Prospects for genetic intervention in man. *Science* 170:1279–1283.
DeKöning, J., Dooren, L. J., van Bekkum, D. W., van Rood, J. J., Dickie, K. A., and Radl, J. 1969. Transplantation of bone-marrow cells and fetal thymus in an infant with lymphocytopenic immunological deficiency. *Lancet* 1:1223–1226.
DiGeorge, A. M. 1968. Congenital absence of thymus and its immunologic consequences: Concurrence with congenital hypoparathyroidism. *Birth Defects Original Article Series* 4:116–123. (Bergsma, D., ed.) The National Foundation, New York.
Donahue, R. P., Bias, W. B., Renwick, J. H., and McKusick, V. A. 1968. Probable assignment of the Duffy blood group locus to chromosome 1 in man. *Proc. Nat. Acad. Sci. USA* 61:949–955.
Dronamraju, K. R. 1960. Hypertrichosis of the pinna of the human ear, Y-linked pedigrees. *J. Genet.* 57:230–244.
Eaton, B. R., Morton, J. A., Pickles, M. M., and White, K. E. 1968. A new antibody anti-Yt^a, characterizing a blood group of high incidence. *Brit. J. Haemat.* 2:333–341.
Edelman, G. M. 1970. The structure and function of antibodies. *Sci. Amer.* 223:34–42.
Edelman, G. M., and Gally, J. A. 1967. Somatic recombination of duplicated genes: An hypothesis on the origin of antibody diversity. *Proc. Nat. Acad. Sci. USA* 57:353–358.
Edwards, J. A., and Gale, R. P. 1972. Camptobrachydactyly, a new autosomal dominant trait with two probable homozygotes. *Amer. J. Hum. Genet.* 24:464–474.
Edwards, J. H. 1960. The simulation of Mendelism. *Acta Genet.* (Basel) 10:63–70.
Edwards, J. H., Harnden, D. G., Cameron, A. H., Crosse, V. M., and Wolff, O. H. 1960. A new trisomic syndrome. *Lancet* 1:787–790.
Edwards, R. G., and Fowler, R. E. 1970. Human embryos in the laboratory. *Sci. Amer.* 223:44–54.
Emery, A. E. H., ed. 1973. *Antenatal Diagnosis of Genetic Disease.* Churchill Livingstone, Edinburgh and London.
Epstein, C. J., and Motulsky, A. G. 1965. Evolutionary origins of human proteins. *Progr. Med. Genet.* 4:85–127.
Fallis, N., Lasagna, L., and Tetreault, L. 1962. Gustatory thresholds in patients with hypertension. *Nature* 196:74–75.
Federman, D. D. 1967. *Abnormal Sexual Development. A Genetic and Endocrine Approach to Differential Diagnosis.* W. B. Saunders Company, Philadelphia.
Feinstein, R. N., Seaholm, J. E., Howard, J. B., and Russell, W. L. 1964. Acatalasemic mice. *Proc. Nat. Acad. Sci. USA* 52:661–662.
Fialkow, P. J. 1969. Genetic aspects of autoimmunity. *Progr. Med. Genet.* 6:117–167.
Fisher, R. A. 1918. The correlation between relatives on the supposition of Mendelian inheritance. *Trans. Roy. Soc. Edinburgh* 52:399–433.
Fisher, R. A. 1944. *Statistical Methods for Research Workers.* Oliver and Boyd, Edinburgh.
Fogh-Anderson, P. 1942. Inheritance of hare-lip and cleft-palate. *Op. Dom. Biol. Hered. Hum Kbh.* 4. Munksgaard, Copenhagen.
Ford, E. B. 1940. Polymorphism and taxonomy. In: Huxley, J. ed. *Systematics,* Clarendon Press, Oxford, pp. 493–513.
Fraser, F. C. 1961. Genetics and congenital malformations. *Progr. Med. Genet.* 1:38–80.
Fraser, F. C. 1963. Taking the family history. *Amer. J. Med.* 34:585–593.
Fraser, F. C. 1970. The genetics of cleft lip and cleft palate. *Amer. J. Hum. Genet.* 22:336–352.
Friedmann, T. 1971. Prenatal diagnosis of genetic disease. *Sci. Amer.* 225:34–42.

Lederberg, J. 1960. A view of genetics. *Science* 131:269–276. (Reprint of Nobel prize lecture, 1959.)
Lehmann, H., and Carrell, R. W. 1969. Variations in the structure of human haemoglobin: with particular reference to the unstable haemoglobins. *Brit. Med. Bull* 25:14–23.
Lejeune, J., Gautier, M., and Turpin, R. 1959. Étude des chromosomes somatiques de neuf enfants mongoliens. *C.R. Acad. Sci.* (Paris) 248:1721–1722.
Lenz, W. 1963. *Medical Genetics*. University of Chicago Press, Chicago.
Levine, P., Robinson, E., Celano, M., Briggs, O., and Falkinburg, L. 1955. Gene interaction resulting in suppression of blood group substance B. *Blood* 10:1100–1108.
Li, C. C. 1963. Genetic aspects of consanguinity. *Amer. J. Med.* 34:702–714.
Liley, A. W. 1963. Intrauterine transfusion of foetus in haemolytic disease. *Brit. Med. J.* 5365:1107–1109.
Lyon, M. F. 1961. Gene action in the X-chromosome of the mouse. (Mus musculus L.) *Nature* 190:372–373.
Lyon, M. F. 1962. Sex chromatin and gene action in the mammalian X-chromosome. *Amer. J. Hum. Genet.* 14:135–148.
Lyon, M. F., and Hawkes, S. G. 1970. X-linked gene for testicular feminization in the mouse. *Nature* 225:1217–1219.
MacDiarmid, W. D., Lee, G. R., Cartwright, G. E., and Wintrobe, M. M. 1967. X-inactivation in an unusual X-linked anemia and the Xg^a blood group. *Clin. Res.* 15:132. (Abstract).
McKusick, V. A. 1964. *On the X Chromosome of Man*. Amer. Inst. Bio. Sciences, Washington.
McKusick, V. A. 1969. *Human Genetics*. 2nd ed. Prentice-Hall, Englewood Cliffs, New Jersey. (A Study Guide to this book was published in 1972.)
McKusick, V. A. 1970. *Human genetics*. Ann. Rev. Genet. 4:1–46. (Roman, H. L., ed.) Palo Alto, California.
McKusick, V. A. 1971. *Mendelian Inheritance in Man:* Catalogs of Autosomal Dominant, Autosomal Recessive and X-linked Phenotypes. 3rd ed. The Johns Hopkins Press, Baltimore.
McKusick, V. A., Howell, R. R., Hussels, J. E., Neufeld, E. F., and Stevenson, R. E. 1972. Allelism, non-allelism, and genetic compounds among the mucopolysaccharidoses. *Lancet* 1:993–996.
Mann, D. L., Rogentine, G. N., Jr., Fahey, J. L., and Nathenson, S. G. 1969. Molecular heterogeneity of human lymphoid (HL-A) alloantigens. *Science* 163:1460–1462.
Mann, J. D., Cahan, A., Gelb, A. G., Fisher, N., Hamper, J., Tippett, P., Sanger, R., and Race, R. R. 1962. A sex-linked blood group. *Lancet* 1:8–10.
Markert, C. L. 1963. Lactate dehydrogenase isozymes: Dissociation and recombination of subunits. *Science* 140:1329–1330.
Markert, C. L. 1964. Biochemical events during differentiation. In: Bearn, A. G., ed. *Differentiation and Development*. Little, Brown and Co., Boston, pp. 81–84.
Markert, C. L., and Ursprung, H. 1971. *Developmental Genetics*. Prentice-Hall, Englewood Cliffs, New Jersey.
Maynard-Smith, S., Penrose, L. S., and Smith, C. A. B. 1961. *Mathematical Tables for Research Workers in Human Genetics*. Churchill, London.
Mayr, E. 1963. *Animal Species and Evolution*. Harvard University Press, Cambridge, Massachusetts.
Mee, A. D., and Evans, D. B. 1970. Antilymphocyte-serum preparations in treatment of renal-allograft rejection. *Lancet* 2:16–19.
Mendel, G. 1865. Experiments in plant hybridization. Translation. In: Peters, J. A., ed. *Classic Papers in Genetics*. Prentice-Hall, Englewood Cliffs, New Jersey, pp. 1–20.
Meselson, M., and Stahl, F. W. 1958. The replication of DNA in *Escherichia coli*. *Proc. Nat. Acad. Sci. USA* 44:671–682.
Mettler, L. E., and Gregg, T. G. 1969. *Population Genetics and Evolution*, Prentice-Hall Inc., Englewood Cliffs, New Jersey.
Migeon, B. R., and Miller, C. S. 1968. Human-mouse somatic cell hybrids with single human chromosome (Group E): Link with thymidine kinase activity. *Science* 162: 1005–1006.
Miller, O. J., and Beatty, B. R. 1969. Portrait of a gene. *J. Cell Physiol.* 74, Suppl. 1:225–232.
Miller, O. J., Breg, W. R., Schmickel, R. D., and Tretter, W. 1961. A family with an

XXXXY male, a leukemic male and two 21-trisomic mongoloid females. *Lancet* 2: 78–79.

Milunsky, A., Littlefield, J. W., Kanfer, H. N., Kolodny, E. H., Shih, V. E., and Atkins, L. 1970. Prenatal genetic diagnosis. *New Eng. J. Med.* 283:Part I, 1370–1381; Part II, 1441–1447; Part III, 1498–1504.

Mohr, J. 1951. A search for linkage between the Lutheran blood group and other hereditary characters. *Acta Path. Microbiol. Scand.* 28:207–210.

Moller, G. 1971. Immunocompetent cells in graft rejection. *Transplant. Proc.* 3:15–20.

Mollison, P. L. 1972. *Blood Transfusion in Clinical Medicine,* 5th ed. Blackwell Scientific Publications, Oxford.

Moore, K. L. 1966. *The Sex Chromatin.* W. B. Saunders Company, Philadelphia.

Moore, K. L., and Barr, M. L. 1955. Smears from the oral mucosa in the detection of chromosomal sex. *Lancet* 2:57–58.

Morgan, W. T. J., and Watkins, W. M. 1969. Genetic and biochemical aspects of human blood group A-, B-, H-, Le^a- and Le^b-specificity. *Brit. Med. Bull.* 25:30–34.

Morton, N. E. 1959. Genetic tests under incomplete ascertainment. *Amer. J. Hum. Genet.* 11:1–16.

Morton, N. E. 1959. Empirical risks in consanguineous marriages: Birth weight, gestation time and measurements of infants. *Amer. J. Hum. Genet.* 10:344–349.

Morton, N. E. 1972. The future of human population genetics. *Progr. Med. Genet.* 8:103–124.

Motulsky, A. G., Fraser, G. R., and Felsenstein, J. 1971. Public health and long-term genetic implications of intrauterine diagnosis and selective abortion. *Birth Defects Original Articles Series* 7:22–32.

Mourant, A. E. 1946. A "new" human blood group of frequent occurrence. *Nature* 158:237.

Murphy, E. A. 1969. Considerations in genetic counseling. *Amer. Acad. General Practice* 40:102–110.

Murphy, E. A., and Mutalik, G. S. 1969. The application of Bayesian methods in genetic counseling. *Hum. Hered.* 19:126–151.

Murray, J. E., Barnes, B. A., and Atkinson, J. C. 1971. Eighth report of the human kidney transplantation registry. *Transplantation* 11:328–337.

Nadler, H. L. 1968. Patterns of enzyme development using cultivated human fetal cells derived from amniotic fluid. *Biochem. Genet.* 2:119–126.

Nadler, H. L. 1969. Prenatal detection of genetic defects. *J. Pediat.* 74:132–143.

Nance, W. E. 1964. Genetic tests with a sex-linked marker: Glucose-6-phosphate dehydrogenase. *Symp. Quant. Biol.* 29:415–425.

Nance, W. E. 1968. *Genetic Studies of Human Serum and Erythrocyte Polymorphisms. Glucose-6-Phosphate Dehydrogenase, Haptoglobin, Hemoglobin, Transfusion, Lactate Dehydrogenase and Catalase.* Ph.D. Thesis, University of Wisconsin.

Nance, W. E., and Uchida, I. 1964. Turner's syndrome, twinning and an unusual variant of glucose-6-phosphate dehydrogenase. *Amer. J. Hum. Genet.* 16:380–392.

Neel, J. V. 1949. The inheritance of sickle cell anemia. *Science* 110:64–66.

Neel, J. V., and Rusk, M. L. 1963. Polydactyly of the second metatarsal with associated defects of the feet: A new, simply inherited skeletal deformity. *Amer. J. Hum. Genet.* 15:288–291.

Neel, J. V., and Schull, W. J. 1954. *Human Heredity.* University of Chicago Press, Chicago.

Neel, J. V., and Schull, W. J. 1962. The effect of inbreeding on mortality and morbidity in two Japanese cities. *Proc. Nat. Acad. Sci. USA* 48:573–582.

Nelson, M. M., and Forfar, J. O. 1969. Congenital abnormalities at birth: Their association in the same patient. *Develop. Med. Child. Neurol.* 11:3–16.

Newcombe, H. B. 1964. Epidemiological Studies: Discussion. In: Fishbein, M., ed. *Congenital Malformations: Papers and Discussions Presented at the Second International Conference on Congenital Malformations.* International Medical Congress, New York, pp. 306–313.

Nichols, W. W. 1963. Relationships of viruses, chromosomes, and carcinogenesis. *Hereditas* 50:53–80.

Nora, J. J. 1968. Multifactorial inheritance hypothesis for the etiology of congenital heart diseases. The genetic-environmental interaction. *Circulation* 38:604–617.

Oakes, W. R., and Lushbaugh, C. C. 1952. Course of testicular injury following accidental exposure to nuclear radiations; report of a case. *Radiology* 59:737–743.

O'Connor, B. 1963. Toward a healthier heritage. In: Fishbein, M. ed. *Congenital Defects.*

First Inter-American Conference on Congenital Defects. J. B. Lippincott, Philadelphia, pp. 6–11.

Ohno, S. 1967. *Sex Chromosomes and Sex-Linked Genes.* Springer-Verlag New York, Inc., New York.

Ohno, S., Klinger, H. P., and Atkin, N. B. 1962. Human oogenesis. *Cytogenetics* 1:42–51.

Old, L. J., Boyse, E. A., and Stockert, E. 1964. Typing of mouse leukemias by serological methods. *Nature* 201:777–779.

Partington, M., Craig, A. W., and Jackson, H. 1962. The effect of radiation on spermatogenic cells in the rat and mouse. *Brit. J. Radiol.* 35:713–718.

Pauling, L., Itano, H. A., Singer, S. J., and Wells, I. C. 1949. Sickle cell anemia, a molecular disease. *Science* 110:543–548.

Penrose, L. S. 1948. The problem of anticipation in pedigrees of dystrophia myotonica. *Ann. Eugen.* 14:125–132.

Penrose, L. S. 1951. Measurement of pleiotropic effects in phenylketonuria. *Ann. Eugen.* 16:134–141.

Penrose, L. S. 1963. Fingerprints, palms and chromosomes. *Nature* 197:933–938.

Penrose, L. S. 1968. *Memorandum on Dermatoglyphic Nomenclature.* Birth Defects Original Article Series Vol. IV, No. 3. The National Foundation, New York.

Penrose, L. S. 1969. Dermatoglyphics. *Sci. Amer.* 221:72–84.

Penrose, L. S., and Smith, G. F. 1966. *Down's Anomaly.* J. & A. Churchill, London.

Perutz, M. F. 1965. Structure and function of haemoglobin. I. A tentative atomic model of horse oxyhaemoglobin. *J. Molec. Biol.* 13:646–668.

Porter, I. H. 1968. *Heredity and Disease.* McGraw-Hill Book Company. New York and Toronto.

Poswillo, D. E., Sopher, D., and Mitchell, S. 1972. Induction of foetal malformations with "blighted" potato: A preliminary report. *Nature* 239:462–464.

Race, R. R., and Sanger, R. 1968. *Blood Groups in Man,* 5th ed. Blackwell Scientific Publications, Oxford.

Race, R. R., and Sanger, R. 1969. Xg and sex-chromosome abnormalities. *Brit. Med. Bull.* 25:99–103.

Rake, M. O., Williams, R., Freeman, T., and McFarlane, A. S. 1970. Protein synthesis by the liver after transplantation. *Lancet* 2:341–342.

Rapaport, F. A., ed. 1971. Proceedings of the Third International Congress of the Transplantation Society. *Transplant. Proc.* 3:1.

Reed, T. E. 1969. Caucasian genes in American Negroes. *Science* 165:762–768.

Reed, T. E., and Chandler, J. H. 1958. Huntington's chorea in Michigan. *Amer. J. Hum. Genet.* 10:201–225.

Reed, T. E., Borgaonkar, D. S., Conneally, P. M., Yu, P. L., Nance, W. E., and Christian, J. C. 1970. Dermatoglyphic nomogram for the diagnosis of Down's syndrome. *J. Pediat.* 77:1024–1032.

Renwick, J. H. 1971. The mapping of human chromosomes. In: *Ann. Rev. Genet.* 5:81–120. (Roman, H. L., ed.) Palo Alto, California.

Renwick, J. H. 1972. Hypothesis: Anencephaly and spina bifida are usually preventable by avoidance of a specific but unidentified substance present in certain potato tubers. *Brit. J. Prev. Soc. Med.* 26:67–88.

Rhodes, A. J. 1961. Virus infections and congenital malformations. In: Fishbein, M., ed. *Congenital Malformations: Papers and Discussions Presented at the First International Conference on Congenital Malformations.* J. B. Lippincott, Philadelphia, pp. 106–116.

Ris, H., and Kubai, D. F. 1970. Chromosome structure. *Ann. Rev. Genet.* 4:263–294. (Roman, H. L., ed.) Palo Alto, California.

Roberts, J. A. F. 1970. *An Introduction to Medical Genetics.* 5th ed. Oxford University Press Inc., New York and London.

Robson, E. B., Sutherland, O., and Harris, H. 1966. Evidence for linkage between the transferrin locus (Tf) and the serum cholinesterase locus (E_1) in man. *Ann. Hum. Genet.* 29:325–336.

Roitt, I. M., Greaves, M. F., Torrigiani, G., Brostoff, J., and Playfair, J. H. L. 1969. The cellular basis of immunological responses. *Lancet* 2:367–371.

Ropartz, C., Lenoir, J., and Rivat, L. 1961. A new inheritable property of human sera: The InV factor. *Nature* 189:586.

Rosenberg, L. E. 1967. Genetic heterogeneity in cystinuria. In: Nyhan, W. L., ed. *Amino Acid Metabolism and Genetic Variation.* McGraw-Hill, New York.

Rosenberg, L. E. 1974. Inborn errors of metabolism. In: Bondy, P. K., and Rosenberg, L. E. *Duncan's Diseases of Metabolism.* 7th ed. W. B. Saunders Company, Philadelphia.

Rosenberg, L. E., and Scriver, C. R. 1969. Disorders of amino acid metabolism. In: Bondy, P. K., ed. *Duncan's Diseases of Metabolism.* 6th ed. W. B. Saunders Company, Philadelphia.

Ruddle, F. H., Chapman, V. M., Chen, T. R., and Klebe, R. J. 1970. Genetic analysis with man-mouse somatic cell hybrids. *Nature* 227:248–257.

Ruddle, F., Ricciuti, F., McMorris, F. A., Tischfield, J., Creagan, R., Darlington, G., and Chen, T. 1972. Somatic cell genetic assignment of peptidase C and the Rh linkage group to chromosome A-1 in man. *Science* 176:1429–1431.

Russell, E. S. 1954. Review of the pleiotropic effects of *W*-series genes on growth and differentiation. In: Rudnick, D., ed. *Aspects of Synthesis and Order in Growth.* Princeton University Press, Princeton, pp. 113–126.

Sanger, R., and Race, R. R. 1958. The Lutheran-secretor linkage in man: Support for Mohr's findings. *Heredity* 12:513–520.

Sank, D. 1963. Genetic aspects of early total deafness. In: Ranier, J. D., et al., eds. *Family and Mental Health Problems in a Deaf Population.* New York State Psychiatric Insitute, New York.

Schachter, H., Michaels, M. A., Tilley, C. A., Crookston, M. C., and Crookston, J. H. 1973. Qualitative differences in the *N*-acetyl-D galactosaminyl transferases produced by human A^1 and A^2 genes. *Proc. Nat. Acad. Sci. USA* 70:220–224.

Scheinfeld, A., and Schachter, J. 1961. Bio-social effects on twinning incidence: I. Intergroup and generation differences in the United States in twinning incidence and MZ:DZ ratios. In: *Proc. II Int. Congr. Hum. Genet.* 1:300–302. (Gedda, L., ed.) Instituto G. Mendel, Rome.

Schull, W. J., and Neel, J. V. 1958. Radiation and the sex ratio in man. *Science* 128:343–348.

Seegmiller, J. E., Rosenbloom, F. M., and Kelley, W. N. 1967. Enzyme defect associated with a sex-linked human neurological disorder and excessive purine synthesis. *Science* 115:1682–1684.

Simpson, N. E. 1964. Multifactorial inheritance: A possible hypothesis for diabetes. *Diabetes* 13:462–471.

Simpson, N. E. 1968. Diabetes in the families of diabetics. *Canad. Med. Assoc. J.* 98:427–432.

Smith, D. W. 1970. *Recognizable Patterns of Human Malformation.* W. B. Saunders Company, Philadelphia.

Smith, D. W., and Wilson, A. A. 1973. *The Child with Down's Syndrome (Mongolism).* W. B. Saunders Company, Philadelphia.

Smithies, O. 1955. Zone electrophoresis in starch gels: Group variations in the serum proteins of normal human adults. *Biochem. J.* 61:629–641.

Smithies, O. 1957. Variations in human serum β-globulins. *Nature* 180:1482–1483.

Smithies, O. 1967. Antibody variability. *Science* 157:267–273.

Smithies, O., Connell, G. E., and Dixon, G. H. 1962. Chromosomal rearrangements and the evolution of haptoglobin genes. *Nature* 196:232–236.

Snell, G. D. 1964. The terminology of tissue transplantation. *Transplantation* 2:655–657.

Snell, G. D., Cherry, M., and Demant, P. 1971. Evidence that H-2 private specificities can be arranged in two mutually exclusive systems possibly homologous with two subsystems of HL-A. *Transplant. Proc.* 3:183–186.

Srb, A. M., Owen, R. D., and Edgar, R. S. 1965. *General Genetics.* 2nd ed. W. H. Freeman and Co., Publishers, San Francisco.

Stanbury, J. B., Wyngaarden, J. B., and Fredrickson, D. S., eds. 1972. *The Metabolic Basis of Inherited Disease.* 3rd ed. McGraw-Hill Book Company, New York.

Stanners, C. P., and Till, J. E. 1960. DNA synthesis in individual L-strain mouse cells. *Biochim. Biophys. Acta* 37:406–419.

Steinberg, A. G., and Bearn, A. G., eds. 1961–1973. *Progress in Medical Genetics.* Grune and Stratton, Inc., New York, Vols. 1–9.

Stephens, F. E., and Tyler, F. H. 1951. Studies in disorders of muscle. V. The inheritance of childhood progressive muscular dystrophy in 33 kindreds. *Amer. J. Hum. Genet.* 3:111–125.

Stern, C. 1957. The problem of complete Y-linkage in man. *Amer. J. Hum. Genet.* 9:147–166.

Stern, C. 1973. *Principles of Human Genetics*. 3rd ed. W. H. Freeman and Co., Publishers, San Francisco.
Stevenson, A. C., and Cheeseman, E. A. 1956. Hereditary deaf mutism, with particular reference to Northern Ireland. *Ann. Hum. Genet.* 20:177–231.
Stevenson, A. C., Davison, B. C. C., and Oakes, M. W. 1970. *Genetic Counselling*. J. B. Lippincott, Philadelphia.
Sutton, H. E. 1961. *Genes, Enzymes, and Inherited Diseases*. Holt, Rinehart, & Winston, New York.
Sutton, H. E. 1965. *An Introduction to Human Genetics*. Holt, Rinehart & Winston, New York.
Swanson, C. P., Merz, T., and Young, W. J. 1967. *Cytogenetics*. Prentice-Hall, Inc., Englewood Cliffs, New Jersey.
Swanson, J., Polesky, H. F., Tippett, P., and Sanger, R. 1965. A "new" blood group antigen, Do^a. *Nature* 206:313.
Takahara, S. 1952. Progressive oral gangrene probably due to lack of catalase in the blood (acatalasemia): Report of nine cases. *Lancet* 2:1101–1104.
Taylor, J. H. 1963. The replication and organization of DNA in chromosomes. In: Taylor, J. H., ed. *Molecular Genetics*. Academic Press, New York, Part I, pp. 65–111.
Thompson, M. W. 1965. Genetic consequences of heteropyknosis of an X-chromosome. *Canad. J. Genet. Cytol.* 7:202–213.
Tjio, J. H., and Levan, A. 1956. The chromosome number in man. *Hereditas* 42:1–6.
Uchida, I. A., and Lin, C. C. 1972. Identification of triploid genome by fluorescence microscopy. *Science* 176:304–305.
Uchida, I. A., and Soltan, H. C. 1963. Evaluation of dermatoglyphics in medical genetics. *Pediat. Clin. N. Amer.* 10:409–422.
Waardenburg, P. J. 1932. *Das menschliche Auge und seine Erbanlagen*. Nijoff, The Hague.
Walker, N. F. 1958. The use of dermal configurations in the diagnosis of mongolism. *Pediat. Clin. N. Amer.* 5:531–543.
Walker, N. F., Carr, D. H., Sergovich, F. R., Barr, M. L., and Soltan, H. C. 1963. Trisomy-21 and 13-15/21 translocation chromosome patterns in related mongol defectives. *J. Ment. Defic. Res.* 7:150–163.
Walker, W., Murray, S., and Russell, J. K. 1957. Stillbirth due to haemolytic disease of the newborn. *J. Obstet. Gynec. Brit. Empire* 64:573–581.
Wallace, B. 1968. *Topics in Population Genetics*. W. W. Norton & Company, Inc., New York.
Walton, J. N., and Nattrass, F. J. 1954. On the classification, natural history and treatment of the myopathies. *Brain* 77:169–231.
Warkany, J. 1971. *Congenital Malformations: Notes and Comments*. Year Book Medical Publishers, Inc., Chicago.
Watson, J. D. 1968. *The Double Helix*. Atheneum Publishers, New York.
Watson, J. D. 1970. *Molecular Biology of the Gene*. 2nd ed. W. A. Benjamin Inc., New York.
Watson, J. D., and Crick, F. H. C. 1953. Molecular structure of nucleic acids—a structure for dexoryribose nucleic acid. *Nature* 171:737–738.
White, C., and Wyshak, G. 1964. Inheritance in human dizygotic twinning. *New Eng. J. Med.* 271:1003–1005.
Whittinghill, M. 1965. *Human Genetics and Its Foundations*. Reinhold Publishing Corporation, New York.
WHO Technical Report Series No. 438. 1970. *Genetic Factors in Congenital Malformations*. Report of a WHO Scientific Group. World Health Organization, Geneva.
Wiener, A. S., Unger, L. J., Cohen, L., and Feldman, J. 1956. Type-specific cold auto-antibodies as a cause of acquired hemolytic anemia and hemolytic transfusion reactions: Biological test with bovine red cells. *Ann. Int. Med.* 44:221–240.
Witkop, C. J. 1957. Hereditary defects in enamel and dentin. *Acta Genet.* (Basel) 7:236–239.
Witkop, C. J. 1971. Albinism. In: *Advances in Human Genetics* 2:61–142. (Harris, H., and Hirschhorn, K., eds.) Plenum Publishing Company, New York.
Woodruff, M. F. A. 1971. Antilymphocytic serum and its mode of action. *Transplant. Proc.* 3:34–47.
Wynne-Davies, R. 1964. Family studies and the cause of congenital club foot. *J. Bone Joint Surg.* 46:445–464.

ANSWERS TO PROBLEMS

CHAPTER TWO

1. (a) A and A' (b-1) At first meiotic division (b-2) At second meiotic division.
2. $\frac{1}{4}AB$, $\frac{1}{4}A'B$, $\frac{1}{4}AB'$, $\frac{1}{4}A'B'$.
3. (a) 2^3 (b) 2^5 (c) 2^{23}.
4. (a) Three
   (b) $ABCD$ $ABC'D$
   $A'BCD$ $A'BC'D$
   $AB'CD$ $AB'C'D$
   A′B′CD A′B′C′D
   (c) $ABCD$ and $A'B'C'D$ are parental combinations.
   (d) The other six are recombinations.

CHAPTER THREE

1. TCG; TTA; AGA; GCT; CCA.
2. (a) UGU (b) ACA (c) See Table 3–1.

CHAPTER FOUR

1. (a) $C/c \times C/c$
   (b) $C/C \times C/C$ curly × curly
   $C/C \times C/c$ curly × curly
   $C/C \times c/c$ curly × straight
   $C/c \times C/c$ curly × curly
   $C/c \times c/c$ curly × straight
2. (a) $\frac{1}{2}$ (b) $(\frac{1}{2})^3$ (c) $(\frac{1}{2})^3$
3. (a) $B/b\ d/d \times b/b\ D/d$
   (b) $\frac{1}{4}\ B/b\ D/d$ brachydactyly and dentinogenesis imperfecta
   $\frac{1}{4}\ B/b\ d/d$ brachydactyly and normal teeth
   $\frac{1}{4}\ b/b\ D/d$ normal fingers and dentinogenesis imperfecta
   $\frac{1}{4}\ b/b\ d/d$ normal fingers and normal teeth

4. (a) $Pk/pk \times Pk/pk$ carrier × carrier
$Pk/pk \times pk/pk$ carrier × affected
$pk/pk \times pk/pk$ affected × affected (pk is symbol for PKU gene)
(b) $Pk/pk \times Pk/pk$
(c) $1/4$
(d) $2/3$

5. (a) Pk/pk man
$1/8$ chance his first cousin is also Pk/pk.
If so, $1/4$ chance first child is affected.
$1/8 \times 1/4 = 1/32$
(b) $1/4$, since probability the wife is a carrier is now 1.

6. (a) $1/2$ (b) $1/4 \times 1/2 = 1/8$ (c) 0.

7. (a) $a/a\ D/D \times A/A\ d/d$.
(b) Only $A/a\ D/d$, normal.
(c) $A/a\ D/d \times A/a\ D/d$.
Progeny:

$A/A\ D/D$	$A/A\ D/d$	$A/a\ D/d$	$A/a\ D/d$
$A/A\ D/d$	$A/A\ d/d$	$A/a\ D/d$	$A/a\ d/d$
$A/a\ D/D$	$A/a\ D/d$	$a/a\ D/D$	$a/a\ D/d$
$A/a\ D/d$	$A/a\ d/d$	$a/a\ D/d$	$a/a\ d/d$

i.e., $9/16$ $A{-}D{-}$ (normal): $3/16$ $a/a\ D{-}$ (albino with normal hearing): $3/16$ $A{-}\ d/d$ (deaf, with normal pigmentation): $1/16$ $a/a\ d/d$ (albino and deaf).

8. (a) $1/2$ chance wife is a carrier.
If so, $1/2$ chance she transmits the color blindness gene to the son.
(1) $1/2 \times 1/2 = 1/4$ (2) $1/2 \times 1/2 \times 1/2 = 1/8$
(b) Same as (1) above.
(c) Male-to-male transmission does not occur.

9. Half the sons of all his daughters, i.e., one-fourth of his grandsons.

10. (a) Dominant inheritance.
(b) Autosomal dominant.
(c) X-linked dominant.

11. (a) $C/c\ a/a \times c/c\ A/a$.
(b) $1/4$ $C/c\ A/a$ curly hair, normal pigmentation
$1/4$ $c/c\ A/a$ straight hair, normal pigmentation
$1/4$ $C/c\ a/a$ curly hair, albino
$1/4$ $c/c\ a/a$ straight hair, albino.

12. Parents $Xcb\ Y\ a/a \times XXcb\ A/a$; child $Xcb\ Xcb\ a/a$.

13. $Bb\ a/a\ Xcb\ Y \times bb\ A/A\ XX$
Children:
$Bb\ A/a\ XXcb$ female with brachydactyly
$bb\ A/a\ XXcb$ female with normal phenotype
$Bb\ A/a\ XY$ male with brachydactyly
$bb\ A/a\ XY$ male with normal phenotype.

CHAPTER FIVE

1. $E_1{}^a/E_1{}^a\ W/w \times E_1{}^u/E_1{}^a\ w/w$.
Progeny: $\frac{1}{4}\ E_1{}^u/E_1{}^a\ W/w$ Intermediate cholinesterase, carrier Wilson's disease
$\frac{1}{4}\ E_1{}^u/E_1{}^a\ w/w$ Intermediate cholinesterase, Wilson's disease
$\frac{1}{4}\ E_1{}^a/E_1{}^a\ W/w$ Atypical cholinesterase, carrier Wilson's disease
$\frac{1}{4}\ E_1{}^a/E_1{}^a\ w/w$ Atypical cholinesterase, Wilson's disease.

2. (a) ½ (b) None.

3. All have sickle cell trait. (Some homozygotes now survive to reproductive age.)

4. ⅔ chance each is heterozygous
Risk = $\frac{2}{3} \times \frac{2}{3} \times \frac{1}{4} = \frac{1}{9}$

5. $E_1{}^a/E_1{}^a\beta^A/\beta^A \times E_1{}^u/E_1{}^u\beta^A/\beta^S$
Progeny: $\frac{1}{2}\ E_1{}^u/E_1{}^a\beta^A/\beta^A$ Intermediate cholinesterase, normal hemoglobin
$\frac{1}{2}\ E_1{}^u/E_1{}^a\beta^A/\beta^S$ Intermediate cholinesterase, sickle cell trait.

6. $\alpha^A/\alpha^{Ho-2}\,\beta^A/\beta^S \times \alpha^A/\alpha^A\,\beta^A/\beta^A$
Progeny: $\frac{1}{4}\ \alpha^A/\alpha^A\beta^A/\beta^A$ Normal
$\frac{1}{4}\ \alpha^A/\alpha^{Ho-2}\beta^A/\beta^A$ Heterozygous for Hopkins-2 hemoglobin
$\frac{1}{4}\ \alpha^A/\alpha^A\beta^A/\beta^S$ Sickle cell trait
$\frac{1}{4}\ \alpha^A/\alpha^{Ho-2}\beta^A/\beta^S$ Heterozygous for Hopkins-2 hemoglobin, and with sickle cell trait.

7. $E_1{}^u/E_1{}^a\ D/d\,\beta^A/\beta^S$
Gametes: $E_1{}^uD\beta^A$ $E_1{}^aD\beta^A$
$E_1{}^ud\beta^A$ $E_1{}^ad\beta^A$
$E_1{}^uD\beta^S$ $E_1{}^aD\beta^S$
$E_1{}^ud\beta^S$ $E_1{}^ad\beta^S$

8. $\beta^A/\beta^S \times \beta^A/\beta^C$
One-fourth of progeny have β^A/β^A, thus have normal hemoglobin.

CHAPTER SIX

1. (a) A/B/O (b) A/A/O or B/B/O

2. (a) Theoretically, she could produce the following gametes:
(1) Normal, with 23 chromosomes.
(2) Balanced, with 22 chromosomes including a DqGq.

(3) Unbalanced, with 23 chromosomes including a 21 and a DqGq.
(4) Unbalanced and deficient, with 22 chromosomes and lacking a 21.
(5) Unbalanced, with 23 chromosomes including a DqGq and an extra D.
(6) Unbalanced and deficient, with 22 chromosomes and lacking one D.

(Types 5 and 6 have not been observed, and can be ignored in your answer. See text for explanation.)

(b) One-third, since 5 and 6 can be ignored and 4 is usually inviable.

3. (a) She has the following chromosomes: 14, DqGq and a pair of 21's. She might produce the following gametes, with respect to these chromosomes: (i) 14 + 21, (ii) DqGq + 21, (iii) 14 + 21 + 21, (iv) DqGq. (This assumes that DqGq is 14.21.)

(b) (i) Normal karyotype and phenotype.
(ii) Translocation Down's syndrome.
(iii) Trisomy 21 Down's.
(iv) Translocation carrier with normal phenotype.

CHAPTER SEVEN

1. Apparently from the mother, since the father has given her an X chromosome.

2. (a) X*cb*X*cb*Y
(b) Father is XY; mother is XX*cb*
Nondisjunction probably occurred in the mother at the second meiotic division.

3. (a) Theoretically, X and XX.
(b) XX XXX
XY XXY

CHAPTER EIGHT

1. (a) C3H, C57BL/6, C3B6F_1
(b) C3B6F_1

2. MZ twin, DZ twin, sib or parent; uncle; cousin.

CHAPTER NINE

In these answers the symbol *R* is used for any Rh positive combination and *r* for Rh negative.

1. R/r or r/r.
2. The father might be A/A, A/O, B/B, B/O, or A/B, but *not* O/O.
3. (a) No (b) No (c) No (d) His N blood group excludes him.
4. O child — O × O parents
   A child — A × O parents
   B child — AB × O parents
   AB child — A × AB parents
5. (a) Wife R/r; her father R/R or R/r, her mother r/r.
   Husband r/r; his father R/r, his mother r/r.
   (b) Nil.
6. (a) A or B (b) Segregate (c) A and B are alleles.
7. (a) $A/O\ R/r$; $A/O\ r/r$; $B/O\ R/r$; $B/O\ r/r$. (b) Assort independently.
   (c) A and R are not alleles.
8. $\frac{1}{4}$ Hp 1-1; $\frac{1}{2}$ Hp 2-1; $\frac{1}{4}$ Hp 2-2.

CHAPTER ELEVEN

1. (a) $(\frac{1}{2})^6 = \frac{1}{64}$ (b) $20\ (\frac{1}{2})^3\ (\frac{1}{2})^3 = \frac{5}{16}$
2. (a) $\frac{1}{2}$ (b) $(\frac{1}{2})^4 = \frac{1}{16}$ (c) $1 - (\frac{1}{2})^4 = \frac{15}{16}$
3. (a) $\frac{1}{4}$ (b) $(\frac{3}{4})^5 = \frac{243}{1024}$ (c) $10\ (\frac{3}{4})^3\ (\frac{1}{4})^2 = \frac{270}{1024}$
4. (a) $(\frac{1}{2})^3 = \frac{1}{8}$ (b) $(\frac{1}{2})^3 = \frac{1}{8}$ (c) $3\ (\frac{1}{2})\ (\frac{1}{2})^2 = \frac{3}{8}$
5. (a) $\frac{1}{20{,}000}$ (b) $\frac{1}{100}$; $\frac{2}{100}$ (c) $\frac{1}{10{,}000}$; $\frac{1}{100{,}000{,}000}$
6. 25 per cent M, 50 per cent MN, 25 per cent N
7. (1) Engineering (2) Arts (3) Nursing
   (Class 1 has been selected by the presence of at least one male per sibship, Class 3 by the presence of at least one female. Class 2 has an approximately normal sex ratio.)
8. (a) 57.1 per cent (b) 62.5 per cent.
9. (a) Frequency of $a = 0.1$, $A = 0.9$.
   (b) Same.
   (c) $(0.18)^2 = 0.0324$.
10. 36 per cent group A, 64 per cent group O.
11. Recombinant: son with normal color vision and G6PD deficiency.
    Non-recombinants: other three sons.
    Linkage is in coupling (both genes received from father).

12. (a) Probably in repulsion.
(b) Probably not, unless all four are recombinants.

13. Sons might receive the following combinations:

$Xg^a\ cb\ h$ } most likely (parental combinations)
$Xg\ +^{cb}\ +^h$ }
$Xg^a\ +^{cb}\ +^h$
$Xg\ cb\ h$
$Xg^a\ cb\ +^h$
$Xg\ +^{cb}\ +^h$
$Xg^a\ +^{cb}\ h$ } least likely (double crossovers)
$Xg\ cb\ +^h$ }

(This answer assumes that the order of the genes on the chromosome is as shown.)

14. $1/8$

15. $3/8$

CHAPTER TWELVE

1. 4 in 40,000 (1×10^{-4})

2. The frequency will rise, because selection has been relaxed and mutation will continue.

3. Any dominant, whether autosomal or X-linked.

CHAPTER THIRTEEN

1. Probability MZ = 98.2 per cent; if haptoglobin types differ, twins must be DZ.

2. $1.00 - 2(0.33) = 0.34$ (34 per cent).

NAME INDEX

Refer also to the References, page 369.

INDEX